AF581756

Climate Change and Air Pollution Effects on Forest Ecosystems

Climate Change and Air Pollution Effects on Forest Ecosystems

Editors

Ovidiu Badea
Alessandra De Marco
Pierre Sicard
Mihai A. Tanase

MDPI • Basel • Beijing • Wuhan • Barcelona • Belgrade • Manchester • Tokyo • Cluj • Tianjin

Editors
Ovidiu Badea
National Institute for Research and Development in Forestry "Marin Dracea"
"Transilvania" University of Brasov
Romania

Alessandra De Marco
Italian National Agency for New Technologies, Energy and Sustainable Economic Development (ENEA)
Italy

Pierre Sicard
ARGANS
France

Mihai A. Tanase
University of Alcal'a
Spain
National Institute for Research and Development in Forestry "Marin Dracea"
Spain

Editorial Office
MDPI
St. Alban-Anlage 66
4052 Basel, Switzerland

This is a reprint of articles from the Special Issue published online in the open access journal *Forests* (ISSN 1999-4907) (available at: https://www.mdpi.com/journal/forests/special_issues/climate_pollution_ecosystem).

For citation purposes, cite each article independently as indicated on the article page online and as indicated below:

LastName, A.A.; LastName, B.B.; LastName, C.C. Article Title. *Journal Name* **Year**, *Volume Number*, Page Range.

ISBN 978-3-0365-2666-9 (Hbk)
ISBN 978-3-0365-2667-6 (PDF)

Cover image courtesy of Șerban Mihai Chivulescu

Contents

About the Editors vii

Ovidiu Badea
Climate Change and Air Pollution Effect on Forest Ecosystems
Reprinted from: *Forests* **2021**, *12*, 1642, doi:10.3390/f12121642 1

Serban Chivulescu, Juan García-Duro, Diana Pitar, Ștefan Leca and Ovidiu Badea
Past and Future of Temperate Forests State under Climate Change Effects in the Romanian Southern Carpathians
Reprinted from: *Forests* **2021**, *12*, 885, doi:10.3390/f12070885 5

Juan García-Duro, Albert Ciceu, Serban Chivulescu, Ovidiu Badea, Mihai A. Tanase and Cristina Aponte
Shifts in Forest Species Composition and Abundance under Climate Change Scenarios in Southern Carpathian Romanian Temperate Forests
Reprinted from: *Forests* **2021**, *12*, 1434, doi:10.3390/f12111434 23

Andrei Popa and Ionel Popa
Photosynthesis Traits of Pioneer Broadleaves Species from Tailing Dumps in Călimani Mountains (Eastern Carpathians)
Reprinted from: *Forests* **2021**, *12*, 658, doi:10.3390/f12060658 41

Gheorghe-Marian Tudoran, Avram Cicșa, Albert Ciceu and Alexandru-Claudiu Dobre
Growth Relationships in Silver Fir Stands at Their Lower-Altitude Limit in Romania
Reprinted from: *Forests* **2021**, *12*, 439, doi:10.3390/f12040439 55

Cristian Gheorghe Sidor, Radu Vlad, Ionel Popa, Anca Semeniuc, Ecaterina Apostol and Ovidiu Badea
Impact of Industrial Pollution on Radial Growth of Conifers in a Former Mining Area in the Eastern Carpathians (Northern Romania)
Reprinted from: *Forests* **2021**, *12*, 640, doi:10.3390/f12050640 71

Areum Kim, Bongsoon Lim, Jaewon Seol, Chihong Lim, Younghan You, Wansup Lee and Changseok Lee
Diagnostic Assessment and Restoration Plan for Damaged Forest around the Seokpo Zinc Smelter, Central Eastern Korea
Reprinted from: *Forests* **2021**, *12*, 663, doi:10.3390/f12060663 83

Nicu Constantin Tudose, Mirabela Marin, Sorin Cheval, Cezar Ungurean, Serban Octavian Davidescu, Oana Nicoleta Tudose, Alin Lucian Mihalache and Adriana Agafia Davidescu
SWAT Model Adaptability to a Small Mountainous Forested Watershed in Central Romania
Reprinted from: *Forests* **2021**, *12*, 860, doi:10.3390/f12070860 109

Georgeta Mihai, Alin Madalin Alexandru, Emanuel Stoica and Marius Victor Birsan
Intraspecific Growth Response to Drought of *Abies alba* in the Southeastern Carpathians
Reprinted from: *Forests* **2021**, *12*, 387, doi:10.3390/f12040387 127

Gabriela Elena Baciu, Carmen Elena Dobrotă and Ecaterina Nicoleta Apostol
Valuing Forest Ecosystem Services. Why Is an Integrative Approach Needed?
Reprinted from: *Forests* **2021**, *12*, 677, doi:10.3390/f12060677 151

Alexandru Claudiu Dobre, Ionuț-Silviu Pascu, Ștefan Leca, Juan Garcia-Duro, Carmen-Elena Dobrota, Gheorghe Marian Tudoran and Ovidiu Badea
Applications of TLS and ALS in Evaluating Forest Ecosystem Services: A Southern Carpathians Case Study
Reprinted from: *Forests* **2021**, *12*, 1269, doi:10.3390/f12091269 . . . . **169**

About the Editors

Ovidiu Badea has a PhD în Forestry and is a senior scientist and member of the Romanian Academy. He is a forest ecologist and specializes in dendrometrics, growth and forest ecosystems monitoring. He is an employee of the National Institute for Research and Development in Forestry (INCDS). Marin Drăcea is a PhD advisor at Transylvania University of Brasov and has significant experience in climate change, air pollution and other stress factors effects on forests. Throughout his research career, he has led more than 45 competitively funded projects and published 58 ISI indexed journal articles, and the H-index according to WOS is 15. He has made significant contributions in the implementation of the Romanian National Strategies for Research, Forestry and Environment. Internationally, he developed an important collaboration with European institutions for use in monitoring the effects of climate change and air pollution on forest condition (ICP-Forests of UN/ECE) as head of the National Focal Center (NFC) since 1991. He is a representative for Romania in the IUFRO International Council, and since 2019, he has been the coordinator of the IUFRO Research WP 08.04.01—Detection and monitoring (between 2013 and 2019, he was the deputy of this IUFRO research group).

Alessandra De Marco is a biologist with a PhD in Bio-systematic and Plant Ecophysiology. She is a researcher at the Italian National Agency for New Technologies, Energy and Sustainable Economic Development (ENEA). She has been involved in IUFRO since 2014; from October 2019 she has been the coordinator of the RG 8.04.00 of Division 8 of the IUFRO. Since 2018, she has been the vice-chair of the Working Group on Effects under the LRTAP convention. She is a National Responsible for the implementation of ecosystem monitoring in the NEC Directive. She is involved in many national or European projects as Principal Investigator (FO3REST, MOTTLES, MITIMPACT, AIRFRESH, VEG-GAP, MODERN NEC). She is a member of the editorial board of *Environmental Pollution* and *Helyon*; she is an editor of Sustainability and Frontiers in *Forest* and *Global Change*. She is a member of the scientific committee of several meetings and is author of more than 100 scientific papers, with a Google h index of 37. Her main interest fields are: climate change and air pollution interactions and their synergistic impacts on natural and anthropogenic ecosystems; integrated assessment modeling to estimate beneficial effects of policies and measures to reduce air pollution; the modeling of air quality and its impacts on ecosystems.

Pierre Sicard, with a PhD in Atmospheric Chemistry, is working on air pollution and climate change impacts on forests ecosystems to reduce the risk for plant ecosystems. Currently, he is leading a European project AIRFRESH (2020-2025) to quantify and map the environmental and socio-economic benefits provided by urban trees at the city scale. He is involved in numerous national and EU-funded projects as coordinator (e.g., FO3REST, AIRFRESH) or Principal Investigator and steering committee (e.g., MOTTLES). He is very active in communication: he is Deputy Coordinator of the RG 8.04.00 "Air Pollution & Climate Change" under the International Union of Forest Research Organizations (IUFRO); he is involved in an UNECE Expert Panel on Clean Air in Cities and is active in the EU Clean Air Forum; he is a member of the Editorial Board of journals (e.g. *Environmental Research, Forests* and the *Atmosphere and Climate* journal); he is a member of the scientific committee of meetings and has published >80 papers, with an h-index of 32. He is also involved in a Regional Expert Group on Climate in the Provence-Alpes-Côte d'Azur region.

Mihai A. Tanase has a PhD in Spatial Planning and Environment, an MSc in Environmental Management, and a Diploma in Silviculture, which he attained at the University of Zaragoza (2010), CIHEAM (2006), and the University of Suceava (1999), respectively. He was a visiting scholar at CESBIO, Gamma Remote Sensing, University of Maryland and Boston University and was a postdoctoral fellow at the University of Melbourne. His research is focused on the use of remote sensing technologies for landscape monitoring, with an emphasis on forests. Dr. Tanase has been involved in 20+ national and international projects, has published over 65 scientific articles in JRC-indexed journals and conferences, has presented at 20+ national and international conferences, co-chaired sessions at two international conferences, was guest editor for three Special Issues and was part of the PLOS ONE journal editorial board. He led ten national and international projects and has supervised or co-supervised six master's and three PhD students.

MDPI

Editorial

Climate Change and Air Pollution Effect on Forest Ecosystems

Ovidiu Badea [1,2]

[1] National Institute for Research and Development in Forestry "Marin Dracea", 128 Bvd. Eroilor, 077190 Voluntari, Romania; ovidiu.badea63@gmail.com; Tel.: +40-744650981
[2] Faculty of Silviculture and Forest Engineering, "Transilvania" University of Brasov, 500123 Brasov, Romania

Climate change, air pollution, urbanization, globalization, demographic changes and changing consumption patterns affect forests and their social, cultural, ecological and economic functions, resulting in consequences for the social value of forests and for people's livelihoods, health and quality of life. These consequences are more acutely felt in regions where people are directly dependent on the environmental services provided by forests. Additionally, these consequences rapidly affect growing urban populations, as forests and trees make important contributions to urban resilience and human health and wellbeing.

Achieving a better understanding of the drivers of the changing relationship between forests and people is a major challenge of forest research and a prerequisite for the development of more sustainable relationships between forests and society. Additionally, scientific transdisciplinary and interdisciplinary research, at different regional levels, is needed in order to contribute to the successful adaptation of forests to climate change, and to the strengthening of tree health, resistance and resilience. At the same time, all scientific results are a precondition for maintaining and improving the potential to mitigate the effects of climate change and air pollution on the natural environment.

Climate change and air pollution have large negative impacts on physiological processes and functions at both an individual tree level and on the whole forest ecosystem. Our ability to take urgent measures to combat climate change and its impact on forest ecosystems, and conserve forest biodiversity, depends upon knowing the latest scientific results on the status of forest ecosystems.

At present, climate and air quality monitoring in forests around the globe is performed in different networks, by different organizations. Unfortunately, there are a lot of gaps in our knowledge concerning the detection and monitoring of the effects on forest ecosystems. There is a need to better understand the interactions and fluxes at an ecosystem level, and to understand how different pollutants and climate effects are reflected or transferred in quantifiable ecosystem variables, in both the short and long term. For the detection and monitoring of air pollution actions in the climate change context to be relevant, there is a need for better science–policy interactions. Using Earth Observation data for processing, validation and analysis, new technical developments may provide us with new results in air pollution investigations.

This Special Issue, "Effects of Climate Change and Air Pollution on Forest Ecosystems", includes 10 peer-reviewed contributions, dedicated to increasing the visibility of forest science in the European and global change research policy, and developing the link between forest science and practices in a changing environment and society. The topics addressed in these scientific articles refer to a range of themes, including: the promotion of adaptive management concepts; methods and techniques of restoring forest ecosystems; monitoring the forest's condition under climate change, atmospheric pollution and other biotic and abiotic stressors; the conservation of nature and forest-protected areas; forest genetic resource conservation; forest pests and diseases; and the value of the forest ecosystem services in the context of climate change, for the sustainable, adaptive, management of forests.

Citation: Badea, O. Climate Change and Air Pollution Effect on Forest Ecosystems. *Forests* **2021**, *12*, 1642. https://doi.org/10.3390/f12121642

Received: 23 November 2021
Accepted: 24 November 2021
Published: 26 November 2021

Nowadays, climate change and biodiversity losses are major challenges of global society. Forests, which cover one-third of the Earth's land surface, are an immense and renewable source of ecosystem services (ES). Understanding forest species interactions and their responses to past climates, as well as the different concepts of management, is critical in foreseeing forests' responses to future conditions, and in creating optimal strategies for climate change mitigation and adaptation. Thus, forest management, which is one of the main factors modifying forest structure and succession, can be used to promote resilient mixed forests, which are expected to accumulate a higher biomass quantity under intense climate change, and contribute to climate change mitigation and adaptation [1]. Additionally, research results suggest that climate change will alter the forests' composition and species abundance, with some forests being particularly vulnerable to climate change, e.g., F. Sylvatica forests in the Southern Carpathians. As far as productivity and forest composition changes are concerned, management practices should accommodate these new conditions, in order to mitigate the impacts of climate change [2]. In many cases, human activities change the condition of natural vegetation, leading to disturbances such as the degradation of vegetation, the erosion of soil, a decline in land productivity and even a reduction in ecosystem services. Gaining a better understanding of natural colonization with a pioneer woody species, for example by studying primary natural succession, can offer valuable knowledge about the species that are most adapted to these particular environmental conditions [3]. For the sustainable management of forests, knowledge on the increments of the main dendrometric characteristics of trees (diameter, height and volume) and the relationships among them, can contribute to adaptations in silvicultural work, with the purpose of reducing the risks generated by environmental factors at the stand level. Thus, the existence of stable stand structures is the main condition for an adaptive forest management [4].

Climate change and anthropic activities have given rise to serious environmental problems, and in an increasing number of ecosystems human influences are harming biodiversity and their functions. Through compiling the results of diagnostic assessments of damaged forest ecosystems by air pollution and reference information collected from intact natural forests, restoration plans have shown that ecological restoration is required urgently, as the extent of vegetation damage and soil acidification is very severe. However, tree growth recovery has been observed when the environmental condition has improved due to a significant reduction in air pollution [5,6]. Future climate change projections also underline the importance of hydrological assessments to investigate watershed behavior under climate-related risks and the endangered economic objectives, where it is necessary to intervene in protected forest areas. In this way, a hydrological model, the Soil and Water Assessment Tool, was built and tested in order to support decision makers in conceiving sustainable watershed management; thus, it has also contributed to guides that prioritize the most suitable measures to increase small river basin resilience, and ensure the water demand under climate change [7]. In order to ensure the benefits of forests in the future through the conservation and sustainable use of the forest tree species, silvicultural practices and forest adaptive management should increase, and maintain high genetic diversity and resilience within forest stands. One of the adaptive measures could be the selection, transfer and planting of highly productive and drought-resilient forest reproductive material in reforestation programs (assisted migration) [8].

To emphasize and maximize the ecological, social and economic benefits of forests, suitable assessment methods are required. Active remote-sensing technology, with proven advantages and characteristic limitations, can represent the foundation of multiple approaches, aiming to quantify the capacity of the forest ecosystem to provide services [9]. Representing an immense opportunity to mitigate climate change through carbon sequestration, soil stabilization and natural disaster mitigation, an integrative approach for valuing and assessing forest ES is needed, taking into account the many interdependent factors involving ES and their associated values, as well as the current challenges that people face [10].

All the scientific contributions to forest science in this Special Issue, "Effects of Climate Change and Air Pollution on Forest Ecosystems", will have an important role in promoting sustainable forest management based on mitigating the effects of climate change and air pollution on forests, and their adaptation to a changing environment and society in the global context.

This information will bring new knowledge concerning forests' conditions and their ecosystem service values in the context of climate change, air pollution and other biotic and abiotic factors. Additionally, they will promote adaptive management concepts, methods and techniques of restoring forest ecosystems based on nature and forest genetic resources conservation, for the sustainable and responsible adaptive management of forests.

Acknowledgments: We thank all the Guest Editors (representatives of IUFRO RG 08.04—Impacts of air pollution and climate change on forest ecosystems), reviewers and authors for their very fruitful work.

Conflicts of Interest: The author declares no conflict of interest.

References

1. Chivulescu, S.; García-Duro, J.; Pitar, D.; Leca, Ș.; Badea, O. Past and Future of Temperate Forests State under Climate Change Effects in the Romanian Southern Carpathians. *Forests* **2021**, *12*, 885. [CrossRef]
2. García-Duro, J.; Ciceu, A.; Chivulescu, S.; Badea, O.; Tanase, M.A.; Aponte, C. Shifts in Forest Species Composition and Abundance under Climate Change Scenarios in Southern Carpathian Romanian Temperate Forests. *Forests* **2021**, *12*, 1434. [CrossRef]
3. Popa, A.; Popa, I. Photosynthesis Traits of Pioneer Broadleaves Species from Tailing Dumps in Călimani Mountains (Eastern Carpathians). *Forests* **2021**, *12*, 658. [CrossRef]
4. Tudoran, G.-M.; Cicșa, A.; Ciceu, A.; Dobre, A.-C. Growth Relationships in Silver Fir Stands at Their Lower-Altitude Limit in Romania. *Forests* **2021**, *12*, 439. [CrossRef]
5. Sidor, C.G.; Vlad, R.; Popa, I.; Semeniuc, A.; Apostol, E.; Badea, O. Impact of Industrial Pollution on Radial Growth of Conifers in a Former Mining Area in the Eastern Carpathians (Northern Romania). *Forests* **2021**, *12*, 640. [CrossRef]
6. Kim, A.R.; Lim, B.S.; Seol, J.; Lim, C.H.; You, Y.H.; Lee, W.S.; Lee, C.S. Diagnostic Assessment and Restoration Plan for Damaged Forest around the Seokpo Zinc Smelter, Central Eastern Korea. *Forests* **2021**, *12*, 663. [CrossRef]
7. Tudose, N.C.; Marin, M.; Cheval, S.; Ungurean, C.; Davidescu, S.O.; Tudose, O.N.; Mihalache, A.L.; Davidescu, A.A. SWAT Model Adaptability to a Small Mountainous Forested Watershed in Central Romania. *Forests* **2021**, *12*, 860. [CrossRef]
8. Mihai, G.; Alexandru, A.M.; Stoica, E.; Birsan, M.V. Intraspecific Growth Response to Drought of Abies alba in the Southeastern Carpathians. *Forests* **2021**, *12*, 387. [CrossRef]
9. Dobre, A.C.; Pascu, I.-S.; Leca, Ș.; Garcia-Duro, J.; Dobrota, C.-E.; Tudoran, G.M.; Badea, O. Applications of TLS and ALS in Evaluating Forest Ecosystem Services: A Southern Carpathians Case Study. *Forests* **2021**, *12*, 1269. [CrossRef]
10. Baciu, G.E.; Dobrotă, C.E.; Apostol, E.N. Valuing Forest Ecosystem Services. Why Is an Integrative Approach Needed? *Forests* **2021**, *12*, 677. [CrossRef]

 MDPI

Article

Past and Future of Temperate Forests State under Climate Change Effects in the Romanian Southern Carpathians

Serban Chivulescu [1], Juan García-Duro [1,*], Diana Pitar [1], Ștefan Leca [1] and Ovidiu Badea [1,2]

[1] National Institute for Research and Development in Forestry "Marin Drăcea", 077191 Voluntari, Romania; serban.chivulescu@icas.ro (S.C.); diana.silaghi@icas.ro (D.P.); stefan.leca@icas.ro (Ș.L.); obadea@icas.ro (O.B.)
[2] Faculty of Silviculture and Forest Engineering, "Transilvania" University of Brașov, 500123 Brașov, Romania
* Correspondence: juan.garcia.duro@icas.ro

Abstract: Research Highlights: Carpathian forests hold high ecological and economic value while generating conservation concerns, with some of these forests being among the few remaining temperate virgin forests in Europe. Carpathian forests partially lost their original integrity due to their management. Climate change has also gradually contributed to forest changes due to its modification of the environmental conditions. Background and Objectives: Understanding trees' responses to past climates and forms of management is critical in foreseeing the responses of forests to future conditions. This study aims (1) to determine the sensitivity of Carpathian forests to past climates using dendrochronological records and (2) to describe the effects that climate change and management will have on the attributes of Carpathian forests, with a particular focus on the different response of pure and mixed forests. Materials and Methods: To this end, we first analysed the past climate-induced growth change in a dendrochronological reference series generated for virgin forests in the Romanian Curvature Carpathians and then used the obtained information to calibrate spatially explicit forest Landis-II models for the same region. The model was used to project forest change under four climate change scenarios, from mild to extreme. Results: The dendrochronological analysis revealed a climate-driven increase in forest growth over time. Landis-II model simulations also indicate that the amount of aboveground forest biomass will tend to increase with climate change. Conclusions: There are differences in the response of pure and mixed forests. Therefore, suitable forest management is required when forests change with the climate.

Keywords: temperate forests; climate change effects; Southern Carpathian forest management; forest growth; forest biomass; virgin forests

Citation: Chivulescu, S.; García-Duro, J.; Pitar, D.; Leca, Ș.; Badea, O. Past and Future of Temperate Forests State under Climate Change Effects in the Romanian Southern Carpathians. *Forests* **2021**, *12*, 885. https://doi.org/10.3390/f12070885

Academic Editor: Steve Chhin

Received: 21 May 2021
Accepted: 4 July 2021
Published: 7 July 2021

1. Introduction

Forests have a specific structure and function that determine their capacity to provide multiple ecosystem services, including C (carbon) sequestration [1,2], which has a main role in atmospheric CO_2 (carbon dioxide) balance and, therefore, climate change mitigation strategies and agreements. Environmental conditions, particularly with regard to the soil and climate, play a significant role in forest structure development, forest succession, productivity, etc. Since the climate is continuously changing [3,4], forests also do [5,6]. Even when climate extremes do not have strong effects at the local level, forest changes are forced by direct [6] and indirect [7] climate change effects [8].

However, in terms of structure and composition, forests are often characterized by a high diversity and spatial heterogeneity [9,10]. Carpathian temperate forests, for instance, host both pure and mixed altitudinal forests [11,12], which have very different ecological preferences and properties [11,13,14]. Mixed stands proved to have high resistance and resilience and, often, a high level of productivity, compared to pure stands [15–19]. On the other hand, even-aged monospecific stands instead are less complex functionally and structurally complex [20]. Functional diversity, particularly in mixed forests, provides redundancy [21,22], resistance [15,16,22] and resilience [16,17] to different disturbances. Having a relatively high forest integrity compared to other temperate regions in the world [10,18,19], the mixed forests in the Carpathians are usually constituted by uneven-aged stands with high structural and functional diversity [9,23,24].

Characterized by their high stability and adaptability to climate change and stress factors [25] and by their high productivity [26], the mixed forest management model of the Carpathians might be a useful reference for other managed forests [23,27] in the region. Forest management, including its effect on forest structure and functioning, can be used to promote long-term forest resilience and resistance [22,28,29], contributing to C sequestration, as part of the atmospheric CO_2 balance and climate change mitigation and adaptation strategies and specific programs [22,30]. However, a careful assessment of the forest's state is required.

Even though mixed stands benefit from functional diversity [31], they are also sensitive to management and changes in environmental conditions [31]. Proven drivers of mixed forest change include temperature rises [16] and drought [17]. Functional diversity guarantees greater stability at the stand and forest levels and can increase stress resilience and productivity under drought events [16], but climate change can trigger modifications in species abundance [16,32,33] and can, therefore, alter the structure and functioning of mixed forests, as well as the services they provide [1]. There is still high uncertainty regarding the local impacts of climate change and their potential changes in temperate forests [30], either pure or mixed, as well as their capacity for stocking biomass under regular management. There is also the need to reconstruct and to reanalyse past changes [8] in the context of climate change effects in order to explore future changes.

The complexity of the ecosystem processes and the need for reliable forecasts of climate change mitigation and adaptation actions make the use of a model approach that integrates plant–soil–climate relationships necessary. Landis-II models have been proven to be robust in simulating ecophysiological processes at the cohort level, including natural disturbances and management at the landscape spatial scale [34]. Landis-II models, coupled with the PnET model [35], cover local climate effects on vegetation development [35,36], and they are suitable tools for climate change forecasting [36]. However, their implementation is complex, because many parameters are intercorrelated, and many data inputs must be measured locally [36]. In addition, their calibration is complex and potentially affected by operator decisions [36]. Error propagation can affect the entire model, and the calibrated model can rarely be exported to other contexts, either geographical or temporal. Therefore, new alternatives for models' implementation are required, and this may be achieved by integrating ecological models with machine learning (ML) tools [37–39] and by utilizing the best data available, such as those from long-term ecological research databases [24], e.g., dendrochronological records.

It is well known that tree ring width is significantly influenced by climate variation [40], and therefore, tree ring cores provide a historical record of the climate variability (dendroclimatology) [41], forest growth and age structure, disturbances, site conditions (dendroecology) [42], etc. Climate reconstruction is commonly achieved through the study of the climate signals found in the tree rings [40–43] of open-grown trees [14,40,44], because they are more exposed to climate events and because isolated trees are less affected by competition and other ecological processes [14,45]. Non-isolated trees, mainly in virgin mixed forests, are the best option for tackling intra- and inter-specific interactions. Carpathian virgin forests, internationally recognized for their conservation value, for research and for the services they provide [22,25,46], together with other old-grown forests in the Curvature Carpathians, give an opportunity to deepen our understanding of the roles of mixed vs. pure forests role and their response to long-term processes [23,40–43], such as climate change [8,44,47], as well as their effects on biodiversity, productivity and C cycling [9,22,29]. Understanding species interactions and responses to past climates and forms of management is critical in foreseeing forest responses to future conditions and in creating optimal strategies for climate change mitigation and adaptation. Within this context, this study aims (1) to determine the sensitivity of Carpathian forests to past climates using dendrochronological records and (2) to describe the future effects of forest management and climate change on Carpathian Forest biomass, with a particular focus on the different response of pure and mixed forests.

2. Materials and Methods

2.1. Study Area

The research area is located in the Penteleu Mountains within the Penteleu Forest District (Figure 1). It is a very representative area of Eastern Europe temperate forests specific to the Romanian Curvature Carpathians, at altitudes between 400 and 1800 m.a.s.l. The average annual temperature ranges between 4 °C and 6 °C. The minimum and maximum absolute temperatures recorded are −33.5 °C and +38 °C. The frost period ranges from 120 days at 800 m.a.s.l. to 220 days at the highest peaks. The average annual precipitation is 830 mm, mainly concentrated in summertime. The prevailing winds come from the NW, W and NE directions, with a 6–7 m·s^{-1} annual average speed in high-altitude areas and 2.5–3.5 m·s^{-1} in low-altitude areas.

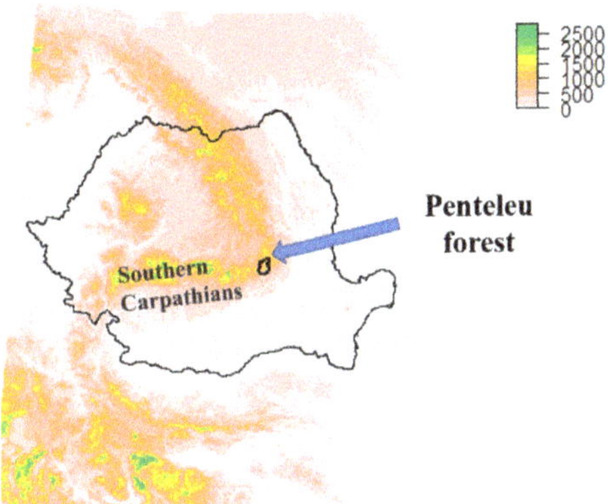

Figure 1. Study area location in Southern Carpathians.

Most of the forest stands included in the research area are temperate montane mixed conifer–broadleaved forests. The main species in terms of timber volume are *Fagus sylvatica* L. (beech—34.6%), *Picea abies* (L.) H. Karst. (Norway spruce—38.1%) and *Abies alba* Mill. (silver fir—22.6%). Other deciduous and softwood species provide less than 4.7% of the overall volume [48]. There is no historical evidence that extreme events (such as windthrows, drought, insect damage, fires, etc.) had a relevant role in the development of the Penteleu forest over time.

Penteleu forest, similar to most of the Romanian Carpathian forests, suffered the effects of profound changes in its management. Being initially private property owned by local landowners and local communities, logging at the beginning of the 20th century was carried out at the pleasure of the owners and in accordance with market demand. However, forest integrity was not heavily affected, regardless of the massive harvests that occurred around a forest railway. Within this period, conifer stands were clear-cut and artificially regenerated afterwards. Mixed stands were subjected to selective cutting, extracting only conifers with a diameter greater than 30 cm, while beech was not extracted for commercial purposes, because it was considered, at that time, a species of low economic value. The Romanian forest nationalization of 1948 was an important step toward obtaining quality timber in maximum quantities, while, at the same time, satisfying the protection functions. Later, in 1975, the forest protection status generated by a change in the forest legislation (especially functional zoning of forests in particular) contributed to forest conservation. Nowadays, protected forests, in which no silvicultural prescriptions or only special conservation prescriptions are allowed, occupy 39% of the total area [48]. The remaining areas are only eligible for selection cuttings (99.6%) and clear-cut prescriptions in a very small amount (0.4%). As a result, around 60% of the forest is older than 60 years old, with almost 20% of it being more than 100 years old [48], including over 200 hectares of virgin and quasi-virgin forests [49,50]. Within the context of temperate European forests, an important fraction of the Penteleu forest district area is occupied by relatively old forest.

The historical management of the different stands in Penteleu forest was reconstructed for the modelling approach following the ecological conditions and actual stand age structure and composition. In general, given the historical management changes and the information contained in the management plans [48], there is a high level of confidence in the historical management of the vast majority of the stands. An estimative historical management of Penteleu forest can be found in Supplementary Material S1.

The climate variables taken from van Oldenborgh et al.'s [51] multi-model mean climate change CMIP5 scenarios (https://climexp.knmi.nl/start.cgi; accessed on 1 July 2020) and additional climate series [43,52] for the region 25–26 E 45–46 N were statistically downscaled to 50 m resolution following Zorita et al. [53] using WorldClim 1.4 and 2.0 [54]. The downscaled monthly climate variables are the Tmin (monthly average of minimum temperature), Tmax (monthly average of maximum temperatures), PET (average monthly evapotranspiration), Prec (average monthly precipitation) and BAL (average monthly water balance). Spectral analysis of the climate variables is provided in Supplementary Material S2. Monthly radiation was also taken from WorldClim [54]. These climate products with past observations and future climate change forecasts until 2100 were used to analyse climate signal in dendrochronological series and in Landis-II simulations.

2.2. Dendrochronological Series and Climate Signal

In order to understand the principles of forests' structure and functioning, a deep analysis of the past and present structure and functionality of virgin forests was performed in a 1 ha control plot inside the Penteleu forest. This control plot (circular, with a 56.41 m radius) was installed and measured in 2014 in a representative temperate montane conifer–broadleaved mixed–virgin forest stand. The stand is unevenly aged, and the reference plot includes 439 trees from three species (beech—63%, Norway spruce—26% and silver fir—11%) [48]. Being unaffected by human activity, we assumed that this stand is optimal for detecting climate signals on tree growth and species interactions.

Tree spatial distribution, breast height diameter (DBH) and height (h) were measured for all living trees over 8 cm DBH. The aboveground tree volume (v) of each tree was calculated with the following regression equation: $\log v = b_0 + b_1 \cdot \log d + b_2 \cdot \log^2 d + b_3 \cdot \log h + b_4 \cdot \log^2 h$, where b_0, b_1, b_2, b_3 and b_4 are the species' national regression coefficients [55]. The overall stand volume measured was 803 m^3. Tree aboveground biomass was computed using species wood density [55].

To build the reference dendrochronological series of silver fir, beech and Norway spruce, 406 radial core samples of living trees collected from the control plot were analysed and processed. The radial growth was calculated with CooRecorder 7.4 image analysis techniques [47,56]. Measurement checking and cross-dating was performed with COFECHA [42,57], and the growth series standardization was carried out with ASTRAN-win [58]. The age structure was intensively analysed in an exploratory analysis to ascertain how stand development might have affected tree growth. DBH, measured through tree ring radial growth, was analysed by linear mixed models and analysis of variance in R software [59].

The statistical analysis of the dendrochronological series was performed with dplR [60], treeclim [61] and waveslim [62] packages in R software [59], and wavelet coherence among species dendrochronological series was explored with biwavelet [63]. The overall and moving correlations between climate variables (i.e., Tmin, Tmax, PET, Prec, BAL) and tree ring growth for the main species were calculated by treeclim dcc response and correlation function analysis [61]. In order to find interactions among species in mixed stands, the moving correlation matrices were inspected. This was completed with a factorial analysis of the of the species' moving.

In addition, since there was a net change in the slope, the 1 ha plot was divided into 2 sectors during the fieldwork in order to explore and compare the relative species abundance in both sectors as a function of the local conditions. Being representative of 2 ecoregions, the reconstruction of the tree aboveground biomass in the 2 sectors was later used in the calibration of the Landis-II models.

2.3. Climate Change Projections

Landis-II [34,64], coupled with PnET models [35,65,66], was developed to simulate forest succession and the effect of climate [66] and management [67] under different climate change scenarios [66]. We built Landis-II models for the Penteleu forest district, aiming to assess management effects and the climate change impacts on Carpathian forest ecoregions, i.e., forests with homogeneous local conditions.

The model parameterization was performed following Gustafson [53] recommendations, using data extracted from scientific and technical works [12,68–71]. Since the full dataset had missing data issues, inconsistency, etc., the values were not used directly in the Landis-II model, but as input data for Landis-II calibration using the genetic algorithms (GA) [39,72] machine learning (ML) technique, with GA R package [38,39,59]. For calibration purposes, the GA fitness function for a given set of chromosome values used the difference in the biomass estimated for the 1 ha plot sectors (using the dendrochronological series) and the biomass predicted by the Landis-II models for those sectors (introduced in the model as fully independent ecoregions). In this, the supervised species calibration was bypassed, and species parameter values that are valid for the whole study area were set. Only the last period of the dendrochronological series was taken into account in order to minimize the effects of the stand dynamics effects on the calibration process.

After the calibration, the 50 m × 50 m resolution landscape was segmented in ecoregions combining the site type, forest types and relative productivity class in the management plans [48]. Soil properties for the ecoregions' parameterization were also extracted from the management plan [48]. Overall, 22 ecoregions, covering an altitudinal gradient and a wide range of site properties, were established. Every ecoregion was provided for monthly photosynthetically active radiation (PAR) from WorldClim 2.0 radiation [54], downscaled CMIP5 Tmax, Tmin and Prec data (https://climexp.knmi.nl/start.cgi; accessed on 1 July 2020), and atmospheric CO_2 concentrations for the 1500–2100 period [3] under the 2.6, 4.5, 6.0 and 8.5 greenhouse gas (GHG) atmospheric concentration RCP scenarios [3].

The 1685 forest stands in the Penteleu forest district were aggregated into 568 management system types, from which 1369 treatment prescriptions (silvicultural regeneration interventions) were sequentially combined and implemented over time at the stand and management system levels, according to historical changes in forest management, species composition and abundance, silvicultural system and age structure. All of the spatial information was managed in R software [59] spatial libraries [73–76].

The silvicultural systems implemented in Landis-II are as follows: (i) strict protection (mostly mixed uneven-aged forests; all virgin forests, quasi-virgin forests and some pure stands are covered by this management type); (ii) old selection forests (predominantly mixed uneven-aged forests); (iii) selection forests (natural composition, often mixed stands, with an uneven-aged structure); (iv) quasi-selection systems (transformation to selection forests 60–70 years; natural composition with relative even to relative uneven-aged structures); (v) group selection system (even aged with pure and mixed composition); (vi) clear-cut system (mostly even-age pure stands and some mixed stands with non-native species).

The timeframe covered by the Landis-II models comprises four main management types and periods: (i) virgin forests and traditional selection forests (1900–1940); (ii) salvage logging (1941–1973); (iii) selection forests (1973–2020); (iv) increased protection of forests (2021–2100).

Disturbances (e.g., windthrows) were modelled through harvest extension, because (a) management plans show evidence of their low relevance and (b) the salvage logging, which is commonly practiced in disturbed stands, has similar effects to unplanned clearcuttings.

The Landis-II models, calibrated and validated for the entire research area, were computed for past conditions in the 1900–2010 period and for the 2010–2100 period under the RCP 2.6, 4.5, 6.0 and 8.5 climate change scenarios. Biomass and, implicitly, C stock, capitalizing on the energy and mass fluxes within the modelling approach, were the main target variables in this study. The ecoregions' biomass over time and the biomass in the natural distribution areas of pure and mixed forests were fitted and analysed using linear mixed models and an analysis of variance. Three key periods (1901–1910, 2001–2010 and 2091–2100) were targeted and compared. Comparisons between those periods were performed for every ecoregion and climate change scenario using Benjamini–Hochberg FDR correction [77]. The familywise and post hoc comparisons were carried out with nlme [78] and multcomp packages [79] in R software [59].

3. Results

3.1. Forest Sensitivity to Past Climate

Older trees in the 1 ha plot mixed–virgin forest stand are more than 350 years old and the maximum tree DBH is 110 cm. Exploratory analysis of the age structure and recruitment suggest that, despite the truncated diameter-based sampling, the age structure follows a geometric distribution, with a strong predominance of young age classes. This age structure has been quite stable over time, emphasized by the small peaks of recruitment detected over the last 300 years.

The analysis of the climate signal in a mixed–virgin forest stand indicates that tree radial growth depends on climate and interspecific relationships. Tree DBH, reconstructed from radial core samples collected in the Penteleu 1 ha plot, increased approximately linearly over time (Figure 2) at the stand level. The mean growth of the central 90% (0.05–0.95 quantiles) of trees in the plot ranged from 0.9 to 4.2 cm·year^{-1}. At the tree level, radial growth has the general tendency to be constant over time, with a few exceptions. Radial growth changed strongly over time for a reduced number of trees: some trees with intense radial growth in the early stages of development gradually reduced their growth rate when they reached physiological maturity, while other trees with reduced radial growth in the early stages suddenly increased their growth rate in later stages. The dendrochronological dataset analysis showed a non-significant effect of the tree age on the tree ring width (p-value > 0.05).

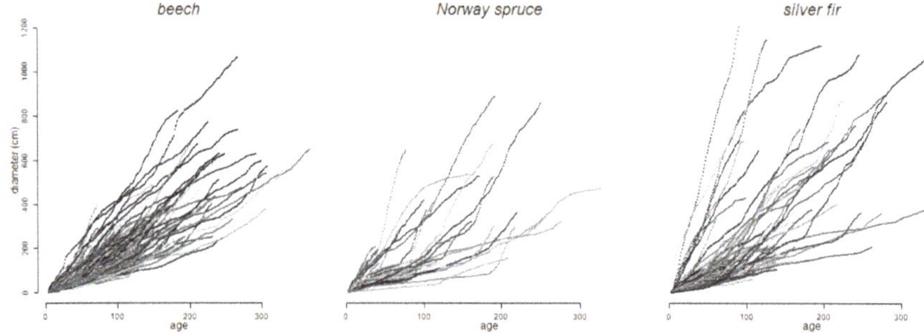

Figure 2. DBH–age relationship in Penteleu 1 ha mixed–virgin forest plot.

All three biwavelets among pairs of species (Figure 3) contain regions with significant wavelet coherence among dendrochronological series of pairs of species. Anti-phase relationships dominate in the beech–silver fir biwavelet, while beech–Norway spruce and silver fir–Norway spruce biwavelets tend to be in phase when there is a lag in the phase.

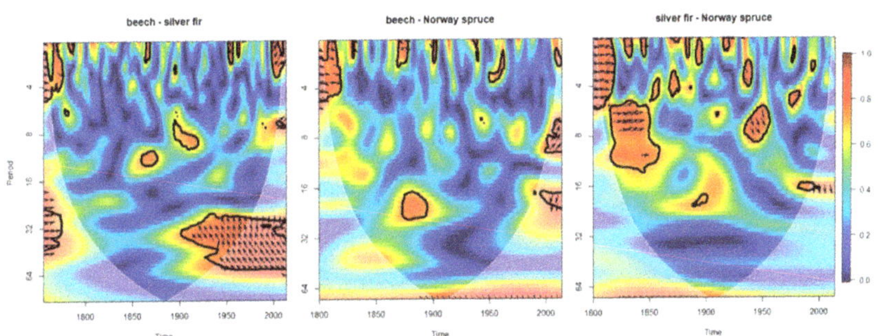

Figure 3. Dendrochronological biwavelet between species, after 100 Monte Carlo randomizations. Arrows direction indicate that both dendrochronological series are in phase (pointing to the **right**) or anti-phase (pointing to the **left**), the black contour indicates $\leq$0.05 significance, and the transparency, the cone of influence.

Climate has a significant effect on tree ring growth. At the stand level, it was summarised the relationship between some of the main climate variables (i.e., Tmin, Tmax, PET, Prec and BAL) and the radial growth extracted from tree cores through moving correlations (details can be found in Supplementary Material S3), and then assessed the similarities between the responses of the three main species in the study area (silver fir, beech and Norway spruce) to the climate variables (Figure 4). The results showed synergic and antagonistic responses to the climate. Silver fir showed a similar trend to the Norway spruce, with the first not being as significant as the second. Silver fir tended to show negative correlations with climate variables in periods in which beech and Norway spruce growth were favoured. Beginning with the end of the 19th century and until the beginning of the 20th century, a consistently positive growth rate was recorded for beech, correlated with higher-than-average summer temperatures. Beech's correlation with September Prec and BAL varied over time, from a negative correlation to a positive one. Beech growth also had a positive correlation with late spring Prec. Norway spruce, in general, tended to show negative correlations with climate variables in the same periods in which beech was favoured by the climate. Beech moving correlations around 1750 showed significant negative correlations with September Prec and BAL, as in 1840. Additionally, around 1840, positive growth was associated with spring Prec and BAL. Later, around 1900, positive correlations were detected with spring Tmin, Tmax and PET. Around 1965, growth was positively correlated with September Prec and BAL and negatively correlated with July and September Tmin. Silver fir growth correlated negatively with spring Tmin and PET and positively with late-summer Prec and BAL around 1970. During the 19th century, silver fir growth was negatively correlated with May Tmin, Tmax and PET, although it was positively correlated with September and April Tmin and Tmax and also with May Prec and BAL. At the beginning of the 20th century, silver fir growth was negatively correlated with late-summer Tmin, Tmax and Prec and positively correlated with April Prec and BAL. Later, occasional positive significant positive correlations were detected with Tmin, Tmax and PET, while negative correlations were detected with July and August Prec and August BAL. Finally, no correlations were found for the 21st century. Norway spruce growth showed positive significant correlations with September Tmin, Tmax and PET around 1825, though negative correlations were found with April Prec and BAL. Later, until 1990, positive correlations tended to occur with spring and summer Tmin, Tmax and PET, particularly around 1875, 1925 and 1970. Significant positive correlations with April and June Prec were detected in the last decades of the 19th century and at the beginning of the 20th century. Finally, around 1945 and 1985, Norway spruce growth was eventually negatively correlated with late-summer Prec and BAL.

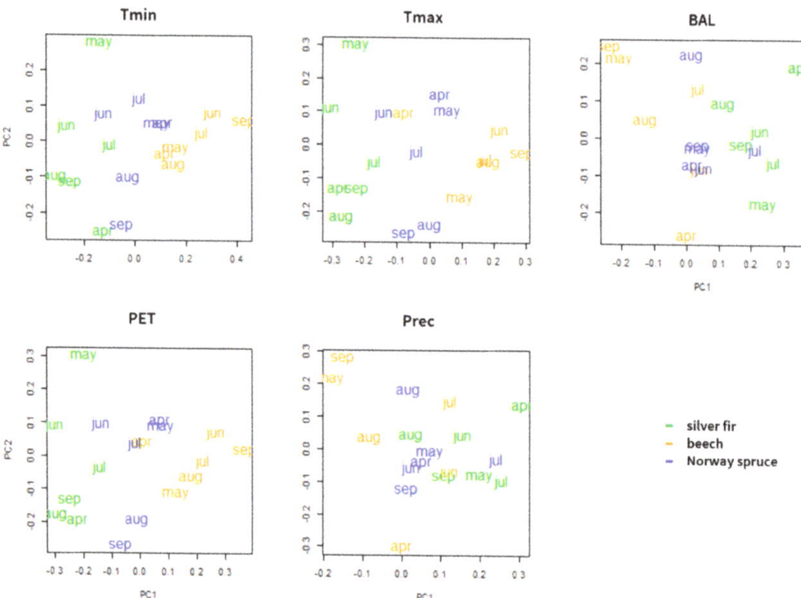

Figure 4. Saturation of the first and second components of the factorial analysis of beech, silver fir and Norway spruce moving correlations between tree ring growth and monthly climate variables (from April to September, respectively: Apr, May, Jun, Jul, Aug, Sep), here sorted by climate variable (i.e., Tmin, Tmax, BAL, PET, Prec).

Thus, the first and second components of the factorial analysis of the moving correlation matrices were built based on the antagonistic behaviour of beech and silver fir. Norway spruce displayed an intermediate behaviour, and consequently, less weight was placed on the construction of the first and second components.

Tree biomass reconstructed from tree cores showed that, at the stand level, around one third of the current biomass in the Penteleu 1 ha plot was produced more than 150 years ago and approximately half was formed in the last 50 years (Figure 5a). Regarding species biomass, before 1850, most of it belonged to beech. After 1900, only half of the biomass found was attributed to beech, with the remaining biomass being provided by conifers, particularly silver fir.

Due to the plot's heterogeneity and microtopography, we compared the biomass accumulation on steep and flat terrain. Before 1850, the accumulated tree biomass was higher in the flattest sector of the plot compared to the steepest one. After 1950, the biomass per hectare in the steepest sector (Figure 5b) was equal to the biomass in the flat sector. As extracted from the dendrochronological series, beech and silver fir are the dominant and codominant species in both sectors, with silver fir being dominant in the steep sector and beech in the flat. Norway spruce, having the lowest contribution in terms of biomass, accumulated more than 1.5 times the biomass per hectare in the flat sector, as compared to the amount of biomass measured in the steep sector.

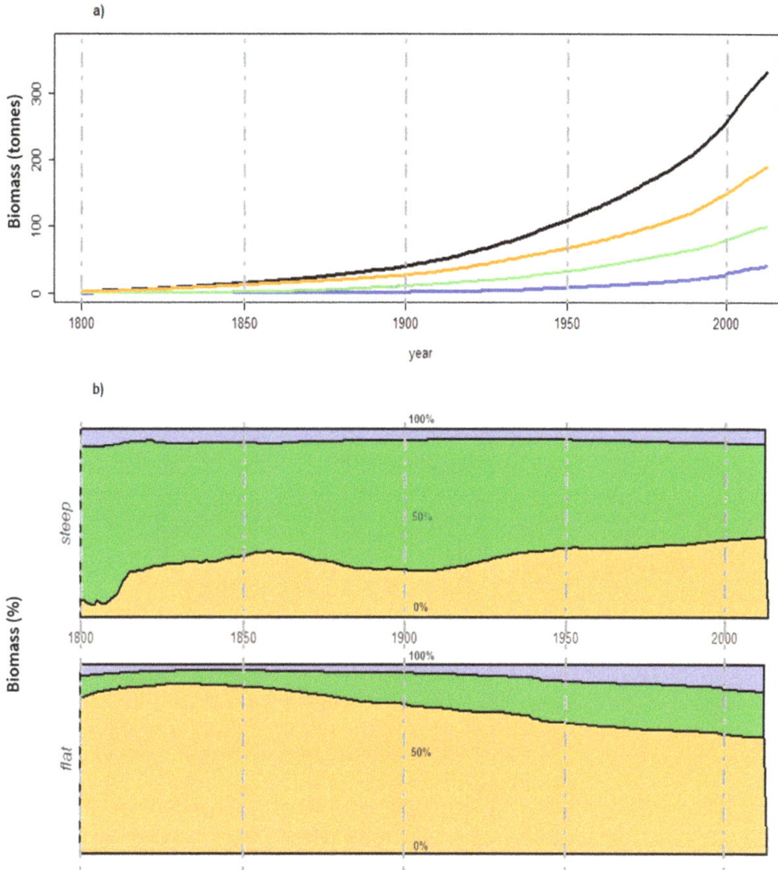

Figure 5. Biomass (tonnes) reconstruction based on tree rings (**a**) at the stand level (continuous) and (**b**) the relative abundance of the main species for the steepest (dotted) and flattest (continuous) sectors in the Penteleu 1 ha plot (black), beech (orange), silver fir (green), Norway spruce (blue).

3.2. Aboveground Living Biomass under Climate Change Scenarios

Landis-II simulations show that the forest biomass changed from 1900 to 2010 in almost all of the ecoregions (Figure 6). Some ecoregions showed a moderate increase in biomass throughout the period. Other ecoregions hardly changed, while the remaining few displayed a reduced amount of biomass. Biomass reductions in the ecoregions were often associated with the harvesting process.

Harvesting caused very strong biomass reductions in some ecoregions, particularly through clearcutting during the 1941–1973 period. However, biomass reductions were also found in other periods, harvest types and management systems, such as selection cuttings in shelterwood systems. The rise in the level of forest protection, together with the aging of the forest, contributed to a reduction in biomass loss and to moderate biomass accumulation in recent decades. In ecoregions with null or very low harvest intensity, the amount of biomass tends to increase.

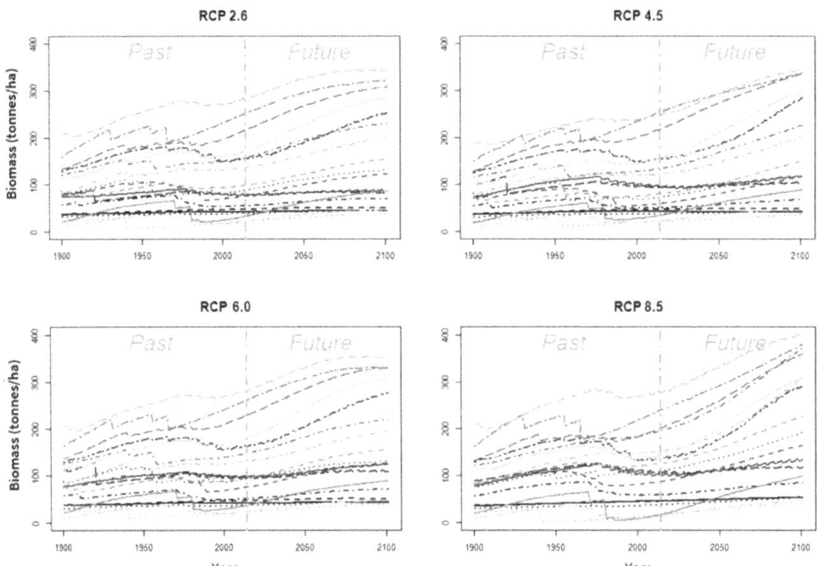

Figure 6. Average quantity of biomass (tonnes/ha) in the 22 ecoregions (different line styles) under the four climate change scenarios: RCP 2.6, RCP 4.5, RCP 6.0 and RCP 8.5.

Management, which, in the past, caused strong reductions in biomass through intense harvest, will contribute to biomass accumulation in the future through protection and low harvest pressure. The Landis-II projections show that, from 2010 to 2100, under a more protective management, the amount of biomass in most of the ecoregions will increase, particularly after 2025.

Apart from harvest effects, the different climate change scenarios (RCP 2.6 to RCP 8.5) had a significant role on forest biomass accumulation after 2010 (p-value ≤ 0.05). This increase in biomass was particularly noticeable in periods where harvesting was less intense. The intensity of the growth differed strongly among the ecoregions. After 2010, the accumulation of biomass is a result of a combination of both local ecoregion conditions, management and climate change scenarios. There is no common response to climate scenarios for all ecoregions, even though intense climate change scenarios, particularly RCP 8.5, tend to increase biomass accumulation. Indeed, some ecoregions have the lowest accumulated biomass under intermediate climate change scenarios (RCP 4.5 and RCP 6.0) and the highest under RCP 8.5. The amount of accumulated biomass barely changed over time in some ecoregions.

The ecoregions, aggregated according to their domination by pure and mixed forests (Figure 7), had different biomass accumulated at the beginning of the study period and over time. In 1900, when salvage logging had not begun, the biomass per hectare in pure stands was higher than in mixed stands. Around 2010, the amount of biomass was still higher in pure forest ecoregions. However, in the last period of the simulations (up to 2100), the biomass per hectare in mixed forests was similar to or even higher than that of pure forests, particularly in RCP 8.5, where the biomass per hectare in mixed forests broadly surpasses that of pure forests.

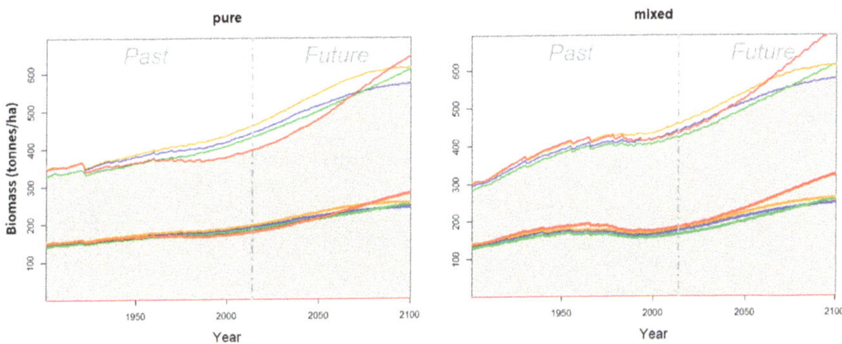

Figure 7. Average quantity of biomass (tonnes/ha—width lines) in pure and mixed forests from 1990 to 2100 and confidence intervals (shaded areas).

4. Discussion

This study shows that, even though virgin forests have a proven stability, both tree radial growth and biomass are affected by climate change and species interactions, thus providing long-term faithful records that are unaffected by management. In addition, it explores the reconstruction of the forest stand aboveground biomass in the past, projection in the future under different climate change scenarios and the C sequestration, as part of the atmospheric CO_2 balance.

Tree DBH reconstructed from Penteleu dendrochronological series was approximately linear, as also found in other similar studies [8,31,80]. Thus, radial growth is not particularly affected by tree age. Radial growth and, thus, the DBH–age relationship, is conditioned by the environmentally limiting conditions [80] and by biological stress. Additionally, the change in the growth rates of individual trees often depends on their interactions with neighbouring trees (e.g., facilitation and competition) and other processes (e.g., mortality) [11,81], meaning that they are likely the cause of the changes found in the measured trees in Penteleu.

We found that small differences in the ecological preferences, local conditions and climate lead to several spatiotemporal patterns in the biomass accumulation over time. In mixed–virgin forest stands, intra- and inter-specific interactions have a strong impact on tree growth, while also having further consequences at the stand level. Such interactions differ in relation to local conditions and species dominance and abundance, which can lead to modifications even at short distances. The results showed that there are differences between growth rates of some trees in early vs. late stages, which (presumably) strongly depend on the physiological activity and dynamics of the in-stand competition and mortality [11,81] of the species. In the Penteleu virgin forest 1 ha plot, the steepest areas are dominated by silver fir, which is less susceptible to warmer and drier conditions than beech [13], while flat areas are occupied by beech, which is also accompanied with a higher abundance of Norway spruce. Having a higher abundance in the flattest area, beech, compared to silver fir and Norway spruce shows better resistance to different stress factors and also has a significant growth in the understory [23,81]. Due to withstand such conditions (including drought), beech growth usually has a low variability [11] compared to coniferous species. These heterogeneous responses to the climate are produced in mixed forests even when the climate conditions are favourable for all three species.

The dendrochronological series analysis proved that silver fir and beech tend to have opposite behaviours in response to climate in mixed stands, while Norway spruce, displaying intermediate behaviour, tends to negatively correlate with the climate when beech correlations are positive. Having different ecological preferences and life traits, the beech–silver fir association in mixed stands is advantageous in terms of their ability to endure drought [13,17,31,33,82] and also because of their efficient light use [83,84]. There is also evidence that beech–Norway spruce mixed stands perform better than beech pure stands [13,33] due to the complementarity of the two species.

The overall biomass, capitalizing all ecological processes, is homogeneously distributed over the 1 ha plot, highlighting the stability and species complementarity in mixed forests, as other studies have suggested [13,18,19,33]. However, despite their high efficiency in terms of the use of resources, there are concerns on the response of mixed and pure forests to climate change (e.g., Piovesan et al. [32]). It was emphasized that pure beech forests have already been affected by climate-change-induced drought in recent decades [32], as Norway spruce shows high drought resistance when compared to beech [17,82,85]. Additionally, silver fir is expected to be less susceptible to warmer and drier conditions than beech [13].

Since tree ring measurements include the effects of past conditions, such types of long-term data can be used in a modelling approach to analyse past changes and to foresee, anticipate and manage future changes [86,87]. However, the complexity, variability and information gaps contained in biological records [87,88] often require solutions such as ML methods [37,86]. GA [38,39] provided a suitable mechanism to speed up and objectivize the modelling approach and was successfully used in this study for Landis-II model calibration in order to simulate landscape changes and for forest change forecasting under different climate change scenarios, which proved to be robust.

The Landis-II models demonstrate that management had a strong impact on forest structure and functioning over time, in agreement with official records and statistics [48]. Forest resilience to management in terms of relative biomass compared to the pre-harvest level is high at the ecoregion level. Biomass commonly recovers in less than 50 years, although impacts on other forest attributes, e.g., diversity, last much longer [48]. Forest resistance to management is very low, as denoted by both (a) the impact of punctual harvest events at the ecoregion level and (b) sustained biomass stock changes due to the change in the silvicultural system, following stakeholders and policymakers' decisions in the past and expectations after 2025, when even more protective management will be implemented.

Biomass change in the different ecological regions is not only related to management, but also to climate change effects. The most important impacts of climate change in the Carpathians will be caused by rising temperature and CO_2 levels. Precipitation will be similar to the current conditions in some areas and may even slightly increase [4]; therefore, drought events are not expected to have a significant role, other than in a few ecological regions and marginal areas of the Carpathians [4]. Temperate forests will benefit from relative stability and better conditions for growth [2,4], contrary to Mediterranean forests, where droughts and fires will be the main drivers of change [89].

Some areas in the Carpathians will endure harsher conditions [4], and our results indicate that impacts will also differ at the local scale, with some ecoregions being more affected than others.

Uncertainty of climate change impacts and characteristics on forest ecosystems is relatively high [30,90], and our results show that temperate forest resistance to climate change is relatively low; however, forest biomass is predicted to be relatively similar under different climate change scenarios. Despite the findings of Nabuurs [2], projections from our Landis-II model also show that the Penteleu forest's C sequestration capacity is still far from saturation, and the aboveground biomass will continue to grow at least until the end of the current century, even when signs of deceleration appear at the end of the period in some ecoregions.

It is important to note that this C pool-growing trend is also related to a rise in the level of forest protection provided, forest aging and the decreasing harvest pressure throughout the period. Thus, the management of temperate forests has the potential to increase carbon sinks and mitigate climate change [1].

On the other hand, adaptation strategies can focus on sustainable management. The general interest is to promote C sequestration and management should be oriented toward forest adaptation to climate change, by reducing potential negative climate change impacts, e.g., drought events [2,89], etc.

In this regard, management can be used to promote mixed forests, which reach higher biomass accumulation under intense climate change scenarios, and thus, to maximize C sequestration. Mixed stands and their particularities contribute to forest resistance and resilience at the landscape level, in comparison to pure stands [15–17].

5. Conclusions

The dendrochronological series in mixed stands not only contain climate signal, but species interactions, and they are useful for past events and projections. Beech, Norway spruce and silver fir, whose radial growth was not dependent of tree age, have complementary responses in mixed stands.

PnET-Landis-II models, GA calibrated using dendrochronological records, applied to the Penteleu forest, which hosts virgin and managed stands, brings to light (i) the relevance of virgin forests in vegetation long-term monitoring and as a reference data source for all forests in the Carpathians, (ii) the similarities and differences in the structure and functioning of mixed and pure forest stands and their adaptative management and (iii) the effect of climate on tree growth and climate change impact on forests, in which mitigation and adaptation strategies, programs and actions must be implemented.

Landis-II models provide valuable insights into the medium- to long-term succession for forest management and forest planning. They provide a good approach to climate change impacts, which are becoming increasingly accentuated. Landis-II simulations revealed that aboveground forest biomass changed in almost all ecological regions in the Carpathian Curvature, often associated with harvesting. Forest projections demonstrate that climate change in the Carpathian Curvature will promote forest aboveground biomass and C sequestration. Climate change will modify species and cohort interactions, leading to changes in the forest structure and functioning. Climate change impacts will depend on climate change intensity and local conditions and management. Among forest ecosystems, mixed forests are capable of higher C sequestration and higher biomass accumulation under intense climate change.

In this study, management, which is one of the main factors modifying forest structure and succession, can be used to promote resilient mixed forests, which are expected to accumulate higher biomass quantity under intense climate change, and thus, contributes to climate change mitigation and adaptation.

Supplementary Materials: The following are available online at https://www.mdpi.com/article/10.3390/f12070885/s1, Figure S1: Estimative historical management, Figure S2: Spectral analysis of climate variables (i.e., Tmin, Tmax, PET, Prec, BAL), Figure S3: Moving correlations between beech (*F. sylvatica* L.), Norway spruce (*P. abies* (L.) H. Karst.) and silver fir (*A. alba* Mill.) tree ring growth and climate variables (i.e., Tmin, Tmax, PET, Prec, BAL).

Author Contributions: Conceptualization, J.G.-D. and S.C.; methodology, J.G.-D., S.C. and Ș.L.; software, J.G.-D. and S.C.; validation, J.G.-D., S.C., D.P. and O.B.; formal analysis, J.G.-D.; investigation, J.G.-D. and S.C.; resources, S.C. and J.G.-D.; data curation, J.G.-D.; writing—original draft preparation, J.G.-D. and S.C.; writing—review and editing, J.G.-D., S.C., D.P., Ș.L. and O.B.; visualization, J.G.-D.; supervision, S.C. and O.B.; project administration, S.C. and O.B.; funding acquisition, S.C. and O.B. All authors have read and agreed to the published version of the manuscript.

Funding: This research was funded by the BIOSERV Programme, Project IDs PN19070103, PN19070102 and EO-ROFORMON project, ID P_37_651/SMIS 105058.

Institutional Review Board Statement: Not applicable.

Informed Consent Statement: Not applicable.

Conflicts of Interest: The authors declare no conflict of interest.

References

1. Dymond, C.C.; Beukema, S.; Nitschke, C.R.; Coates, K.D.; Scheller, R.M. Carbon Sequestration in Managed Temperate Coniferous Forests under Climate Change. *Biogeosciences* **2016**, *13*, 1933–1947. [CrossRef]
2. Nabuurs, G.-J.; Lindner, M.; Verkerk, P.J.; Gunia, K.; Deda, P.; Michalak, R.; Grassi, G. First Signs of Carbon Sink Saturation in European Forest Biomass. *Nat. Clim. Chang.* **2013**, *3*, 792–796. [CrossRef]
3. van Vuuren, D.P.; Edmonds, J.; Kainuma, M.; Riahi, K.; Thomson, A.; Hibbard, K.; Hurtt, G.C.; Kram, T.; Krey, V.; Lamarque, J.-F.; et al. The Representative Concentration Pathways: An Overview. *Clim. Chang.* **2011**, *109*, 5–31. [CrossRef]
4. Micu, D.M.; Dumitrescu, A.; Cheval, S.; Birsan, M.-V. *Climate of the Romanian Carpathians*; Springer Atmospheric Sciences; Springer International Publishing: Cham, Switzerlands, 2015; ISBN 978-3-319-02885-9.
5. Huntley, B.; Baxter, R. Vegetation Ecology and Global Change. In *Vegetation Ecology*; van der Maarel, E., Franklin, J., Eds.; John Wiley & Sons, Ltd.: Oxford, UK, 2013; pp. 509–530. ISBN 978-1-118-45259-2.
6. McDowell, N.G.; Allen, C.D.; Anderson-Teixeira, K.; Aukema, B.H.; Bond-Lamberty, B.; Chini, L.; Clark, J.S.; Dietze, M.; Grossiord, C.; Hanbury-Brown, A.; et al. Pervasive Shifts in Forest Dynamics in a Changing World. *Science* **2020**, *368*, eaaz9463. [CrossRef]
7. Seidl, R.; Thom, D.; Kautz, M.; Martin-Benito, D.; Peltoniemi, M.; Vacchiano, G.; Wild, J.; Ascoli, D.; Petr, M.; Honkaniemi, J.; et al. Forest Disturbances under Climate Change. *Nat. Clim. Chang.* **2017**, *7*, 395. [CrossRef]
8. Pretzsch, H. The Course of Tree Growth. Theory and Reality. *For. Ecol. Manag.* **2020**, *478*, 118508. [CrossRef]
9. Chivulescu, S.; Leca, S.; Silaghi, D.; Cristea, V. Structural Biodiversity and Dead Wood in Virgin Forests from Eastern Carpathians. *J. Agric. For.* **2018**, *64*. [CrossRef]
10. Grantham, H.S.; Duncan, A.; Evans, T.D.; Jones, K.R.; Beyer, H.L.; Schuster, R.; Walston, J.; Ray, J.C.; Robinson, J.G.; Callow, M.; et al. Anthropogenic Modification of Forests Means Only 40% of Remaining Forests Have High Ecosystem Integrity. *Nat. Commun.* **2020**, *11*, 5978. [CrossRef]
11. Vrška, T.; Adam, D.; Hort, L.; Kolář, T.; Janík, D. European Beech (*Fagus Sylvatica* L.) and Silver Fir (*Abies Alba* Mill.) Rotation in the Carpathians—A Developmental Cycle or a Linear Trend Induced by Man? *For. Ecol. Manag.* **2009**, *258*, 347–356. [CrossRef]
12. Şofletea, N.; Curtu, L. *Dendrologie*; Pentru Viaţă: Braşov, Romania, 2001; Volume II, ISBN 978-973-99456-1-5.
13. Bosela, M.; Lukac, M.; Castagneri, D.; Sedmák, R.; Biber, P.; Carrer, M.; Konôpka, B.; Nola, P.; Nagel, T.A.; Popa, I.; et al. Contrasting Effects of Environmental Change on the Radial Growth of Co-Occurring Beech and Fir Trees across Europe. *Sci. Total Environ.* **2018**, *615*, 1460–1469. [CrossRef]
14. Primicia, I.; Camarero, J.J.; Janda, P.; Čada, V.; Morrissey, R.C.; Trotsiuk, V.; Bače, R.; Teodosiu, M.; Svoboda, M. Age, Competition, Disturbance and Elevation Effects on Tree and Stand Growth Response of Primary Picea Abies Forest to Climate. *For. Ecol. Manag.* **2015**, *354*, 77–86. [CrossRef]
15. Griess, V.C.; Acevedo, R.; Härtl, F.; Staupendahl, K.; Knoke, T. Does Mixing Tree Species Enhance Stand Resistance against Natural Hazards? A Case Study for Spruce. *For. Ecol. Manag.* **2012**, *267*, 284–296. [CrossRef]
16. Pretzsch, H.; Schütze, G.; Uhl, E. Resistance of European Tree Species to Drought Stress in Mixed *versus* Pure Forests: Evidence of Stress Release by Inter-Specific Facilitation: Drought Stress Release by Inter-Specific Facilitation. *Plant Biol.* **2013**, *15*, 483–495. [CrossRef] [PubMed]
17. Magh, R.-K.; Bonn, B.; Grote, R.; Burzlaff, T.; Pfautsch, S.; Rennenberg, H. Drought Superimposes the Positive Effect of Silver Fir on Water Relations of European Beech in Mature Forest Stands. *Forests* **2019**, *10*, 897. [CrossRef]
18. MacPherson, D.M.; Lieffers, V.J.; Blenis, P.V. Productivity of Aspen Stands with and without a Spruce Understory in Alberta's Boreal Mixedwood Forests. *For. Chron.* **2001**, *77*, 351–356. [CrossRef]
19. Man, R.; Lieffers, V.J. Are Mixtures of Aspen and White Spruce More Productive than Single Species Stands? *For. Chron.* **1999**, *75*, 505–513. [CrossRef]
20. Laiho, O.; Lahde, E.; Pukkala, T. Uneven- vs Even-Aged Management in Finnish Boreal Forests. *Forestry* **2011**, *84*, 547–556. [CrossRef]
21. Barsoum, N.; Coote, L.; Eycott, A.E.; Fuller, L.; Kiewitt, A.; Davies, R.G. Diversity, Functional Structure and Functional Redundancy of Woodland Plant Communities: How Do Mixed Tree Species Plantations Compare with Monocultures? *For. Ecol. Manag.* **2016**, *382*, 244–256. [CrossRef]
22. Petritan, I.C.; Commarmot, B.; Hobi, M.L.; Petritan, A.M.; Bigler, C.; Abrudan, I.V.; Rigling, A. Structural Patterns of Beech and Silver Fir Suggest Stability and Resilience of the Virgin Forest Sinca in the Southern Carpathians, Romania. *For. Ecol. Manag.* **2015**, *356*, 184–195. [CrossRef]
23. Chivulescu, S.; Ciceu, A.; Leca, S.; Apostol, B.; Popescu, O.; Badea, O. Development Phases and Structural Characteristics of the Penteleu-Viforta Virgin Forest in the Curvature Carpathians. *IForest Biogeosci. For.* **2020**, *13*, 389–395. [CrossRef]
24. Mirtl, M. Introducing the Next Generation of Ecosystem Research in Europe: LTER-Europe's Multi-Functional and Multi-Scale Approach. In *Long-Term Ecological Research*; Müller, F., Baessler, C., Schubert, H., Klotz, S., Eds.; Springer: Dordrecht, The Netherlands, 2010; pp. 75–93, ISBN 978-90-481-8781-2.
25. Parviainen, J. Virgin and Natural Forests in the Temperate Zone of Europe. *For. Snow Landsc. Res.* **2005**, *79*, 9–18.
26. Dănescu, A.; Albrecht, A.T.; Bauhus, J. Structural Diversity Promotes Productivity of Mixed, Uneven-Aged Forests in Southwestern Germany. *Oecologia* **2016**, *182*, 319–333. [CrossRef] [PubMed]

27. Vacek, S.; Vacek, Z.; Podrázský, V.; Bílek, L.; Bulušek, D.; Štefančík, I.; Remeš, J.; Štícha, V.; Ambrož, R. Structural Diversity of Autochthonous Beech Forests in Broumovské Stěny National Nature Reserve, Czech Republic. *Austrian J. For. Sci.* **2014**, *131*, 191–214.
28. Duveneck, M.J.; Scheller, R.M. Measuring and Managing Resistance and Resilience under Climate Change in Northern Great Lake Forests (USA). *Landsc. Ecol.* **2016**, *31*, 669–686. [CrossRef]
29. Curovic, M.; Spalevic, V.; Sestras, P.; Motta, R.; Dan, C.; Garbarino, M.; Vitali, M.; Urbinati, C. Structural and Ecological Characteristics of Mixed Broadleaved Old-Growth Forest (Biogradska Gora-Montenegro). *Turk. J. Agric. For.* **2020**, *44*, 428–438. [CrossRef]
30. Bouriaud, L.; Bouriaud, O.; Elkin, C.; Temperli, C.; Reyer, C.; Duduman, G.; Barnoaiea, I.; Nichiforel, L.; Zimmermann, N.; Bugmann, H. Age-Class Disequilibrium as an Opportunity for Adaptive Forest Management in the Carpathian Mountains, Romania. *Reg. Environ. Chang.* **2015**, *15*, 1557–1568. [CrossRef]
31. Pretzsch, H.; Hilmers, T.; Uhl, E.; Bielak, K.; Bosela, M.; del Rio, M.; Dobor, L.; Forrester, D.I.; Nagel, T.A.; Pach, M.; et al. European Beech Stem Diameter Grows Better in Mixed than in Mono-Specific Stands at the Edge of Its Distribution in Mountain Forests. *Eur. J. For. Res.* **2020**. [CrossRef]
32. Piovesan, G.; Biondi, F.; Filippo, A.D.; Alessandrini, A.; Maugeri, M. Drought-Driven Growth Reduction in Old Beech (*Fagus Sylvatica* L.) Forests of the Central Apennines, Italy. *Glob. Chang. Biol.* **2008**, *14*, 1265–1281. [CrossRef]
33. Schwarz, J.A.; Bauhus, J. Benefits of Mixtures on Growth Performance of Silver Fir (Abies Alba) and European Beech (*Fagus Sylvatica*) Increase with Tree Size Without Reducing Drought Tolerance. *Front. For. Glob. Chang.* **2019**, *2*, 79. [CrossRef]
34. Scheller, R.M.; Domingo, J.B.; Sturtevant, B.R.; Williams, J.S.; Rudy, A.; Gustafson, E.J.; Mladenoff, D.J. Design, Development, and Application of LANDIS-II, a Spatial Landscape Simulation Model with Flexible Temporal and Spatial Resolution. *Ecol. Model.* **2007**, *201*, 409–419. [CrossRef]
35. Aber, J.D.; Ollinger, S.V.; Federer, C.A.; Reich, P.B.; Goulden, M.L.; Sicklighter, D.W.; Melillo, J.M.; Lathrop, R.G. Predicting the Effects of Climate Change on Water Yield and Forest Production in the Northeastern United States. *Clim. Res.* **1995**, *5*, 207–222. [CrossRef]
36. Gustafson, E.J.; Sturtevant, B.R.; de Bruijn, A.M.G.; Lichti, N.; Jacobs, D.F.; Kashian, D.M.; Miranda, B.R.; Townsend, P.A. Forecasting Effects of Tree Species Reintroduction Strategies on Carbon Stocks in a Future without Historical Analog. *Glob. Chang. Biol.* **2018**, *24*, 5500–5517. [CrossRef] [PubMed]
37. Bishop, C.M. *Pattern Recognition and Machine Learning*; Information Science and Statistics; Springer: New York, NY, USA, 2006; ISBN 978-0-387-31073-2.
38. Scrucca, L. On Some Extensions to GA Package: Hybrid Optimisation, Parallelisation and Islands Evolution. *R J.* **2017**, *9*, 187–206. [CrossRef]
39. Scrucca, L. GA: A Package for Genetic Algorithms in R. *J. Stat. Softw.* **2013**, *53*, 1–37. [CrossRef]
40. Fritts, H.C. *Tree Rings and Climate*; Academic Press: London, UK; New York, NY, USA, 1976; ISBN 978-0-12-268450-0.
41. Briffa, K.R.; Jones, P.D.; Wigley, T.M.; Pilcher, J.R.; Baillie, M.G.L. Climate Reconstruction from Tree Rings: Part 1, Basic Methodology and Preliminary Results for England. *J. Climatol.* **1983**, *3*, 233–242. [CrossRef]
42. Cook, E.R.; Johnson, A.H.; Blasing, T.J. Forest Decline: Modeling the Effect of Climate in Tree Rings12. *Tree Physiol.* **1987**, *3*, 27–40. [CrossRef]
43. Pauling, A.; Luterbacher, J.; Casty, C.; Wanner, H. Five Hundred Years of Gridded High-Resolution Precipitation Reconstructions over Europe and the Connection to Large-Scale Circulation. *Clim. Dyn.* **2006**, *26*, 387–405. [CrossRef]
44. Popa, I. *Fundamente Metodologice Şi Aplicaţii de Dendrocronologie*; Editura Tehnică Silvică, 2004.
45. Hasenauer, H. Dimensional Relationships of Open-Grown Trees in Austria. *For. Ecol. Manag.* **1997**, *96*, 197–206. [CrossRef]
46. Kuemmerle, T.; Chaskovskyy, O.; Knorn, J.; Radeloff, V.C.; Kruhlov, I.; Keeton, W.S.; Hostert, P. Forest Cover Change and Illegal Logging in the Ukrainian Carpathians in the Transition Period from 1988 to 2007. *Remote Sens. Environ.* **2009**, *113*, 1194–1207. [CrossRef]
47. Sidor, C.G.; Popa, I.; Vlad, R.; Roibu, C.C. *Rețeaua Națională de Serii Dendrocronologice Pentru Pinul Silvestru (Pinus Sylvestris) Din România-PIDECRO*; Editura Silvica: Voluntari, Romania, 2020; ISBN 978-606-8020-73-0.
48. ICAS. *Amenajamentul. O.S. Penteleu*; Regia Naţională a pădurilor Romsilva; Institutul de Cercetări şi amenajări Silvice: Bucureşti, Romania, 2012.
49. INCDS. *Amenajamentul. O.S. Penteleu*; Regia Naţională a pădurilor Romsilva; Institutul de Cercetări şi amenajări Silvice: Bucureşti, Romania, 2021.
50. Cristea, V.C. Dendrometrics and Auxological Identification and Characterization of Forests with Natural and Semi-Natural Structure from Penteleu Massif, Transilvania. Ph.D. Thesis, University of Brașov, Brașov, Romania, 2019.
51. Van Oldenborgh, G.J.; Drijfhout, S.; van Ulden, A.; Haarsma, R.; Sterl, A.; Severijns, C.; Hazeleger, W.; Dijkstra, H. Western Europe Is Warming Much Faster than Expected. *Clim. Past* **2009**, *5*, 1–12. [CrossRef]
52. Luterbacher, J. European Seasonal and Annual Temperature Variability, Trends, and Extremes since 1500. *Science* **2004**, *303*, 1499–1503. [CrossRef]
53. Zorita, E.; von Storch, H. The Analog Method as a Simple Statistical Downscaling Technique: Comparison with More Complicated Methods. *J. Clim.* **1999**, *12*, 2474–2489. [CrossRef]

54. Fick, S.E.; Hijmans, R.J. WorldClim 2: New 1-Km Spatial Resolution Climate Surfaces for Global Land Areas. *Int. J. Climatol.* **2017**, *37*, 4302–4315. [CrossRef]
55. Giurgiu, V.; Decei, I.; Drăghiciu, D. *Metode Şi Tabele Dendrometrice [Methods and Dendrometric Tables]*; Ceres Publishing House, Bucharest: Bucharest, Romania, 2004; ISBN 973-40-0639-8.
56. Larson, L.A. Online User Manual for Cybis CooRecorder and CDendro Programs. Available online: http://www.cybis.se/forfun/dendro/ (accessed on 1 July 2020).
57. Holmes, R.L. Computer-Assisted Quality Control in Tree-Ring Dating and Measurement. *Tree-Ring Bull.* **1983**, *3*, 69–78.
58. Cook, B.D.; Krusic, P.J. *ARSTAN v. 4.1d: A Tree-Ring Standardization Program Based on Detrending and Autoregressive Time Series Modeling, with Interactive Graphics*; Tree-Ring Laboratory, Lamont-Doherty Earth Observatory of Columbia University: Palisades, NY, USA, 2005.
59. R Core Team. *R: A Language and Environment for Statistical Computing*; R Foundation for Statistical Computing: Vienna, Austria, 2018.
60. Bunn, A.G. A Dendrochronology Program Library in R (DplR). *Dendrochronologia* **2008**, *26*, 115–124. [CrossRef]
61. Zang, C.; Biondi, F. Treeclim: An R Package for the Numerical Calibration of Proxy-Climate Relationships. *Ecography* **2015**, *38*, 431–436. [CrossRef]
62. Whitcher, B. Waveslim: Basic Wavelet Routines for One-, Two-, and Three-Dimensional Signal Processing; R Package Version 1.8.2. Available online: https://cran.r-project.org/web/packages/waveslim/ (accessed on 1 January 2021).
63. Gouhier, T.C.; Grinsted, A.; Simko, V. R Package Biwavelet: Conduct Univariate and Bivariate Wavelet Analyses. Available online: https://cran.r-project.org/web/packages/biwavelet/ (accessed on 1 January 2021).
64. Duveneck, M.J.; Scheller, R.M.; White, M.A. Effects of Alternative Forest Management on Biomass and Species Diversity in the Face of Climate Change in the Northern Great Lakes Region (USA). *Can. J. For. Res.* **2014**, *44*, 700–710. [CrossRef]
65. Scheller, R.M.; Mladenoff, D.J. A Forest Growth and Biomass Module for a Landscape Simulation Model, LANDIS: Design, Validation, and Application. *Model. Disturb. Succession For. Landsc. Using LANDIS* **2004**, *180*, 211–229. [CrossRef]
66. Gustafson, E.J.; De Bruijn, A.M.G.; Pangle, R.E.; Limousin, J.-M.; McDowell, N.G.; Pockman, W.T.; Sturtevant, B.R.; Muss, J.D.; Kubiske, M.E. Integrating Ecophysiology and Forest Landscape Models to Improve Projections of Drought Effects under Climate Change. *Glob. Chang. Biol.* **2015**, *21*, 843–856. [CrossRef]
67. Gustafson, E.J.; Shifley, S.R.; Mladenoff, D.J.; Nimerfro, K.K.; He, H.S. Spatial Simulation of Forest Succession and Timber Harvesting Using LANDIS. *Can. J. For. Res.* **2000**, *30*, 32–43. [CrossRef]
68. Kattge, J.; Díaz, S.; Lavorel, S.; Prentice, I.C.; Leadley, P.; Bönisch, G.; Garnier, E.; Westoby, M.; Reich, P.B.; Wright, I.J.; et al. TRY–A Global Database of Plant Traits. *Glob. Chang. Biol.* **2011**, *17*, 2905–2935. [CrossRef]
69. Vittoz, P.; Engler, R. Seed Dispersal Distances: A Typology Based on Dispersal Modes and Plant Traits. *Bot. Helv.* **2007**, *117*, 109–124. [CrossRef]
70. Giurgiu, V.; Draghiciu, D. *Modele Matematico-Auxologice Si Tabele de Productie Pentru Arborete [Mathematic-Auxologic Models and Yield Tables for Forest Stands]*; Ceres Publishing House: Bucharest, Romania, 2004; ISBN 973-40-0637-1.
71. Tanase, M.A.; Villard, L.; Pitar, D.; Apostol, B.; Petrila, M.; Chivulescu, S.; Leca, S.; Borlaf-Mena, I.; Pascu, I.-S.; Dobre, A.-C.; et al. Synthetic Aperture Radar Sensitivity to Forest Changes: A Simulations-Based Study for the Romanian Forests. *Sci. Total Environ.* **2019**, *689*, 1104–1114. [CrossRef]
72. Schmitt, L.M. Theory of Genetic Algorithms. *Theor. Comput. Sci.* **2001**, *259*, 1–61. [CrossRef]
73. Bivand, R.; Rundel, C. Rgeos: Interface to Geometry Engine-Open Source ('GEOS'). Available online: https://cran.r-project.org/web/packages/rgeos/ (accessed on 1 January 2021).
74. Hijmans, R.J. *Raster: Geographic Data Analysis and Modeling*; R Foundation for Statistical Computing: Vienna, Austria, 2016.
75. Bivand, R.; Keitt, T.; Rowlingson, B. Rgdal: Bindings for the "Geospatial" Data Abstraction Library. Available online: https://cran.r-project.org/web/packages/rgdal/ (accessed on 1 January 2021).
76. Bivand, R.; Lewin-Koh, N. Maptools: Tools for Reading and Handling Spatial Objects. Available online: https://cran.r-project.org/web/packages/maptools/ (accessed on 1 January 2021).
77. White, T.; van der Ende, J.; Nichols, T.E. Beyond Bonferroni Revisited: Concerns over Inflated False Positive Research Findings in the Fields of Conservation Genetics, Biology, and Medicine. *Conserv. Genet.* **2019**, *20*, 927–937. [CrossRef]
78. Pinheiro, J.; Bates, D.; DebRoy, S.; Sarkar, D.; R Core Team. Nlme: Linear and Nonlinear Mixed Effects Models. Available online: https://cran.r-project.org/web/packages/nlme/ (accessed on 1 January 2021).
79. Hothorn, T.; Bretz, F.; Westfall, P. Simultaneous Inference in General Parametric Models. *Biom. J.* **2008**, *50*, 346–363. [CrossRef]
80. Pretzsch, H.; Dieler, J. The Dependency of the Size-Growth Relationship of Norway Spruce (*Picea Abies* [L.] Karst.) and European Beech (*Fagus Sylvatica* [L.]) in Forest Stands on Long-Term Site Conditions, Drought Events, and Ozone Stress. *Trees* **2011**, *25*, 355–369. [CrossRef]
81. Bose, A.K.; Weiskittel, A.; Wagner, R.G. Occurrence, Pattern of Change, and Factors Associated with American Beech-Dominance in Stands of the Northeastern USA Forest. *For. Ecol. Manag.* **2017**, *392*, 202–212. [CrossRef]
82. Čater, M.; Levanič, T. Beech and Silver Fir's Response along the Balkan's Latitudinal Gradient. *Sci. Rep.* **2019**, *9*, 16269. [CrossRef] [PubMed]
83. Costea, C. *Codrul Grădinărit*; Editura Agro-silvica: Bucharest, Romania, 1962.
84. Čater, M.; Kobler, A. Light Response of *Fagus Sylvatica* L. and *Abies Alba* Mill. in Different Categories of Forest Edge–Vertical Abundance in Two Silvicultural Systems. *For. Ecol. Manag.* **2017**, *391*, 417–426. [CrossRef]

85. Töchterle, P.; Yang, F.; Rehschuh, S.; Rehschuh, R.; Ruehr, N.K.; Rennenberg, H.; Dannenmann, M. Hydraulic Water Redistribution by Silver Fir (*Abies Alba* Mill.) Occurring under Severe Soil Drought. *Forests* **2020**, *11*, 162. [CrossRef]
86. Garcia-Duro, J.; Manzoni, L.; Arias, I.; Casal, M.; Cruz, O.; Pesqueira Xose, M.; Munoz, A.; Alvarez, R.; Mariot, L.; Bandini, S.; et al. *Hidden Costs of Modelling Post-Fire Plant Community Assembly Using Cellular Automata*; Springer: Berlin/Heidelberg, Germany, 2018.
87. García-Duro, J.; Cruz, O.; Casal, M.; Reyes, O. Fire as Driver of the Expansion of Paraserianthes Lophantha (Willd.) I. C. Nielsen in SW Europe. *Biol. Invasions* **2019**, *21*, 1427–1438. [CrossRef]
88. García-Duro, J.; Álvarez, R.; Basanta, M.; Casal, M. Aplicación de redes bayesianas ás relacións entre especies vexetais despois de incendio forestal e a súa sensibilidade ó tamaño das unidades de mostraxe. In *BIOapps2016. Encontro Galaico-Portugués de Biometría, con Aplicación ás Ciencias da Saúde, á Ecoloxía e ás Ciencias do Medio Ambiente*; Ginzo, M.J., Pértega, S., Santiago Pérez, M.I., Eds.; SGAPEIO & SPE: Santiago de Compostela, Spain, 2016.
89. Pausas, J.G.; Fernández-Muñoz, S. Fire Regime Changes in the Western Mediterranean Basin: From Fuel-Limited to Drought-Driven Fire Regime. *Clim. Chang.* **2012**, *110*, 215–226. [CrossRef]
90. Hubau, W.; Lewis, S.L.; Phillips, O.L.; Affum-Baffoe, K.; Beeckman, H.; Cuní-Sanchez, A.; Daniels, A.K.; Ewango, C.E.N.; Fauset, S.; Mukinzi, J.M.; et al. Asynchronous Carbon Sink Saturation in African and Amazonian Tropical Forests. *Nature* **2020**, *579*, 80–87. [CrossRef] [PubMed]

 forests

MDPI

Article

Shifts in Forest Species Composition and Abundance under Climate Change Scenarios in Southern Carpathian Romanian Temperate Forests

Juan García-Duro [1,*], Albert Ciceu [1,2], Serban Chivulescu [1], Ovidiu Badea [1,2], Mihai A. Tanase [1,3] and Cristina Aponte [1,4]

[1] National Institute for Research and Development in Forestry "Marin Drăcea", 077191 Voluntari, Ilfov, Romania; albert.ciceu@icas.ro (A.C.); serban.chivulescu@gmail.com (S.C.); ovidiu.badea63@gmail.com (O.B.); mihai@tma.ro (M.A.T.); cristinaaponte@gmail.com (C.A.)
[2] Faculty of Silviculture and Forest Engineering, "Transilvania" University of Brașov, 500123 Brașov, Romania
[3] Department of Geology, Geography and Environment, University of Alcala, 28802 Alcala de Henares, Spain
[4] Department of Environment and Agronomy, Centro Nacional Instituto de Investigación y Tecnología Agraria y Alimentaria, INIA-CSIC, 28040 Madrid, Spain
* Correspondence: juan.garcia.duro@icas.ro

Citation: García-Duro, J.; Ciceu, A.; Chivulescu, S.; Badea, O.; Tanase, M.A.; Aponte, C. Shifts in Forest Species Composition and Abundance under Climate Change Scenarios in Southern Carpathian Romanian Temperate Forests. *Forests* **2021**, *12*, 1434. https://doi.org/10.3390/f12111434

Academic Editor: Daniel J. Johnson

Received: 22 September 2021
Accepted: 18 October 2021
Published: 21 October 2021

Abstract: The structure and functioning of temperate forests are shifting due to changes in climate. Foreseeing the trajectory of such changes is critical to implementing adequate management practices and defining long-term strategies. This study investigated future shifts in temperate forest species composition and abundance expected to occur due to climate change. It also identified the ecological mechanisms underpinning such changes. Using an altitudinal gradient in the Romanian Carpathian temperate forests encompassing several vegetation types, we explored forest change using the Landis-II landscape model coupled with the PnET ecophysiological process model. We specifically assessed the change in biomass, forest production, species composition and natural disturbance impacts under three climate change scenarios, namely, RCP 2.6, 4.5 and 8.5. The results show that, over the short term (15 years), biomass across all forest types in the altitudinal gradient will increase, and species composition will remain unaltered. In contrast, over the medium and long terms (after 2040), changes in species composition will accelerate, with some species spreading (e.g., *Abies alba* Mill.) and others declining (e.g., *Fagus sylvatica* L.), particularly under the most extreme climate change scenario. Some forest types (e.g., *Picea abies* (L.) karst forests) in the Southern Carpathians will notably increase their standing biomass due to climate change, compared to other types, such as *Quercus* forests. Our findings suggest that climate change will alter the forest composition and species abundance, with some forests being particularly vulnerable to climate change, e.g., *F. sylvatica* forests. As far as productivity and forest composition changes are concerned, management practices should accommodate the new conditions in order to mitigate climate change impacts.

Keywords: LANDIS-II; PnET; climate change; Southern Carpathians; forest biomass; production; species composition; species abundance; Romanian temperate forests

1. Introduction

Human-induced climate change is one of the major processes affecting the global environment nowadays [1]. However, the impacts are so complex and diverse that the net effect of climate change on forest systems is still uncertain. For instance, while the increase in atmospheric CO_2 concentrations often leads to greater productivity [2,3], the aridity caused by warming [4,5] and the intensification of disturbances regimes [6] usually have a negative impact on forests' structure and productivity. Recent studies have suggested that despite the positive climate change-associated effects [7,8], the negative ones often prevail [9,10].

Climate change effects, both positive and negative, will trigger quantitative and qualitative changes in forest composition, structure and functioning and will push for species adaptive responses [1,11]. Plant responses to climate change will be species specific [4], though species with similar vital attributes are expected to have close responses [5,12]. Thus, angiosperms and gymnosperms/conifers have different ecophysiological responses to raised CO_2 concentrations, potentially affecting the regional response of each forest type [12].

Temperate forests occupy 1097 M ha worldwide, retain 118.6 Pg of terrestrial carbon [13] and provide 40% of the world's forest harvest [14], thus playing an important role in the overall carbon balance. Shifts in temperate forests' structure, functioning and distribution, such as those driven by climate change and mitigation measures, will affect their economic value [15] and will have notable consequences on their capacity to sequester carbon and to provide other ecosystem services [10,16–19]. Forest carbon stocks and sequestration primarily depend on forest productivity and disturbance regimes [10,14], both of which are expected to change with climate change [6,18]. A study by Nabuurs et al. [19] described the first signs of carbon sink saturation in European forests. Such weakening in C sink/sequestration capacity may be related to aging forests, decreased summer humidity, etc. A lower C sequestration capacity would also be related to the change in disturbance regimes. The main disturbances in temperate forests include drought, insects, windthrows, pathogens and fire [6]. However, disturbances are often interconnected and can generate significant feedback effects [6]. This is, for instance, the case of bark beetle attacks, which are secondary disturbances that tend to occur after drought, harvest or windthrows.

Other studies have already suggested that some temperate forests will eventually decline due to summer droughts [20] and their habitat distribution will be reduced to montane areas due to altitudinal shifts [21]. Some temperate forests will be more intensely affected than others [6,10,16], with vulnerability being related to species composition [16], health status, disturbances regimes [22] and the phytogeographical context [23]. In this context, the evolution of European temperate forests under climate change is largely uncertain, with studies suggesting that conifers will be severely damaged by climate change [16], while beech (*Fagus sylvatica* L.) forests, already affected by drought in recent decades [20], will be more sensitive to climate change. Given the gaps in the knowledge and the relevance of mixed (broadleaf and conifer) forests [24–26], such discrepancies require further research.

This study aimed to unveil the future effects of climate change and its interactions with natural disturbances and land management on the structure and composition of temperate forests by implementing a forest simulation modeling approach. The Southern Carpathian forests were used as a case study as they harbor a large diversity of forest types and are representative of European temperate forests. The forest simulation model LANDIS-II [27] coupled with the PnET model [28,29], which have been highlighted for their capacity to model complex forest dynamics under multiple interacting drivers (e.g., climate, management, pests, windthrows) [30–33], was used to increase our understanding of the potential impact of natural disturbances on temperate forest composition, structure and its productivity. In particular, this study addressed two main questions: (1) What future changes in forest species composition and abundance are expected to occur due to climate change? (2) What are the ecological mechanisms underpinning forest type and species (variable effects—geographical and species dependent) changes?

We hypothesized that species in the lower part of the altitudinal gradient (e.g., oaks) will grow in abundance and those in the high altitudes, particularly conifers, will be constrained, and that the main mechanisms driving the change are warmer climate and increments in droughts and disturbances, all of them modifying species interactions.

2. Materials and Methods

2.1. Study Area

The study area is located in the Fagaraş Mountains, in the Romanian Southern Carpathians (Figure 1). It occupies around 6400 km^2 and comprises a large geographical gradient that extends 110 km north to south and 73 km east to west, covering the main forested lands in the Argeş, Sibiu and Braşov counties. The altitude ranges from 185 to 2544 m.a.s.l, with plain landscapes (c. 12% average slope, elevation < 500 m) on the southern sector and steep and rough topography (c. 32% average slope, elevation > 1000 m) on the high-altitude northern sector. The climate varies along the altitudinal gradient [23]: the annual mean temperature in the highlands is lower than 4 °C, with a winter mean temperature below −5 °C and a summer temperature around 14 °C. The precipitation approaches 1000 mm, with summer rainfall of around 315 mm. The climate in the plains is warmer and drier, with a 10.5 °C annual mean temperature, mean winter temperatures below 0 °C and summer temperatures over 20 °C. The annual precipitation here is about 700 mm, with summer rainfall of around 260 mm. In general, September and October are the driest months, while May and June are the wettest ones. Soil types in the area include protosols, spodozols, cambisols, argiluvisols and hydromorphic soils with pseudogleic properties [34,35].

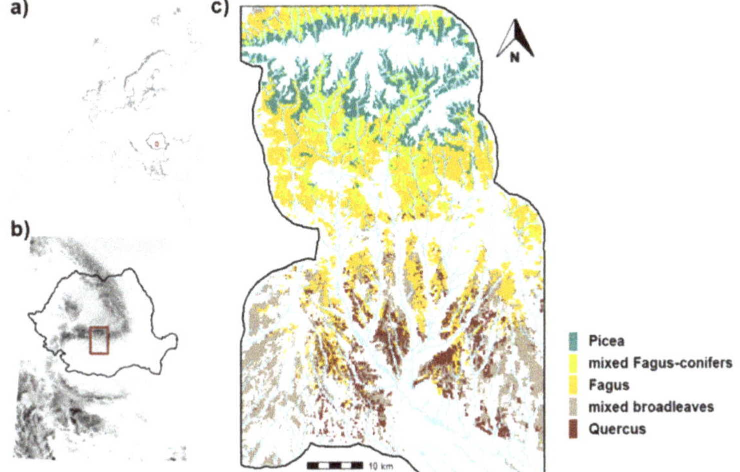

Figure 1. (**a**) Location of the study area within Europe and (**b**) the Carpathians; (**c**) distribution of the five forest types within the study area.

Temperate forests, whose composition changes across the altitudinal belts, cover 52% of the total study area. There are five general forest types: (a) *Picea abies* (L.) karst forests, (b) mixed *Fagus sylvatica* L.–conifer forests, (c) *F. sylvatica* forests, (d) mixed broadleaved forests and (e) *Quercus* forests. *Picea* forests (Norway spruce) occupy the subalpine and montane superior belts. Mixed *Fagus*–conifer forests, mainly *Picea* and *Abies alba* Mill., dominate the intermediate montane belt, with the beech abundance increasing as the altitude decreases. Mixed *Fagus*–conifer forests are substituted by pure *Fagus* forests first and mixed broadleaved forest (*F. sylvatica* and *Carpinus betulus* L.) later, both of them in the inferior montane belt [24,36]. Oaks gradually appear in mixed broadleaved forests in the colline belt and, in low-altitude areas, reach the point where *Quercus* forests dominate the landscape. Within *Quercus* forests, there is an altitudinal transition from *Quercus petraea* (Matt.) Liebl. to *Quercus robur* L., *Quercus cerris* L. and *Quercus frainetto* Ten. in the most southern thermophilus areas [36].

Some of the forested lands are actively managed for timber production, with *F. sylvatica* providing 34% of the overall standing wood volume, *P. abies* 28%, *Quercus* spp. 17%, *A. alba* 5% and other species the remaining 16% [37–46]. Depending on forest species and conditions, different management systems are implemented including tree selection, shelterwood and clearcutting. Clearcutting is practiced only for small areas (lower than 3 ha) in Norway spruce (*P. abies*) and non-natural forest stands [47,48]. The rest of the forested land is either subjected to conservation management to protect forest health (e.g., phytosanitary felling, trees affected by small local windthrows) or is strictly protected with no management actions allowed. The most common natural disturbances in the area include windthrows and insect attacks [49]. Windthrow events mostly affect conifers in the north sector of the study area [50], particularly *Picea* forests, with insect outbreaks (*Hylobius abietis* L., *Ips duplicatus* Sahlberg and *I. typographus* L.) being a common secondary disturbance following windthrows [51] and drought [49,52] in conifer forests.

2.2. Landis-II Model

Landis-II is a collection of spatially explicit forest landscape models [53] that simulate forest change as a function of succession and disturbances [33]. The landscape in Landis-II is defined as a grid of cells (here 200 × 200 m), each of which belongs to an ecoregion (i.e., areas of homogeneous soil and climate) and can contain multiple species cohorts that can be independently killed by disturbances, competition or age-related mortality, as the succession progresses. Landis-II integrates a number of ecological process models through its modular design. Here, we used the PnET succession extension [54] to underpin tree species establishment, growth, mortality and decomposition. This extension embeds elements of the PnET ecophysiology model of [55] and accounts for competition for available light and water. Biomass growth is the result of a number of processes (e.g., photosynthesis, evapotranspiration) controlled by species ecophysiological parameters (e.g., foliar N concentration, photosynthetic rates) given a number of conditions that include precipitation, temperature and atmospheric CO_2 concentration. Climate change and CO_2 enrichment are interwoven as change in environmental parameters over time, making the PnET extension convenient for climate change modeling [30,56,57].

Landis-II interdependent disturbance extensions were used to model the impacts of harvest [31,58], wind [59] and insect outbreaks [60,61]. The harvest module [62] implements prescriptions in different management areas according to a temporal schedule, management systems and stand characteristics, including resource availability, age structure or stand composition. The biological disturbance agent module [63] implements insect impacts based on pest species preferences and resource availability. The wind module [64], which models windthrow events and operates independently of climate, was used to trigger insect outbreaks.

2.3. Landscape Design: Ecoregions and Forest Communities

The initial landscape for Landis-II simulations was built based on the local management plans that were available for the public forests (approximately 60% of the total forest land) [37–46]. Management plans contained the spatial delimitation of forest stands and information of their species composition and cohort ages, with stands ranging from 0.1 to 50 ha, the legal maximum allowed, with a median of 3.25 ha. Stands were classified into 14 ecoregions of homogeneous climate, soil type and tree species abundances (Supplementary Table S1; Supplementary Figure S1). Ecoregions were ascribed to one of the five forest types according to species composition (Figure 1). To constrain the number of communities, i.e., cohorts and species combinations that conform to a forest stand, a total of 223 initial communities, comprising from 1 to 4 species and a range of 1 to 17 age cohorts, were identified in the management plans and assigned, based on their frequency, to the corresponding stand and ecoregion in the simulated landscape. The area of private forest lands, for which information was not available, was delimited based on Corine Land Cover 2012 [65]. To ascribe ecoregions and initial communities in private forests, a random forest algorithm [66] was trained using information from the management plans and environmental variables (DEM, [67], climate [68], soil properties and classification [34] and distance to infrastructure and populations [69,70]). The overall accuracy of the prediction reached 96.7% over the independent testing dataset.

2.4. LANDIS-II Parameterization

The modeled species included *Abies alba* Mill. (silver fir), *Alnus glutinosa* (L.) Gaertn. (European alder), *Alnus incana* (L.) Moench (grey alder), *C. betulus* (European hornbeam), *F. sylvatica* (European beech), *P. abies* (Norway spruce), *Q. cerris* (Turkey oak), *Q. frainetto* Ten. (Hungarian oak), *Q. petraea* (Matt.) Liebl. (sessile oak) and *Q. robur* L. (pedunculate oak). Species contributing less than 2% to the overall wood volume in the study area were not included in the simulation. Tree species parameters required by the model were measured in the study area [71], compiled from unpublished data or obtained from the local Romanian literature [36,72,73] and other international sources [74,75]. Model parameterization was performed following the recommendations of [29]. The three main background disturbances of the studied area were modeled to increase the accuracy of the simulated landscape and were considered a key element in this study.

The wind disturbance extension required the windthrow return interval and severity, which were modeled/introduced based on national historical records [76]. The average wind speed [68] and the area occupied by conifer forests were used to determine the windthrow local impact on each ecoregion. The mortality of age cohorts was based on Popa [76]. It was assumed that wind event occurrence will not be affected by climate change.

Bark beetle outbreaks were linked to windthrows and were followed by harvesting interventions as per standard practices in the study area. Three main bark beetle species were modeled: *I. typographus* and *I. duplicatus*, which target mature *Picea* stands [49,77], and *H. abietis*, which targets stands with a high density of *A. alba* saplings [78]. Pest species preferences (host species and age cohort) and impacts were defined based on expert knowledge, forest health status monitoring [79–81] and national Romanian research studies [49,51,52,82].

Harvesting prescriptions were defined following the national Romanian regulatory framework [47,48,83,84] which establishes that logging cycles for the main species in the study area are between 100 and 200 years for beech, Norway spruce, sessile oak and silver fir, 70 to 160 years for other oaks and up to 80 years for broadleaved softwood species, depending on the silvicultural system, forest function, site productivity and other specific situations. The simulated landscape was divided into management areas, composed of multiple stands, that included unmanaged, strictly protected forest (<1%), protected forest with special conservation prescription practices (less than 10 m^3/ha, approximately 40%), selective logging (20%; with approximately half of the area dedicated to production of mixed beech–conifer, mixed beech–broadleaved and oak forests), shelterwood forests (28%; majority of Fagus, also the remaining mixed beech–conifer, mixed beech–broadleaved and oak forests) and clearcut forests (12%; mostly Norway spruce pure stands). Within each management area, management treatments (stand selection, harvesting periodicity, intensity, targeting species and cycle duration) were implemented based on stand composition and stand age.

2.5. Climate Change Scenarios

Three climate change scenarios were defined (Figure 2) based on the RCP 2.6, 4.5 and 8.5 emission scenarios [85]. For the model spin-up period (1901–2015), we used gridded climate data (precipitation, maximum and minimum temperature) from the Climate Research Unit Time Series (CRU TS) [86] high-resolution dataset. For the simulation period between 2015 and 2100, we used RCP projections from Climatic Data Generator (ClimGen) [87] based on the French National Centre for Meteorological Research CNRM-CN5 climate model [88]. For the simulation period between 2100 and 2140, for which there were no climate projections available, climate data were randomly resampled from the last 3 decades, the last climate normal period, of the corresponding CNRM-CN5 series. Climate series were extended to 2140 because of the species longevity and the long logging cycles used in Romanian forestry.

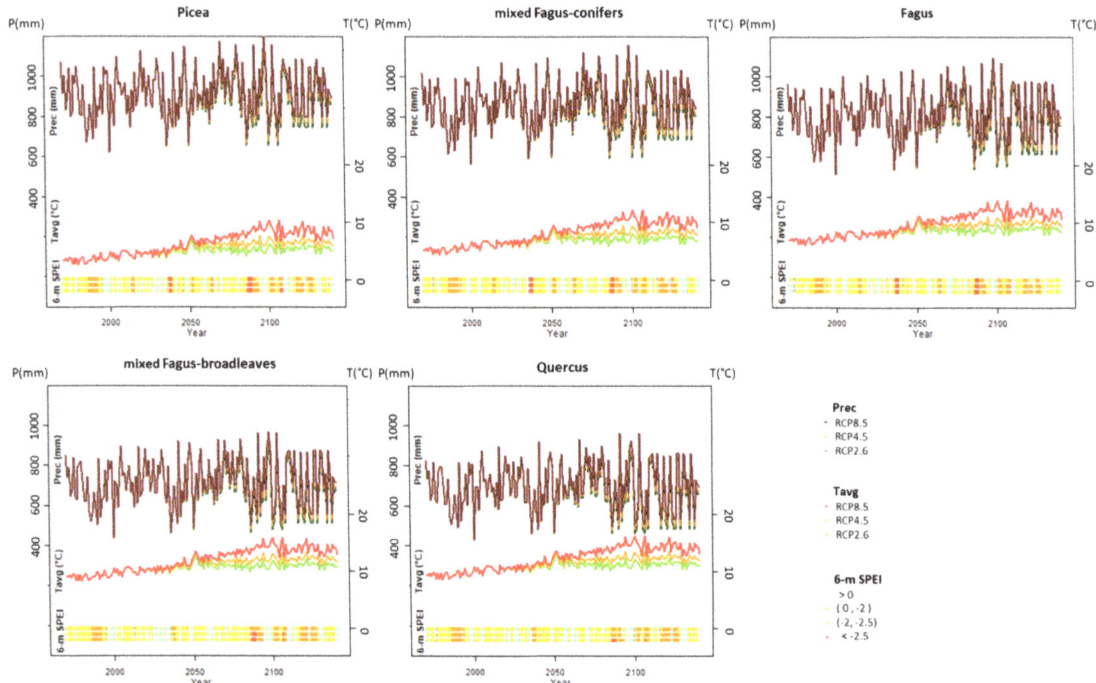

Figure 2. Comparison of climate change scenarios for the period 1970–2140 among the three RCP scenarios (RCP 2.6, RCP 4.5, RCP 8.5) in five main forest types representative of the altitudinal gradient (top to bottom). Precipitation (Prec), average temperature (Tavg) and standardized precipitation evapotranspiration index (SPEI index) (rows top to bottom represent RCP 2.6, RCP 4.5 and RCP 8.5, respectively).

Gridded 0.5° climate data were statistically downscaled following Moreno and Hasenauer [89] using a 30″ resolution gridded climate dataset from WorldClim 1.4 and 2.0 [68]. Mean annual CO_2 concentrations taken from van Vuuren et al. [85] differed among PCR scenarios but were assumed spatially constant. Similar to climate, CO_2 concentrations after 2100 were kept constant at the level of the last climate normal to prevent inconsistencies with resampled climate series. Photosynthetically active radiation (PAR), calculated from WorldClim 2.0 solar radiation [68], changed seasonally and spatially but did not differ across climate change scenarios.

2.6. Model Output and Validation

We simulated changes in forest biomass and species composition over a period of 125 years (2015–2140) under three climate change scenarios (RCP 2.6, RCP 4.5 and RCP 8.5), using 22 replicates of each scenario at a spatial resolution of the 200 m cell size. Model outputs included annual live, dead and harvested biomass throughout the simulation period. Forest net productivity was calculated as the change in biomass accounting for the harvesting. Model outputs for the period 2010–2020, when the management plans were released, were validated using Romanian forest yield tables [72,73] and forest management plans of the study area [37–46]. The simulated initial biomass per hectare (Figure 3) and the extracted timber volume were within the ranges recorded in the management plans for all five forest types. As expected, compared to management, background natural disturbances had minor effects on the current landscape. The simulated windthrow impact was in accordance with volumes estimated from Popa [76] and the area affected by intense windthrows during the period 1986–2016 [79–81]. Insect outbreak occurrence and impact were also in accordance with the area affected by intense insect attacks in the study area in the period 1986–2016 [79–81], and the harvested volumes were in agreement with the management plans [37–46].

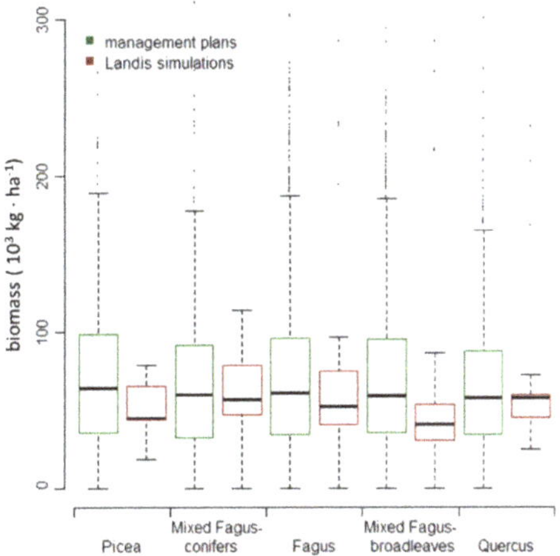

Figure 3. Boxplot diagram comparing the modeled biomass in the five main forest types in the initial Landis-II simulated landscape (year 2015) and the field-measured biomass data recorded from the management plans.

2.7. Statistical Analysis

Biomass results were spatially aggregated for the five main forest types. Differences across scenarios and forest types were assessed by linear mixed models (LME) [90], with changes over time fit to polynomial functions (up to 5 terms) and simulation replicates as a random factor. Natural disturbance outputs (i.e., mean damaged area, number of killed cohorts and severity) and harvested biomass were analyzed for the entire study area. Shifts in forest species composition at the landscape level were analyzed through the change in species biomass, using constrained redundancy analysis [91] with year and climate change scenario as constraining variables.

3. Results

3.1. Disturbances

Windthrows affected a minor percentage of the forest area (1%), with all of the effect located in the Norway spruce forest type. The affected area significantly increased over time ($p \leq 0.05$) such that by 2140, the area damaged was two-fold the area of 2015 (Figure 4a). A similar trend was observed in the severity and the number of cohorts killed by wind, with values increasing over time, and the number of cohorts killed being significantly lower for the RCP 8.5 scenario. These results are in accordance with the aging of Norway spruce forests, as the susceptibility to windthrows increases with tree age.

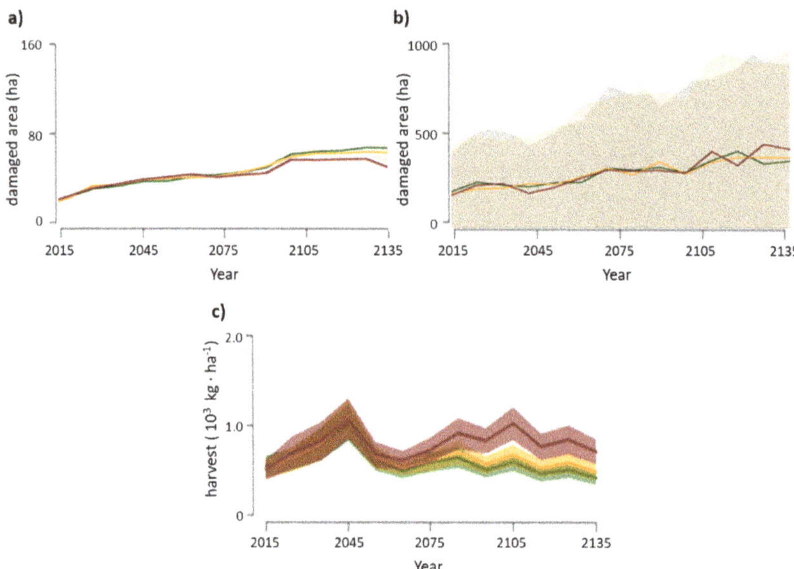

Figure 4. Variation in the average area damaged by (**a**) wind (**b**) and insects and (**c**) the harvested biomass across climate change scenarios. RCP 2.6 (green), RCP 4.5 (orange) and RCP 4.5 (red).

Insect outbreaks annually affected around 200 ha of the forest area (~1%), mostly in the spruce and mixed *Fagus* (beech)–conifer forests. The area and the number of cohorts killed by *Ips* insects, which target mature *P. abies* trees, increased over time ($p < 0.05$), also reflecting a change in the forest age structure. This trend was not observed for *H. abietis* (Figure 4b).

Harvesting occurred mostly in the *Fagus* and *Quercus* forest types, where annual means of 300,000 and 100,000 tons were extracted, respectively, with extraction rates of about 10 and 15% every 10 years (Figure 4c). Harvesting in the spruce and mixed beech–conifer forest types was mostly restricted to conservation works, with rates of extraction below 2%. Differences between climate change scenarios were noted from 2070, with biomass harvested in RCP 8.5 being greater, followed by RCP 4.5, and RCP 2.6, the lowest among all the scenarios. Given that harvesting prescriptions were constant over time, following the national Romanian regulatory framework, this indicates that differences in forest attributes developed over time among climate change scenarios.

3.2. Changes in Biomass across Climate Change Scenarios

Living aboveground biomass (Figure 5) tended to increase in all forest types and climate change scenarios throughout the simulation period, with fluctuations related to disturbance or harvesting events. Comparatively larger increases in biomass were observed for RPC 8.5, ranging from 21% in the *Quercus* forest type to 51% in the mixed beech–broadleaved forest type, than for RCP 4.5 (from 2% to 37%). The lowest change in biomass was observed in RCP 2.6, ranging from a decrease in total biomass (−15%) in the *Quercus* forest type to a 29% increase in the mixed beech–conifer forest type. The effect of climate varied across forest types: mixed beech–broadleaved forests showed the largest differences in the final biomass among scenarios, with RCP 8.5 showing a larger biomass than RCPs 4.5 and 2.6.

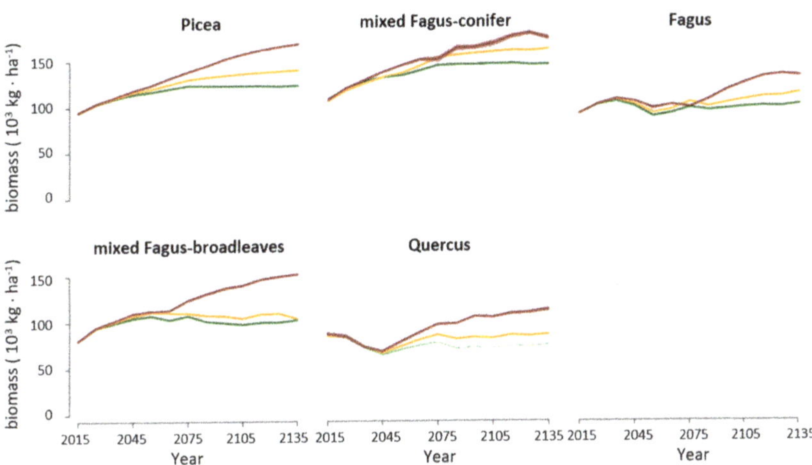

Figure 5. Mean (dashed line) and 95% confidence interval (shaded area) of the living biomass (-10^3 kg·ha^{-1}) measured in the forest belts over time under the three climate change scenarios: RCP 2.6 (green), RCP 4.5 (orange) and RCP 4.5 (red).

The analysis of the biomass at the species level revealed changes in species biomass through the simulation timespan and among climate change scenarios (Figure 6). The five main species, *P. abies*, *A. alba*, *F. sylvatica*, *Q. petraea* and *Q. frainetto*, showed significant biomass increments during the succession (*F. sylvatica* up to 2050) ($p < 0.05$). Of those five, all but *F. sylvatica* showed significant differences in biomass among the scenarios, with the largest biomass under RCP 8.5 ($p < 0.05$). Three species, *C. betulus*, *Q. petraea* and particularly *F. sylvatica*, showed short periods with strong reductions in biomass after 2050, which co-occurred in time with low SPEI values, although they tended to recover later.

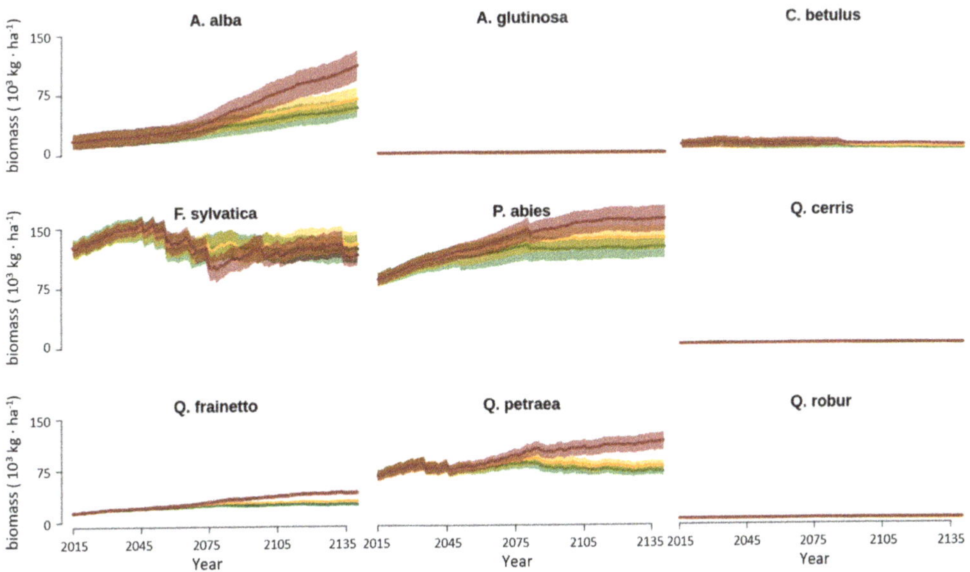

Figure 6. Mean (dashed line) and 95% confidence interval (shaded area) of the species living biomass ($\cdot10^3$ kg·ha^{-1}) measured over time under the three climate change scenarios: RCP 2.6 (green), RCP 4.5 (orange) and RCP 4.5 (red).

The percentage of dead wood biomass relative to the total biomass showed a slight increase in all the forest types over time (Figure 7). Whereas the living biomass of Norway spruce and *Quercus* spp. differed among scenarios, the RCP scenario did not affect the percentage of dead biomass in these forests, which also showed a trend with very small interdecadal variations. Forest types with participation of *F. sylvatica* were subjected to strong variations in dead biomass over time, usually within the range 2.5–5.0%, which is also related to fluctuations in living biomass (Figure 5) and a low 6-month SPEI (Figure 2). The mortality events after drought temporally altered the proportion of live and dead biomass and occurred with a higher intensity in the RCP 8.5 scenario.

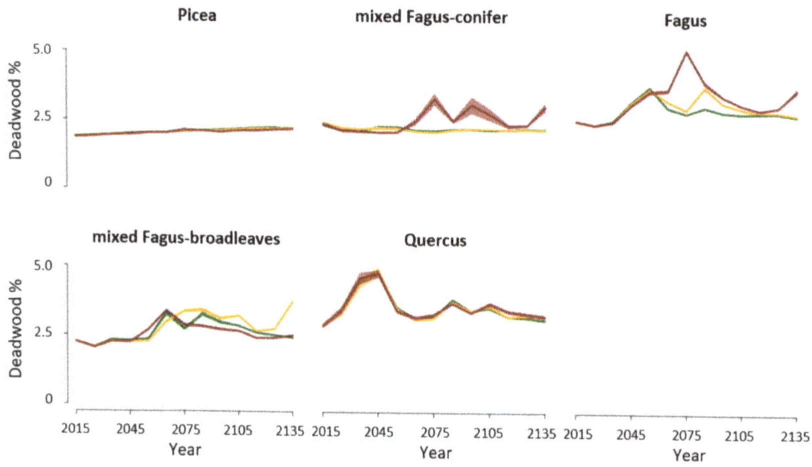

Figure 7. Percentage dead biomass relative to total biomass in the different forest types. Shaded areas indicate 95% confidence interval for every decade and climate change scenario: RCP 2.6 (green), RCP 4.5 (orange) and RCP 4.5 (red).

3.3. Productivity

The main forest types showed different aboveground net productivities, with Norway spruce and mixed forest productivity being typically lower than 15 kg/ha, that of beech pure forests being around that value and that of *Quercus* being usually above it. The aboveground net productivity tended to decrease slightly over time for all forest types. Productivity also differed among climate change scenarios, with the magnitude of the differences depending on forest type (Figure 8). Three forest types, Norway spruce, mixed beech–broadleaved and *Quercus*, showed significantly ($p < 0.05$) higher productivity under RCP 8.5 than under the other two scenarios. Beech forest productivity, despite having high temporal variations, showed significant differences between RCP 8.5 and RCP 4.5 ($p < 0.02$), usually being larger in RCP 8.5. The mixed beech–conifer forest productivity only showed significant differences between the RCP 2.6 and RCP 4.5 scenarios ($p = 0.035$).

3.4. Changes in Landscape Species Composition

Forest species composition varied over time and across climate change scenarios (Figure 9). During the first half of the simulations, the landscape species composition was similar among scenarios; in the following decades, the position of samples started to drift with the displacement increasing over time. This resulted in a significant differentiation in the species composition between RCP 8.5 and the other two scenarios. As shown by the ordination, and in accordance with the biomass analysis, RCP 8.5 was characterized by a greater biomass of *Q. frainetto*, *P. abies* and *A. alba*.

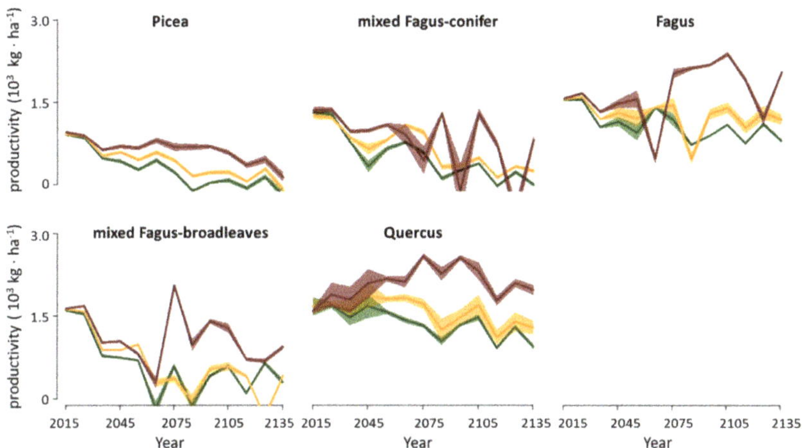

Figure 8. Productivity of the five main forest types (*Picea*, *Fagus*–conifers, *Fagus*, *Fagus*–broadleaved, *Quercus*) for the forecasts for every decade and climate change scenario: RCP 2.6 (green), RCP 4.5 (orange) and RCP 4.5 (red).

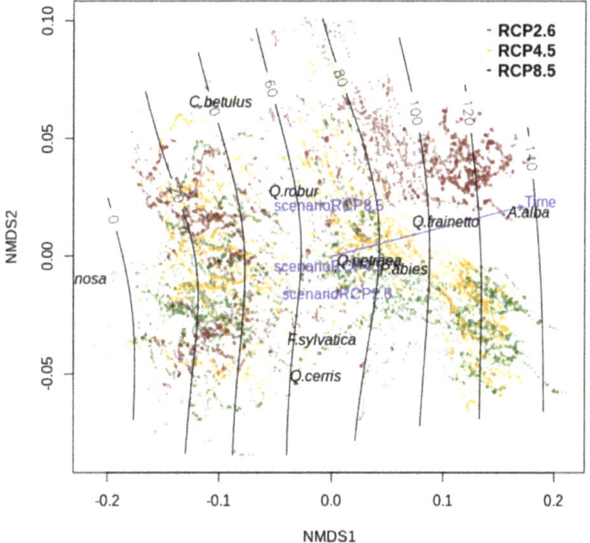

Figure 9. Ordination-based redundancy analysis of species biomass for the whole landscape, every year in the period 2015–2140 and every Landis-II simulation obtained for climate change scenarios RCP 2.6 (green), RCP 4.5 (orange) and RCP 8.5 (red), with the constraining variables year and RCP scenario.

4. Discussion

The simulation model implemented for temperate forests in the Romanian Southern Carpathians showed that climate change produces changes in forest biomass, productivity and species abundance. Such changes were particularly noticeable under RCP 8.5, the most extreme climate change scenario, and rendered an overall increase in the carbon carrying capacity of the studied Carpathian forests.

4.1. Forest Biomass and Productivity

The results of our Landis-II model adapted to Southern Carpathian forests showed an overall increase in forest biomass over time, with significantly greater biomass accumulation in the RCP 8.5 scenario. This is in line with Hubau et al. [10], who found that contrary to some tropical forests, temperate forests maintain a certain capacity to keep stocking carbon. The increased forest biomass accumulation and productivity predicted for the study area with intense climate change point to the raised CO_2 concentration [2,5] and warmer winters [3] as being the main drivers in the Carpathians, where precipitations do not change significantly for most of the area [23]. However, our simulations also indicate that the initial increase in biomass is limited, likely due to increasing drought and the stabilization of climate and atmospheric CO_2 concentrations. This trend suggests that the equilibrium already observed in some tropical forests [10] might occur in temperate forests somewhere at the end of the current century. Indeed, other authors have reached similar conclusions and timing, pointing directly to the proliferation of bark beetles as one of the main natural disturbances limiting carbon sequestration associated with climate change [92]. In the Southern Carpathians, climate change conditions overcompensate the negative effects of the disturbances at the beginning of the period but may not be able to do that after 2040, for most of the forests.

The main forest types studied responded differently to climate change, according to the ecophysiology of the species. Norway spruce and *Quercus* spp. forests had low productivity but higher stability, while mixed beech–conifer forests, beech forests and mixed beech–broadleaved forests were more prone to biomass changes, particularly after 2050. Indeed, low- and medium-altitude forests in the Southern Carpathians are very likely to be more affected by drought, while the precipitation in mountain tops, often covered by Norway spruce forests, might increase slightly [23]. Bouriaud el al. [16] also predicted forest biomass increments for the Romanian Eastern Carpathians but found a decreasing trend for conifer forests. This discrepancy among Bouriaud et al.'s [16] results and ours is likely due to three causes: (a) climate change effects on forests vary in type and intensity at the global, regional and local levels [15,23,93], as do the local conditions and climate change projections of the study areas in the Eastern and Southern Carpathians; (b) the modeling approach, with PnET and Landis-II models also accounting for complex ecophysiological responses to photosynthetically active radiation, atmospheric CO_2 concentration, etc.; and (c) the climate models used [16,94–97]. Indeed, the CNRM-CN5 model [87,88] used here is one of the least limiting climate models regarding water availability in the study area.

The predicted increase in the impact of the main natural disturbances in the Southern Carpathians: windthrows and bark beetle attacks, particularly *Ips* sp., is in agreement with other studies [6,98]. Even though some variations may be caused by harvest and aging forests, the periods with low productivity are likely associated with drought, since the SPEI value tends to decrease within the same periods. Drought events also occur simultaneously with raised mortality and deadwood biomass increments. This is the case of Norway spruce forests and beech forests after 2040. Natural disturbances induce organic carbon release, and hence it is common that disturbance regimes and ecosystem resilience determine forests' carbon sequestration capacity [10,99–102]. A net increase in temperate forest biomass caused by climate change has been forecasted for areas where disturbances and extreme climate events have a limited impact [54,103]. These disturbances usually damage conifer forests [6], and aging forests will likely contribute to it [52]. In our model, *H. abietis* had a minor impact, as a consequence of the low biomass of the cohorts they target, and because the selection harvesting management implemented in mixed beech and silver fir (*A. alba*) stands prevents *H. abietis* outbreaks [104]. The change in wood harvest observed here was also predicted by Bouriaud et al. [16], Ciceu et al. [105] and Chivulescu et al. [33] and concentrated in the intermediate- and low-altitude forests, having consequences for the forest structure. This increment in harvested timber is directly associated with both forest aging and the application of logging cycles [78]. As a result of all previous processes, around the year 2050, forests are expected to reach the carbon sequestration limit. Such an idea is supported by Bouriaud et al.'s [16] and Hubau et al.'s [10] studies. Nabuurs et al. [19] also found evidence of carbon sink saturation in European forests.

4.2. Forest Composition

The Landis-II projections predicted shifts in forest species composition and abundance under the RCP scenarios in Romanian Southern Carpathian temperate forests, with the five main species showing significant biomass increments over time (*F. sylvatica* only up to 2040). Indeed, the reduction in the cohorts killed by windthrows under RCP 8.5 compared to RCP 4.5 and RCP 2.6 is likely due to the success of *A. alba* and the intense modification of the forest structure under RCP 8.5. However, only *F. sylvatica* showed differences among the RCP scenarios, with the highest biomass accumulation under RCP 8.5. Some species, particularly *F. sylvatica*, showed short periods with strong reductions in the biomass after 2050 that co-occurred, and thus are likely associated, with drought conditions. Climate change may alter local environmental conditions, making them no longer suitable for a given species and altering the interspecific competition [18,96], ultimately resulting in a change in species abundance [1]. For instance, Musselman and Fox [14] indicated that drought-tolerant species are expected to succeed at the global scale under the new conditions, and this is likely going to occur for *A. alba* in the Southern Carpathians, as predicted here.

Previous studies have indicated that the intensity of the change in climate is proportional to the magnitude of the impact on forests [16,33]. The intensity of such a change in the Southern Carpathians is expected to be relatively small, as the climate conditions do not strongly depart from the current conditions under the analyzed RCPs. However, some forest types are more vulnerable to changes than others [18,96]. Norway spruce forests at the upper limit will gain from the temperature rise [15], and, as suggested by Landis-II models, at the lower limit, they will also benefit from *F. sylvatica*'s vulnerability to drought. At lower altitudinal belts, conifers coexist with beech. Beech and conifer mixed forests are more successful than pure beech stands [25,26], but Norway spruce grows more and remains relatively unaffected compared to *F. sylvatica* during drought [106–108]. *A. alba* is also less susceptible to warmer and drier conditions than *F. sylvatica* [25]. Thus, at the stand level, the association between *F. sylvatica* and *A. alba* is not only advantageous for enduring drought [25,26,106,107] but also for sunlight use [104,109]. Our results for the Southern Carpathians suggest that climate change will contribute to *A. alba* encroachment in *F. sylvatica* pure forest stands. Indeed, there is strong evidence that beech forests have already been affected by climate change-induced drought in recent decades [20]. Moreover, Broadmeadow et al. [4] recently highlighted beech's vulnerability in temperate forests across Europe.

Conifer and beech forests aside, *Quercus* spp.-dominated forests will partially benefit from the temperature rise and *F. sylvatica* decline. *Quercus* species are relatively drought resistant. According to the Landis-II simulations, and as also supported by previous studies [110], the most thermophile and drought-resistant species in the Southern Carpathians such as *Q. frainetto* [36] will benefit from climate change and increase their biomass during succession, while *Q. petraea* will eventually suffer the impact of drought events.

5. Conclusions

Here, we simulated the changes in the composition and structure of the temperate forests of the Southern Carpathians under three climate change scenarios. Our results indicate that climate change may contribute, overall, to increasing temperate forest productivity and biomass, mainly due to the combination of increasing temperatures and thus extended growing periods in the uplands and only a mild reduction in rainfall across the landscape. Fluctuations in this trend associated with strong biomass reductions and temporary deadwood rise were related to drought periods. However, it is likely that increased drought events may eventually counteract such positive trends.

Climate change selectively affected areas and species, thus contributing to changes in species abundance. *P. abies* benefitted from warmer conditions, whereas forests dominated by *F. sylvatica* were vulnerable to climate change, with drought periods associated with large mortality events. The *A. alba* abundance increased under the new climate conditions at the expense of *Fagus*'s decline. Our results suggest that under the upcoming climate, mixed *F. sylvatica*–conifer (namely Norway spruce and silver fir) stands will have greater resistance and resilience than those of pure *F. sylvatica* stands, a finding that will help in guiding future forest management practices.

Supplementary Materials: The following are available online at https://www.mdpi.com/article/10.3390/f12111434/s1, Figure S1: Distribution of the ecoregions within the study area. Ecoregions represent areas of homogeneous climate, soil and forest types, Table S1: Characteristics of the defined ecoregions within each forest type: main species composition, elevation range, soil attributes and main natural disturbances affecting the ecoregion. Disturbances: It: *I. typografus*; Id: *I. duplicatus*; Ha: *H. abietis* [34,35,111,112].

Author Contributions: Conceptualization, C.A. and J.G.-D.; methodology, C.A. and J.G.-D.; software, J.G.-D.; validation, J.G.-D. and C.A.; formal analysis, J.G.-D.; investigation, J.G.-D. and C.A.; resources, J.G.-D., A.C. and S.C.; data curation, J.G.-D.; writing—original draft preparation, J.G.-D. and C.A.; writing—review and editing, C.A., J.G.-D., A.C., S.C., O.B. and M.A.T.; visualization, J.G.-D.; supervision, C.A.; project administration, M.A.T., C.A. and O.B.; funding acquisition, M.A.T., C.A. and O.B. All authors have read and agreed to the published version of the manuscript.

Funding: This work was supported financially by EO-ROFORMON project, ID P_37_651/SMIS 105058 and PN 19070101.

Institutional Review Board Statement: Not applicable.

Informed Consent Statement: Not applicable.

Data Availability Statement: The data presented in this study are available on request from the corresponding author.

Acknowledgments: The authors acknowledge the contribution of the panel of experts constituted by I. Seceleanu, M. Paraschiv, N. Olenici and D. Chira.

Conflicts of Interest: The authors declare no conflict of interest.

References

1. Huntley, B.; Baxter, R. Vegetation ecology and global change. In *van der Maarel, E & Franklin. Vegetation Ecology*; J Wiley-Blackwell: Hoboken, NJ, USA, 2012; pp. 509–525. ISBN 978-1-4443-3888-1.
2. Cannell, M. UK Conifer Forests May Be Growing Faster in Response to Increased N Deposition, Atmospheric CO2 and Temperature. *Forestry* **1998**, *71*, 277–296. [CrossRef]
3. Richardson, A.D.; Hufkens, K.; Milliman, T.; Aubrecht, D.M.; Furze, M.E.; Seyednasrollah, B.; Krassovski, M.B.; Latimer, J.M.; Nettles, W.R.; Heiderman, R.R.; et al. Ecosystem Warming Extends Vegetation Activity but Heightens Vulnerability to Cold Temperatures. *Nature* **2018**, *560*, 368–371. [CrossRef]
4. Broadmeadow, M.S.J.; Ray, D.; Samuel, C.J.A. Climate Change and the Future for Broadleaved Tree Species in Britain. *For. Int. J. For. Res.* **2005**, *78*, 145–161. [CrossRef]
5. Kirschbaum, M.U.F.; McMillan, A.M.S. Warming and Elevated CO2 Have Opposing Influences on Transpiration. Which Is More Important? *Curr. For. Rep.* **2018**, *4*, 51–71. [CrossRef]
6. Seidl, R.; Thom, D.; Kautz, M.; Martin-Benito, D.; Peltoniemi, M.; Vacchiano, G.; Wild, J.; Ascoli, D.; Petr, M.; Honkaniemi, J.; et al. Forest Disturbances under Climate Change. *Nat. Clim. Chang.* **2017**, *7*, 395. [CrossRef] [PubMed]
7. Pan, Y.; Birdsey, R.A.; Fang, J.; Houghton, R.; Kauppi, P.E.; Kurz, W.A.; Phillips, O.L.; Shvidenko, A.; Lewis, S.L.; Canadell, J.G.; et al. A Large and Persistent Carbon Sink in the World's Forests. *Science* **2011**, *333*, 988–993. [CrossRef]
8. Sitch, S.; Friedlingstein, P.; Gruber, N.; Jones, S.D.; Murray-Tortarolo, G.; Ahlström, A.; Doney, S.C.; Graven, H.; Heinze, C.; Huntingford, C.; et al. Recent Trends and Drivers of Regional Sources and Sinks of Carbon Dioxide. *Biogeosciences* **2015**, *12*, 653–679. [CrossRef]
9. Brienen, R.J.W.; Phillips, O.L.; Feldpausch, T.R.; Gloor, E.; Baker, T.R.; Lloyd, J.; Lopez-Gonzalez, G.; Monteagudo-Mendoza, A.; Malhi, Y.; Lewis, S.L.; et al. Long-Term Decline of the Amazon Carbon Sink. *Nature* **2015**, *519*, 344–348. [CrossRef]

10. Hubau, W.; Lewis, S.L.; Phillips, O.L.; Affum-Baffoe, K.; Beeckman, H.; Cuní-Sanchez, A.; Daniels, A.K.; Ewango, C.E.N.; Fauset, S.; Mukinzi, J.M.; et al. Asynchronous Carbon Sink Saturation in African and Amazonian Tropical Forests. *Nature* **2020**, *579*, 80–87. [CrossRef]
11. Scheller, R.M.; Mladenoff, D.J. A Spatially Interactive Simulation of Climate Change, Harvesting, Wind, and Tree Species Migration and Projected Changes to Forest Composition and Biomass in Northern Wisconsin, USA. *Glob. Chang. Biol.* **2005**, *11*, 307–321. [CrossRef]
12. Klein, T.; Ramon, U. Stomatal Sensitivity to CO_2 Diverges between Angiosperm and Gymnosperm Tree Species. *Funct. Ecol.* **2019**, *33*, 1411–1424. [CrossRef]
13. Pan, Y.; Birdsey, R.A.; Phillips, O.L.; Jackson, R.B. The Structure, Distribution, and Biomass of the World's Forests. *Annu. Rev. Ecol. Evol. Syst.* **2013**, *44*, 593–622. [CrossRef]
14. Musselman, R.C.; Fox, D.G. A Review of the Role of Temperate Forests in the Global CO_2 Balance. *J. Air Waste Manag. Assoc.* **1991**, *41*, 798–807. [CrossRef]
15. IPCC. Climate Change 2014: Impacts, Adaptation and Vulnerability Part. A: Global and Sectoral Aspects. In *Contribution of Working Group II to the Fifth Assessment Report of the Intergovernmental Panel on Climate Change*; Cambridge University Press: Cambridge, UK, 2014.
16. Bouriaud, L.; Bouriaud, O.; Elkin, C.; Temperli, C.; Reyer, C.; Duduman, G.; Barnoaiea, I.; Nichiforel, L.; Zimmermann, N.; Bugmann, H. Age-Class Disequilibrium as an Opportunity for Adaptive Forest Management in the Carpathian Mountains, Romania. *Reg. Environ. Chang.* **2015**, *15*, 1557–1568. [CrossRef]
17. Law, B.E.; Hudiburg, T.W.; Berner, L.T.; Kent, J.J.; Buotte, P.C.; Harmon, M.E. Land Use Strategies to Mitigate Climate Change in Carbon Dense Temperate Forests. *Proc. Natl. Acad. Sci. USA* **2018**, *115*, 3663–3668. [CrossRef]
18. McDowell, N.G.; Allen, C.D.; Anderson-Teixeira, K.; Aukema, B.H.; Bond-Lamberty, B.; Chini, L.; Clark, J.S.; Dietze, M.; Grossiord, C.; Hanbury-Brown, A.; et al. Pervasive Shifts in Forest Dynamics in a Changing World. *Science* **2020**, *368*, eaaz9463. [CrossRef]
19. Nabuurs, G.-J.; Lindner, M.; Verkerk, P.J.; Gunia, K.; Deda, P.; Michalak, R.; Grassi, G. First Signs of Carbon Sink Saturation in European Forest Biomass. *Nat. Clim. Chang.* **2013**, *3*, 792–796. [CrossRef]
20. Piovesan, G.; Biondi, F.; Filippo, A.D.; Alessandrini, A.; Maugeri, M. Drought-Driven Growth Reduction in Old Beech (Fagus Sylvatica, L.) Forests of the Central Apennines, Italy. *Glob. Chang. Biol* **2008**, *14*, 1265–1281. [CrossRef]
21. Rubel, F.; Brugger, K.; Haslinger, K.; Auer, I. The Climate of the European Alps: Shift of Very High Resolution Köppen-Geiger Climate Zones 1800–2100. *Meteorol. Z.* **2017**, *26*, 115–125. [CrossRef]
22. Berland, A.; Shuman, B.; Manson, S.M. Simulated Importance of Dispersal, Disturbance, and Landscape History in Long-Term Ecosystem Change in the Big Woods of Minnesota. *Ecosystems* **2011**, *14*, 398–414. [CrossRef]
23. Micu, D.M.; Dumitrescu, A.; Cheval, S.; Birsan, M.-V. *Climate of the Romanian Carpathians*; Springer Atmospheric Sciences; Springer International Publishing: Cham, Switzerland, 2015; ISBN 978-3-319-02885-9.
24. Beldie, A. *Flora Şi Vegetaţia Munţilor Bucegi*; Academia Republicii Socialiste Romania: Bucharest, Romania, 1967.
25. Bosela, M.; Lukac, M.; Castagneri, D.; Sedmák, R.; Biber, P.; Carrer, M.; Konôpka, B.; Nola, P.; Nagel, T.A.; Popa, I.; et al. Contrasting Effects of Environmental Change on the Radial Growth of Co-Occurring Beech and Fir Trees across Europe. *Sci. Total Environ.* **2018**, *615*, 1460–1469. [CrossRef]
26. Schwarz, J.A.; Bauhus, J. Benefits of Mixtures on Growth Performance of Silver Fir (Abies Alba) and European Beech (Fagus Sylvatica) Increase With Tree Size Without Reducing Drought Tolerance. *Front. Glob. Chang.* **2019**, *2*, 79. [CrossRef]
27. Scheller, R.M.; Domingo, J.B.; Sturtevant, B.R.; Williams, J.S.; Rudy, A.; Gustafson, E.J.; Mladenoff, D.J. Design, Development, and Application of LANDIS-II, a Spatial Landscape Simulation Model with Flexible Temporal and Spatial Resolution. *Ecol. Model.* **2007**, *201*, 409–419. [CrossRef]
28. Aber, J.D.; Federer, C.A. A Generalized, Lumped-Parameter Model of Photosynthesis, Evapotranspiration and Net Primary Production in Temperate and Boreal Forest Ecosystems. *Oecologia* **1992**, *92*, 463–474. [CrossRef]
29. Gustafson, E.J. *PnET-Succession v3.1 Extension User Guide*; USDA Forest Service: Avenue Portland, OR, USA, 2018.
30. Gustafson, E.J.; De Bruijn, A.M.G.; Pangle, R.E.; Limousin, J.-M.; McDowell, N.G.; Pockman, W.T.; Sturtevant, B.R.; Muss, J.D.; Kubiske, M.E. Integrating Ecophysiology and Forest Landscape Models to Improve Projections of Drought Effects under Climate Change. *Glob. Chang. Biol.* **2015**, *21*, 843–856. [CrossRef]
31. Gustafson, E.J.; Shvidenko, A.Z.; Scheller, R.M. Effectiveness of Forest Management Strategies to Mitigate Effects of Global Change in South-Central Siberia. *Can. J. Res.* **2011**, *41*, 1405–1421. [CrossRef]
32. Seidl, R.; Fernandes, P.M.; Fonseca, T.F.; Gillet, F.; Jönsson, A.M.; Merganičová, K.; Netherer, S.; Arpaci, A.; Bontemps, J.-D.; Bugmann, H.; et al. Modelling Natural Disturbances in Forest Ecosystems: A Review. *Ecol. Model.* **2011**, *222*, 903–924. [CrossRef]
33. Chivulescu, S.; García-Duro, J.; Pitar, D.; Leca, Ş.; Badea, O. Past and Future of Temperate Forests State under Climate Change Effects in the Romanian Southern Carpathians. *Forests* **2021**, *12*, 885. [CrossRef]
34. Florea, N.; Balaceanu, V.; Munteanu, I.; Asvadurov, H.; Conea, A.; Oancea, C.; Cernescu, N.; Popovat, M. Harta Solurilor Romaniei, Scara 1: 200.000 (The Soil Map of Romania at the Scale 1: 200,000). *Inst. Geol. -Igfcot Buchar.* **1963**, *1993*, 50.
35. Hengl, T.; Mendes de Jesus, J.; Heuvelink, G.B.M.; Ruiperez Gonzalez, M.; Kilibarda, M.; Blagotić, A.; Shangguan, W.; Wright, M.N.; Geng, X.; Bauer-Marschallinger, B.; et al. SoilGrids250m: Global Gridded Soil Information Based on Machine Learning. *PLoS ONE* **2017**, *12*, e0169748. [CrossRef] [PubMed]
36. Şofletea, N.; Curtu, L. *Dendrologie*; Pentru Viaţă: Braşov, Romania, 2001; Volume 2, ISBN 978-973-99456-1-5.

37. ICAS. *Amenajamentul. O.S. Arpaş.*; Regia Naţională a pădurilor Romsilva. Institutul de Cercetări şi amenajări Silvice: Bucharest, Romania, 2017.
38. ICAS. *Amenajamentul. O.S. Sibiu.*; Regia Naţională a pădurilor Romsilva. Institutul de Cercetări şi amenajări Silvice: Bucharest, Romania, 2016.
39. ICAS. *Amenajamentul. O.S. Topoloveni*; Regia Naţională a pădurilor Romsilva. Institutul de Cercetări şi amenajări Silvice: Bucharest, Romania, 2015.
40. ICAS. *Amenajamentul. O.S. Vidraru*; Regia Naţională a pădurilor Romsilva. Institutul de Cercetări şi amenajări Silvice: Bucharest, Romania, 2015.
41. ICAS. *Amenajamentul. O.S. Domnesti*; Regia Naţională a pădurilor Romsilva. Institutul de Cercetări şi amenajări Silvice: Bucharest, Romania, 2014.
42. ICAS. *Amenajamentul. O.S. Mihăeşti*; Regia Naţională a pădurilor Romsilva. Institutul de Cercetări şi amenajări Silvice: Bucharest, Romania, 2014.
43. ICAS. *Amenajamentul. O.S. Muşăteşti*; Regia Naţională a pădurilor Romsilva. Institutul de Cercetări şi amenajări Silvice: Bucharest, Romania, 2014.
44. ICAS. *Amenajamentul. O.S. Aninoasa*; Regia Naţională a pădurilor Romsilva. Institutul de Cercetări şi amenajări Silvice: Bucharest, Romania, 2014.
45. ICAS. *Amenajamentul. O.S. Cotmeana*; Regia Naţională a pădurilor Romsilva. Institutul de Cercetări şi amenajări Silvice: Bucharest, Romania, 2013.
46. ICAS. *Amenajamentul. O.S. Piteşti*; Regia Naţională a pădurilor Romsilva. Institutul de Cercetări şi amenajări Silvice: Bucharest, Romania, 2013.
47. *Norme Tehnice Pentru Amenajarea Padurilor*; Ministry Apelor Padur Si Protection Mediului: Bucharest, Romania, 2000; p. 160.
48. *Norme Tehnice Privind Alegerea Si Aplicarea Tratamentelor*; Ministry Apelor Padur Si Protection Mediului: Bucharest, Romania, 2000; p. 87.
49. Paraschiv, M.-V. Cercetări Privind Insectele Dăunătoare Molidului Picea Abies (L.) Karst. In *Arboretele Din Munţii Braşovului/Researches Concerning the Damaging Insects of Norway Spruce Picea Abies (L.) Karst. Stands in Braşov Mountains*; Universitatea Transilvania din Braşov: Brașov, Romania, 2012.
50. Bogdan, O.; Niculescu, E. *Riscurile Climatice Din. Romania*; Institutul de Geografie: Bucharest, Romania, 1999; ISBN 973-0-008000-0.
51. Olenici, N.; Duduman, M.-L.; Tulbure, C.; Rotariu, C. *Ips Duplicatus* (Coleoptera, Curculionidae, Scolytinae)-Un Daunator Important al Molidului Din Afara Arealului Natural de Vegetatie. *Rev. Padur.* **2009**, *124*, 17–24.
52. Duduman, M.L.; Isaia, G.; Olenici, N. *Ips Duplicatus* (Sahlberg) (Coleoptera: Curculionidae, Scolytinae) distribution in Romania –preliminary results–. *Bull. Univ. Braşov Ser. II For. Woodlnd. Agric. Food Eng.* **2011**, *4*, 19–26.
53. Scheller, R.M.; Mladenoff, D.J. A Forest Growth and Biomass Module for a Landscape Simulation Model, LANDIS: Design, Validation, and Application. *Ecol. Model.* **2004**, *180*, 211–229. [CrossRef]
54. Duveneck, M.J.; Scheller, R.M.; White, M.A. Effects of Alternative Forest Management on Biomass and Species Diversity in the Face of Climate Change in the Northern Great Lakes Region (USA). *Can. J. Res.* **2014**, *44*, 700–710. [CrossRef]
55. Aber, J.D.; Ollinger, S.V.; Federer, C.A.; Reich, P.B.; Goulden, M.L.; Sicklighter, D.W.; Melillo, J.M.; Lathrop, R.G. Predicting the Effects of Climate Change on Water Yield and Forest Production in the Northeastern United States. *Clim. Res.* **1995**, *5*, 207–222. [CrossRef]
56. Gustafson, E.J.; Kubiske, M.E.; Sturtevant, B.R.; Miranda, B.R. Scaling Aspen-FACE Experimental Results to Century and Landscape Scales. *Landsc. Ecol.* **2013**, *28*, 1785–1800. [CrossRef]
57. Xu, C.; Gertner, G.Z.; Scheller, R.M. Potential Effects of Interaction between CO2 and Temperature on Forest Landscape Response to Global Warming. *Glob. Chang. Biol.* **2007**, *13*, 1469–1483. [CrossRef]
58. Newton, A.C.; Echeverría, C.; Cantarello, E.; Bolados, G. Projecting Impacts of Human Disturbances to Inform Conservation Planning and Management in a Dryland Forest Landscape. *Biol. Conserv.* **2011**, *144*, 1949–1960. [CrossRef]
59. Scheller, R.M.; Hua, D.; Bolstad, P.V.; Birdsey, R.A.; Mladenoff, D.J. The Effects of Forest Harvest Intensity in Combination with Wind Disturbance on Carbon Dynamics in Lake States Mesic Forests. *Ecol. Model.* **2011**, *222*, 144–153. [CrossRef]
60. Kretchun, A.M.; Loudermilk, E.L.; Scheller, R.M.; Hurteau, M.D.; Belmecheri, S. Climate and Bark Beetle Effects on Forest Productivity—Linking Dendroecology with Forest Landscape Modeling. *Can. J. For. Res.* **2016**, *46*, 1026–1034. [CrossRef]
61. Xi, W.; Waldron, J.D.; Lafon, C.W.; Cairns, D.M.; Birt, A.G.; Tchakerian, M.D.; Coulson, R.N.; Klepzig, K.D. Modeling Long-Term Effects of Altered Fire Regimes Following Southern Pine Beetle Outbreaks (North Carolina). *Ecol. Restor.* **2009**, *27*, 24–26. [CrossRef]
62. Gustafson, E.J.; Shifley, S.R.; Mladenoff, D.J.; Nimerfro, K.K.; He, H.S. Spatial Simulation of Forest Succession and Timber Harvesting Using LANDIS. *Can. J. Res.* **2000**, *30*, 32–43. [CrossRef]
63. Sturtevant, B.R.; Gustafson, E.J.; Li, W.; He, H.S. Modeling Biological Disturbances in LANDIS: A Module Description and Demonstration Using Spruce Budworm. *Ecol. Model.* **2004**, *180*, 153–174. [CrossRef]
64. Mladenoff, D.J.; Hong, S.H. Design, behavior and application of LANDIS, an object-oriented model of forest landscape disturbance and succession. In *Spatial Modeling of Forest Landscape Change: Approaches and Applications*; Cambridge University Press: Cambridge, UK, 1999; pp. 125–162.

65. European Environment Agency Corine Land Cover 2012 (Version 18.5.1). 2016. Available online: https://www.eea.europa.eu/data-and-maps/data/external/corine-land-cover-2012 (accessed on 1 January 2019).
66. Liaw, A.; Wiener, M. Classification and Regression by RandomForest. *R News* **2002**, *2*, 18–22.
67. Earth Resources Observation and Science (EROS) Center. *Shuttle Radar Topography Mission (SRTM) 1 Arc-Second Global*; USGS: Reston, VA, USA, 2017.
68. Fick, S.E.; Hijmans, R.J. WorldClim 2: New 1-Km Spatial Resolution Climate Surfaces for Global Land Areas. *Int. J. Climatol.* **2017**, *37*, 4302–4315. [CrossRef]
69. Esch, T.; Heldens, W.; Hirner, A.; Keil, M.; Marconcini, M.; Roth, A.; Zeidler, J.; Dech, S.; Strano, E. Breaking New Ground in Mapping Human Settlements from Space-The Global Urban Footprint. *Isprs J. Photogramm. Remote Sens.* **2017**, *134*, 30–42. [CrossRef]
70. OpenStreetMap contributors Planet. Dump Retrieved. Available online: https://planet.osm.org (accessed on 1 January 2019).
71. Tanase, M.A.; Villard, L.; Pitar, D.; Apostol, B.; Petrila, M.; Chivulescu, S.; Leca, S.; Borlaf-Mena, I.; Pascu, I.-S.; Dobre, A.-C.; et al. Synthetic Aperture Radar Sensitivity to Forest Changes: A Simulations-Based Study for the Romanian Forests. *Sci. Total Environ.* **2019**, *689*, 1104–1114. [CrossRef]
72. Giurgiu, V.; Decei, I.; Drăghiciu, D. *Metode Şi Tabele Dendrometrice [Methods and Dendrometric Tables]*; Ceres Publishing House, Bucharest: Bucharest, Romania, 2004; ISBN 973-40-0639-8.
73. Giurgiu, V.; Draghiciu, D. *Modele Matematico-Auxologice Si Tabele de Productie Pentru Arborete [Mathematic-Auxologic Models and Yield Tables for Forest Stands]*; Ceres Publishing House: Bucharest, Romania, 2004; ISBN 973-40-0637-1.
74. Kattge, J.; Díaz, S.; Lavorel, S.; Prentice, I.C.; Leadley, P.; Bönisch, G.; Garnier, E.; Westoby, M.; Reich, P.B.; Wright, I.J.; et al. TRY–a Global Database of Plant Traits. *Glob. Chang. Biol.* **2011**, *17*, 2905–2935. [CrossRef]
75. Vittoz, P.; Engler, R. Seed Dispersal Distances: A Typology Based on Dispersal Modes and Plant Traits. *Bot. Helv.* **2007**, *117*, 109–124. [CrossRef]
76. Popa, I. *Managementul Riscului La Doborâturi Produse De Vânt*; Editura Tehnică Silvică: Bucharest, Romania, 2007; ISBN 978-973-96001-8-7.
77. Grodzki, W.; Jakuš, R.; Lajzová, E.; Sitková, Z.; Maczka, T.; Škvarenina, J. Effects of Intensive versus No Management Strategies during an Outbreak of the Bark Beetle *Ips Typographus* (L.) (Col.: Curculionidae, Scolytinae) in the Tatra Mts. in Poland and Slovakia. *Ann. Sci.* **2006**, *63*, 55–61. [CrossRef]
78. Fedderwitz, F.; Björklund, N.; Ninkovic, V.; Nordlander, G. Does the Pine Weevil *Hylobius Abietis* Prefer Conifer Seedlings over Other Main Food Sources? *Silva. Fenn.* **2018**, *52*. [CrossRef]
79. Simionescu, A.; Ciornei, C.; Chira, D. *Starea De Sănătate a Pădurilor Din România În Perioada 2001–2010 [Romanian Forest Health Status in the Period 2001–2010]*; Muşatinii: Suceava, Romania, 2013; ISBN 978-606-656-017-7.
80. Simionescu, A.; Mihalciuc, V.; Lupu, D.; Vlăduleasa, A.; Badea, O.; Fulicea, T. *Starea De Sănătate a Pădurilor Din România În Intervalul 1986–2000 [Romanian Forest Health Status in the Period 1986–2000]*; Muşatinii: Suceava, Romania, 2001; ISBN 973-8122-05-8.
81. INCDS, Marin D. *Raport Privind Activitatile Desfasurate in Cadrul Lucrarii Asistenta Tehnica Privind Realizarei Bazei de Date Si Elaborarea Lucrarii Anuale 'Starea de Sanatate a Padurilor Administrate de RNP Romsilva' Cod: 16.12/2018" ['Report about Technical Assistance Regarding the Creation of the Database and the Elaboration of the Annual Paper "The State of Health of the Forests Managed by RNP Romsilva" Code: 16.12/2018']*; Romsilva: Bucharest, Romania, 2018.
82. Olenici, N.; Olenici, V. *Hylobius Abietis* L.-Unele Particularitati Biologice Ecologice Si Comportamentale Si Protectia Culturilor Impotriva Vatamarilor Cauzate de Acesta (I). *Bucov. For.* **1994**, 34–59.
83. *Norme Tehnice Pentru Îngrijirea Si Conducerea Arboretelor*; Ministry Apelor Padur Si Protection Mediului: Bucharest, Romania, 2000; p. 176.
84. *Norme Tehnice Privind Compozitii, Scheme Si Tehnologii de Regenerare a Padurilor Si Impadurire a Terenurilor Degradate*; Ministry Apelor Padur Si Protection Mediului: Bucharest, Romaniadiului, 2000; p. 253.
85. Van Vuuren, D.P.; Edmonds, J.; Kainuma, M.; Riahi, K.; Thomson, A.; Hibbard, K.; Hurtt, G.C.; Kram, T.; Krey, V.; Lamarque, J.-F.; et al. The Representative Concentration Pathways: An Overview. *Clim. Chang.* **2011**, *109*, 5–31. [CrossRef]
86. Harris, I.; Jones, P.D.; Osborn, T.J.; Lister, D.H. Updated High-Resolution Grids of Monthly Climatic Observations–the CRU TS3.10 Dataset. *Int. J. Climatol.* **2014**, *34*, 623–642. [CrossRef]
87. Osborn, T.J.; Wallace, C.J.; Harris, I.C.; Melvin, T.M. Pattern Scaling Using ClimGen: Monthly-Resolution Future Climate Scenarios Including Changes in the Variability of Precipitation. *Clim. Chang.* **2016**, *134*, 353–369. [CrossRef]
88. Voldoire, A.; Sanchez-Gomez, E.; Mélia, D.S.; Decharme, B.; Cassou, C.; Sénési, S.; Valcke, S.; Beau, I.; Alias, A.; Chevallier, M.; et al. The CNRM-CM5. 1 Global Climate Model: Description and Basic Evaluation. *Clim. Dyn.* **2013**, *40*, 2091–2121. [CrossRef]
89. Moreno, A.; Hasenauer, H. Spatial Downscaling of European Climate Data. *Int. J. Climatol.* **2016**, *36*, 1444–1458. [CrossRef]
90. Pinheiro, J.; Bates, D.; DebRoy, S.; Sarkar, D.; R Core Team. *Nlme: Linear and Nonlinear Mixed Effects Models*, R package version; 2017. Available online: https://CRAN.R-project.org/package=nlme (accessed on 1 September 2020).
91. Oksanen, J.; Blanchet, F.G.; Friendly, M.; Kindt, R.; Legendre, P.; McGlinn, D.; Minchin, P.R.; O'Hara, R.B.; Simpson, G.L.; Solymos, P.; et al. *Vegan: Community Ecology Package*. 2020. Available online: https://CRAN.R-project.org/package=vegan (accessed on 1 September 2020).
92. Seidl, R.; Rammer, W.; Jäger, D.; Lexer, M.J. Impact of Bark Beetle (*Ips Typographus* L.) Disturbance on Timber Production and Carbon Sequestration in Different Management Strategies under Climate Change. *For. Ecol. Manag.* **2008**, *256*, 209–220. [CrossRef]

93. Cuculeanu, V.; Tuinea, P.; Bălteanu, D. Climate Change Impacts in Romania: Vulnerability and Adaptation Options. *GeoJournal* **2002**, *57*, 203–209. [CrossRef]
94. Gunton, R.M.; Polce, C.; Kunin, W.E. Predicting Ground Temperatures across European Landscapes. *Methods Ecol. Evol.* **2015**, *6*, 532–542. [CrossRef]
95. Kearney, M.R.; Gillingham, P.K.; Bramer, I.; Duffy, J.P.; Maclean, I.M.D. A Method for Computing Hourly, Historical, Terrain-corrected Microclimate Anywhere on Earth. *Methods Ecol. Evol.* **2020**, *11*, 38–43. [CrossRef]
96. von Arx, G.; Graf Pannatier, E.; Thimonier, A.; Rebetez, M. Microclimate in Forests with Varying Leaf Area Index and Soil Moisture: Potential Implications for Seedling Establishment in a Changing Climate. *J. Ecol.* **2013**, *101*, 1201–1213. [CrossRef]
97. Terando, A.; Youngsteadt, E.; Meineke, E.; Prado, S. Accurate near Surface Air Temperature Measurements Are Necessary to Gauge Large-Scale Ecological Responses to Global Climate Change. *Ecol. Evol.* **2018**, *8*, 5233–5234. [CrossRef] [PubMed]
98. Moreno Rodríguez, J.M.; Cruz Treviño, A.; Martínez Lope, C. *Evaluación Preliminar De Los Impactos En España Por Efecto Del Cambio Climático: Proyecto ECCE-Informe Final*; Centro de Publicaciones, Ministerio de Medio Ambiente: Madrid, Spain, 2005; ISBN 978-84-8320-303-3.
99. Allen, K.A.; Harris, M.P.K.; Marrs, R.H. Matrix Modelling of Prescribed Burning in *Calluna Vulgaris*-Dominated Moorland: Short Burning Rotations Minimize Carbon Loss at Increased Wildfire Frequencies. *J. Appl. Ecol.* **2013**, *50*, 614–624. [CrossRef]
100. Aponte, C.; de Groot, W.J.; Wotton, B.M. Forest Fires and Climate Change: Causes, Consequences and Management Options. *Int. J. Wildland Fire* **2016**, *25*, i–ii. [CrossRef]
101. Bellingham, P.J.; Sparrow, A.D. Resprouting as a Life History Strategy in Woody Plant Communities. *Oikos* **2000**, *89*, 409–416. [CrossRef]
102. Bond, W.J.; Keeley, J.E. Fire as a Global 'Herbivore': The Ecology and Evolution of Flammable Ecosystems. *Trends Ecol. Evol.* **2005**, *20*, 387–394. [CrossRef]
103. Gustafson, E.J.; Shvidenko, A.Z.; Sturtevant, B.R.; Scheller, R.M. Predicting Global Change Effects on Forest Biomass and Composition in South-central Siberia. *Ecol. Appl.* **2010**, *20*, 700–715. [CrossRef]
104. Costea, C. *Codrul Grădinărit*; Editura Agro-silvica: Bucharest, Romania, 1962.
105. Ciceu, A.; Duro, J.G.; Aponte, C.; Pascu, I.S.; Claudiu-Dobre, A.; Zamfira, V.; Leca, Ş.; Pitar, D.; Apostol, B.; Apostol, E.N.; et al. Landis-II simulation model integration in Romania's forest management. *Rev. De Silvic. Si Cineg.* **2020**, *25*, 47–55.
106. Čater, M.; Levanič, T. Beech and Silver Fir's Response along the Balkan's Latitudinal Gradient. *Sci Rep.* **2019**, *9*, 16269. [CrossRef]
107. Magh, R.K.; Bonn, B.; Grote, R.; Burzlaff, T.; Pfautsch, S.; Rennenberg, H. Drought Superimposes the Positive Effect of Silver Fir on Water Relations of European Beech in Mature Forest Stands. *Forests* **2019**, *10*, 897. [CrossRef]
108. Töchterle, P.; Yang, F.; Rehschuh, S.; Rehschuh, R.; Ruehr, N.K.; Rennenberg, H.; Dannenmann, M. Hydraulic Water Redistribution by Silver Fir (*Abies Alba* Mill.) Occurring under Severe Soil Drought. *Forests* **2020**, *11*, 162. [CrossRef]
109. Čater, M.; Kobler, A. Light Response of *Fagus Sylvatica* L. and *Abies Alba* Mill. in Different Categories of Forest Edge–Vertical Abundance in Two Silvicultural Systems. *For. Ecol. Manag.* **2017**, *391*, 417–426. [CrossRef]
110. Ciceu, A.; Popa, I.; Leca, S.; Pitar, D.; Chivulescu, S.; Badea, O. Climate Change Effects on Tree Growth from Romanian Forest Monitoring Level II Plots. *Sci. Total Environ.* **2020**, *698*, 134129. [CrossRef] [PubMed]
111. Hengl, T.; de Jesus, J.M.; MacMillan, R.A.; Batjes, N.H.; Heuvelink, G.B.M.; Ribeiro, E.; Samuel-Rosa, A.; Kempen, B.; Leenaars, J.G.B.; Walsh, M.G.; et al. SoilGrids1km–Global Soil Information Based on Automated Mapping. *PLoS ONE* **2014**, *9*, e105992. [CrossRef] [PubMed]
112. Shangguan, W.; Hengl, T.; Mendes de Jesus, J.; Yuan, H.; Dai, Y. Mapping the Global Depth to Bedrock for Land Surface Modeling. *J. Adv. Modeling Earth Syst.* **2017**, *9*, 65–88. [CrossRef]

MDPI

Article

Photosynthesis Traits of Pioneer Broadleaves Species from Tailing Dumps in Călimani Mountains (Eastern Carpathians)

Andrei Popa [1,2] and Ionel Popa [1,2,*]

[1] Forestry Faculty, Ștefan cel Mare University of Suceava, Universității 13, 720229 Suceava, Romania; popa.andrei.dorna@gmail.com
[2] National Institute for Research and Development in Forestry Marin Drăcea, Calea Bucovinei 73bis, 725100 Câmpulung Moldovenesc, Romania
* Correspondence: popaicas@gmail.com; Tel.: +4-074-465-0967

Abstract: The reforestation and stable ecological restoration of tailings dumps resulting from surface mining activities in the Călimani Mountains represent an ongoing environmental challenge. To assess the suitability of different tree species for restoration efforts, photosynthetic traits were monitored in four broadleaf pioneer species—green alder (*Alnus alnobetula* (Ehrh.) K. Koch), aspen (*Populus tremula* L.), silver birch (*Betula pendula* Roth.), and goat willow (*Salix caprea* L.)—that naturally colonized the tailings dumps. Green alder and birch had the highest photosynthetic rate, followed by aspen and goat willow. Water use efficiency parameters (WUE and iWUE) were the highest for green alder and the lowest for birch, with intermediary values for aspen and goat willow. Green alder also exhibited the highest carboxylation efficiency, followed by birch. During the growing season, net assimilation and carboxylation efficiency exhibited a maximum in late July and a minimum in late June. The key limitation parameters of the photosynthetic process derived from the FvCB model (V_{cmax} and J_{max}) were the highest for green alder and exhibited a maximum in late July, regardless of the species. Based on photosynthetic traits, the green alder—a woody N_2-fixing shrub—is the most well-adapted and photosynthetically efficient species that naturally colonized the tailings dumps in the Călimani Mountains.

Keywords: gas exchange; ecosystem restoration; mountain forests; photosynthesis

Citation: Popa, A.; Popa, I. Photosynthesis Traits of Pioneer Broadleaves Species from Tailing Dumps in Călimani Mountains (Eastern Carpathians). *Forests* **2021**, *12*, 658. https://doi.org/10.3390/f12060658

Academic Editor: Francois Girard

Received: 12 May 2021
Accepted: 21 May 2021
Published: 22 May 2021

1. Introduction

Climate change and anthropic activities have given rise to the most serious environmental problems of the 21st century [1]. Today, in an increasing number of ecosystems, anthropic influences are harming biodiversity and ecosystem functioning. Human activities change the condition of natural vegetation, leading to disturbances such as degradation of vegetation, erosion of soil, decline in land productivity and even reduction of ecosystem services [2].

A large proportion of the Earth's geological resources (metals, minerals, fuel, etc.) are underground; mining activities to access these resources damage the above-ground landscape and have a significant impact on natural ecosystems such as forests, rivers and lakes [3]. Even after the mining activity is complete (especially in the case of surface mining), dump areas remain in the place of former natural ecosystems and have the potential to cause serious environmental problems. Due to social and political pressure to undertake more sustainable development, the restoration of mining areas has gradually become an important phase of mining activities [4]. Multiple restoration solutions are available; of these, soil amendment combined with phytoremediation (reforestation) is the most environmentally friendly method [5].

Due to high soil degradation (lack of nutrients and organic matter, small edaphic volume, high concentration of heavy metals), it is important to choose a mix of species that are well-adapted to the local climate and the unique conditions of the mining habitat [6].

Successful reforestation of these degraded landscapes depends on the adaptability of tree species to degraded soil and the capacity of plant associations to contribute to the restoration of soil proprieties and former environmental conditions [7].

The process of ecosystem reconstruction on degraded mining soils progresses gradually from dump consolidation and soil amendments to pioneering species installation and finally climax tree species establishment [8]. In many cases, human interventions through ecological reconstruction are informed by the natural colonization of pioneer species. The optimal species composition for reforestation is chosen based on criteria such as adaptative capacity (survival rate), growth and biomass production (photosynthesis), and the capacity to cover the land through vegetative or generative regeneration [7].

In mountainous areas (e.g., the Carpathians), pioneer species colonize open soils on degraded or abandoned lands as part of primary succession. These pioneer species include green alder (*Alnus alnobetula* (Ehrh.) K. Koch = *Alnus viridis* (Chaix) DC), aspen (*Populus tremula* L.), silver birch (*Betula pendula* Roth.), goat willow (*Salix caprea* L.) and rowan (*Sorbus aucuparia* L.) [9]. Green alder, an N_2-fixing shrub, plays an important role in mountain ecosystems due to its capacity to stabilize slopes and prevent erosion [10,11]. On the other hand, silver birch is a fast-growing deciduous species that occurs naturally throughout most of Europe and is frequently observed on degraded lands [12,13]. Aspen and goat willow have high ecological amplitudes and are widely distributed throughout the temperate and boreal areas of Europe and Asia [14,15].

Carbon assimilation, as an indicator of species' adaptability to specific habitat conditions, is related to biomass production, CO_2 storage, competitiveness and survival capacity [16]. Gas exchange measurements play a major role in understanding photosynthetic processes [17,18]. The performance of forest species in terms of photosynthetic traits is evaluated through gas exchange, which is measured under controlled experimental conditions or in situ conditions [19,20]. Particular growing conditions specific to mining-degraded soils modify the normal assimilation process [21–23].

To the best of our knowledge, no previous studies have measured gas exchange to analyse the photosynthetic traits of mountain pioneer species growing in former mining areas. However, analyses of photosynthetic parameters under different growing conditions (exposure to ozone, fertilization variants, enriched CO_2, etc.) have been performed for *Populus* spp. [24–26], *Salix* spp. [27–29], *Betula* spp. [30–32], and *Alnus* spp. [33–35].

Our work aimed to quantify the eco-physiological performance of four broadleaf pioneer species that naturally colonized a mountain tailings dump area. The following scientific questions were examined: (i) Which of these four pioneer species is best adapted for use in reforestation processes from the point of view of photosynthetic traits? (ii) What are the seasonal patterns of the photosynthetic traits?

2. Materials and Methods

2.1. Study Site

This study took place in the Călimani Mountains (Eastern Carpathians, Romania; 47°7′24″ N and 25°13′48″ E; 1500 m a.s.l.) on degraded tailings dumps resulting from sulfur surface mining that occurred between 1969 and 1997. During this period, the former natural forest ecosystem (a mixture of Norway spruce and Swiss stone pine) and organic soil were destroyed, and dumps were formed by the successive storage of tailings. In 2007, an ecological reconstruction effort began with the stabilization of the slopes with high slants, alkaline amendment of the soil (debris from former buildings) and afforestation of some areas with Norway spruce (*Picea abies* L., H. Karst.) and green alder. Currently, the upper part of the mining dump is colonized by pioneer forest species (green alder, aspen, goat willow, birch) in combination with herbaceous species, and the area is characterized by a primary succession of vegetation [9,36].

The woody species present on the site comprise species specific to the upper mountain area: green alder, Norway spruce, dwarf mountain pine (*Pinus mugo* Turra), aspen, goat willow and birch. Shrubby vegetation is represented by blueberry (*Vaccinium myr-*

tillus L.), cranberry (*Vaccinium vitis-idaea* L.), bog blueberry (*Vaccinium uliginosum* L.) and rhododendron (*Rhododendron myrtifolium* Schott & Kotschy).

Air temperature (2 m) and soil water content (10 cm depth) were measured using dedicated sensors (HOBO U23-001, Onset Computer Corporation, Bourne, MA, USA and CS650, Campbell Scientific, Logan, UT, USA). The average annual air temperature at the site in 2019 was 4.3 °C and, in the vegetation season, it was 10.7 °C (Figure 1). The average annual precipitation in the study area is 1000–1200 mm. The snow layer is present for an average of 180–200 days each year, the first snowfall can occur in early October, and the typical vegetation season starts in the second decade of May until middle October [37].

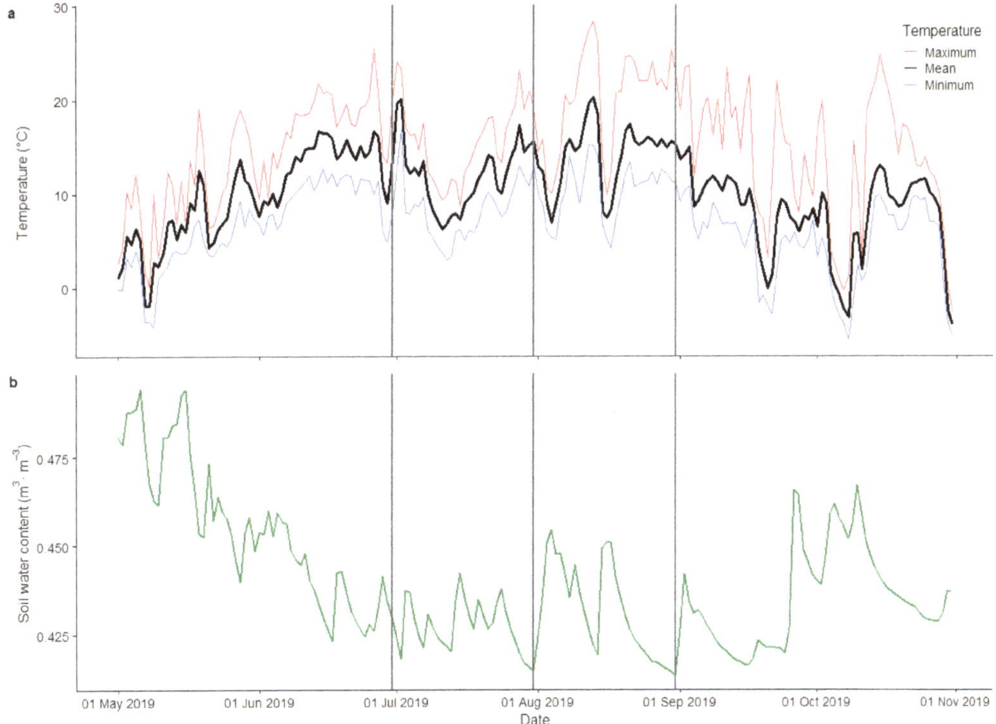

Figure 1. Air temperature (**a**) and soil water content (**b**) variation during the 2019 growing season (vertical lines represent measurements dates).

The tailings dump soils are characterized by high acidity (pH 3.2) and missing organic components [9,36].

2.2. Gas Exchange Measurements

Gas exchange during photosynthesis was measured using a portable photosynthesis system (LI-6800, LI-COR Inc., Lincoln, NE, USA) equipped with a standard infrared gas-exchange analyser (IRGA) and a chamber for broadleaf species (area: 6 cm^2). Instantaneous gas exchange was recorded at three time points in 2019 (29–30 June, 30–31 July and 30–31 August) between 9 am and 3 pm. Measurements, at each time point during the season, were performed on one leaf from five different exemplars of each broadleaf species: green alder, birch, aspen and goat willow. Due to difficult site accessibility (high mountain and degraded land) and the long time required for measurement (multiple dead times for stabilization and between measurements), we extended the day measurement period by

3 h, compared with the standard practice. We planned the experiment, in order to minimize the influence of diurnal variability of photosynthesis, by measuring the first exemplar of each species, followed by the second exemplar of each species and so on. We took into consideration that increasing the number of measurement days (now limited to two days for each time point during the season) could induce more variability because climatic parameters can change significantly (e.g., precipitation or temperature).

All studied species have C3 photosynthetic pathways. The selected trees were saplings, with a mean height of 1.5–2.0 m for silver birch, goat willow and aspen and 1.0 m for green alder. Measurements were performed on different leaves from one time point to another, but from the same exemplars. Selected leaves were completely developed, had no damage and were located on the upper part of the crown, in full exposure to light.

Instrument calibration was performed at the start of each session following the manufacturer's recommendations [38]. To assure measurement accuracy, the gas analyser was matched when the following conditions were met: elapsed time since the last match >10 min, CO_2 measured by the reference analyser have changed by 100 $\mu mol \cdot mol^{-1}$ since the last match, difference between CO_2 reference and CO_2 sample <10 $\mu mol \cdot mol^{-1}$, and difference in H_2O <1 $mmol \cdot mol^{-1}$. While collecting the measurements, the mean temperature and relative humidity in the measurement chamber were 22.3 ± 1.8 °C and $60 \pm 1.2\%$, respectively, the fan speed was 10,000 rpm and the flow rate was 500 $\mu mol \cdot s^{-1}$. The irradiance photosynthetic photon flux density (PPFD) was set to 1000 $\mu mol \cdot m^{-2} \cdot s^{-1}$ and was kept constant for all measurements, with following light composition ratio of 0.9 red and 0.1 blue using the LI-COR 6800 light source.

To obtain response curves for net photosynthetic rate as a function of intercellular CO_2 concentration, a chamber CO_2 gradient consisting of 400, 200, 100, 50, 400, 600, 800, 1000, and 1200 $\mu mol \cdot mol^{-1}$ was used. Response curve measurements were performed on one leaf from five different exemplars for each species. Between each measurement time point during the season, the exemplars were kept the same, but the leaves differ. Steady-state values from each leaf at different CO_2 concentrations were recorded after 2–3 min, an interval which allowed the leaf to adjust to the new environmental conditions (a stability point was reached when the standard deviation for CO_2 and H_2O differences was ≤ 0.1 for 20 s) [38,39].

Light-saturated net photosynthetic rate A ($\mu mol\ CO_2 \cdot m^{-2} \cdot s^{-1}$), transpiration rate E ($mmol\ H_2O \cdot m^{-2} \cdot s^{-1}$), stomatal conductance to water vapour gsw ($mol\ H_2O \cdot m^{-2} \cdot s^{-1}$), and intercellular CO_2 concentration Ci ($\mu mol \cdot mol^{-1}$) and leaf temperature $Tleaf$ (°C) were measured. Using these measured parameters, three efficiency parameters were then calculated: water use efficiency WUE (as A/E, $\mu mol\ CO_2 \cdot m^{-2} \cdot s^{-1} / mmol\ H_2O \cdot m^{-2} \cdot s^{-1}$), intrinsic water use efficiency $iWUE$ (as A/gsw, $\mu mol \cdot mol^{-1}$) and instantaneous carboxylation efficiency A/Ci ($\mu mol\ CO_2 \cdot m^{-2} \cdot s^{-1} / \mu mol \cdot mol^{-1}$) [40]. During the gas exchange measurements, a uniform distribution of photosynthesis and transpiration over the leaf was assumed [41].

2.3. The Farquhar–Von Caemmerer–Berry Model

Data from the response curve of net photosynthetic rate to CO_2 were analysed using the Farquhar–von Caemmerer–Berry model (the FvCB model) [42]. This model was developed to model leaf gas exchange and net photosynthetic rate (A) for C3 plants under any given environmental conditions. It has been used widely in recent decades to summarise the dependence of carbon assimilation rate on intercellular CO_2 concentration (Ci) because of its simple form and the comparable metrics of photosynthetic capacity that are provided. The net photosynthetic rate predicted by the FvCB model is the minimum between the Rubisco limited rate (Ac), the ribulose 1,5-bisphosphate (RuBP)-regeneration or electron (e-) transport limited rate (Aj) and the triose phosphate utilization (TPU) limited rate (Ap) of CO_2 assimilation [43,44]. Considering the three limitation phases, the net photosynthetic rate can be modelled as follows:

$$A = \min(Ac, Aj, Ap) \tag{1}$$

In the Rubisco limited phase, the response of net assimilation (*Ac*) to *Ci* is defined by:

$$Ac = \frac{V_{\text{cmax}} * (C_i - \Gamma_*)}{C_i + K_C * (1 + O/K_O)} - R_d \tag{2}$$

where *Rd* is day respiration (μmol·m^{-2}·s^{-1}), V_{cmax} is the maximum carboxylation rate of Rubisco (μmol CO_2·m^{-2}·s^{-1}), Γ* is the photosynthetic compensation point (μmol·mol^{-1}), K_C is the Michaelis–Menten constant of Rubisco for CO_2 (μmol·mol^{-1}), K_O is the Michaelis–Menten constant of Rubisco for O_2 (μmol·mol^{-1}), and O is the intercellular partial pressure of O_2 (mmol·mol^{-1}) set to 21 KPa.

During the RuBP-regeneration or electron transport, the response of net assimilation (*Aj*) to *Ci* is defined by:

$$Aj = \frac{J * (C_c - \Gamma_*)}{4C_c + 8\Gamma_*} - R_d \tag{3}$$

where *J* is the rate of electron transport (μmol·e^{-1}·m^{-2}·s^{-1}).

In the TPU limited phase, the net assimilation rate (*Ap*) is defined as:

$$Ap = 3T_p - R_d \tag{4}$$

where T_p is the rate of phosphate release in triose phosphate utilization (μmol·m^{-2}·s^{-1}).

The kinetic constants of Rubisco (Γ*, K_C, K_O), which are temperature-dependent, were derived using Arrhenius-type equations using the leaf temperature measurements [45,46]. Traits of photosynthetic capacity (V_{cmax}, J_{max}, TPU) were derived from the FvCB model and were corrected to a temperature of 25 °C [47]. More information and details about A/Ci data fitting can be found in the literature [48–50]. The FvCB model was applied for each leaf, and coefficients were analysed as mean.

2.4. Data Analyses

Differences between species were analysed using ANOVA followed by post hoc Tukey tests [51]. The dependence between photosynthetic parameters was quantified using the Pearson correlation coefficient. Statistical tests were considered significant at the $p < 0.05$ level. FvCB model parameters were estimated using the R package 'plantecophys' [47], and figures were constructed using the packages 'ggplot2' and 'cowplot'. All data processing was done using R 4.0.3 software [52].

3. Results and Discussion

3.1. Photosynthetic Parameters for Deciduous Pioneer Species

To compare the eco-physiological performance of the deciduous pioneer species that naturally colonized the mining dump areas in the Călimani Mountains, the mean values of photosynthetic traits during the 2019 vegetation period were analysed for CO_2 concentration close to the actual environmental concentration (400 μmol·mol^{-1}).

Green alder and birch had the highest net assimilation rate for the entire season among the four species analysed (Table 1). The lowest net assimilation rate was recorded for goat willow. The net assimilation rate for green alder was significantly higher than those of aspen and goat willow, while the net assimilation rate of birch was significantly different only from that of goat willow. Similar values for net assimilation (ranging from 12.8 to 17.3 μmol CO_2·m^{-2}·s^{-1}) have been documented under similar measurement conditions (1400 μmol·m^{-2}·s^{-1} PPFD and 320 μmol·mol^{-1} CO_2 concentration) for the seedlings of different Alnus species [33]. However, our results showed slightly higher net assimilation values for silver birch compared with the range reported in the literature [31,32,53].

Table 1. Mean values (± standard deviation) of photosynthetic parameters for the 2019 growing season, at 400 μmol·mol^{-1} CO_2 concentration and 1000 μmol·m^{-2}·s^{-1} PPFD. Significant differences are shown using letters following the Tukey test ($p < 0.05$).

Parameters	Green Alder	Birch	Aspen	Goat Willow
A (μmol CO_2·m^{-2}·s^{-1})	17.35 ± 3.62 ^{a}	16.62 ± 2.47 ^{ab}	14.43 ± 2.93 ^{bc}	13.31 ± 4.97 ^{c}
E (mmol H_2O·m^{-2}·s^{-1})	2.6 ± 0.78 ^{a}	3.44 ± 0.94 ^{b}	2.94 ± 1.16 ^{ab}	2.75 ± 1.34 ^{ab}
gsw (mol H_2O·m^{-2}·s^{-1})	0.24 ± 0.10 ^{a}	0.31 ± 0.08 ^{a}	0.27 ± 0.10 ^{a}	0.25 ± 0.12 ^{a}
Ci (μmol·mol^{-1})	240.76 ± 26.07 ^{a}	278.21 ± 13.61 ^{b}	272.36 ± 29.13 ^{b}	274.16 ± 24.30 ^{b}
A/*Ci* (μmol CO_2·m^{-2}·s^{-1}/μmol·mol^{-1})	0.07 ± 0.01 ^{a}	0.06 ± 0.01 ^{b}	0.05 ± 0.01 ^{bc}	0.05 ± 0.02 ^{c}
WUE (μmol CO_2·m^{-2}·s^{-1}/mmol H_2O·m^{-2}·s^{-1})	6.92 ± 1.08 ^{a}	5.07 ± 1.09 ^{b}	5.63 ± 2.07 ^{b}	5.49 ± 1.78 ^{b}
iWUE (μmol·mol^{-1})	78.05 ± 18.91 ^{a}	54.92 ± 8.85 ^{b}	60.93 ± 20.73 ^{b}	60.9 ± 18.41 ^{b}

Birch had the highest rates of transpiration and stomatal conductance. The rates of transpiration for birch were significantly higher than those for green alder, but stomatal conductance was not significantly different between species. Similar transpiration rate (3.4 mmol·m^{-2}·s^{-1}) and stomatal conductance (0.4 mol·m^{-2}·s^{-1}) have been observed for mountain birch at the treeline on the Tibetan Plateau [54]. Birch proveniences and leaf types (early vs. late leaves) can induce differences in stomatal conductance [12].

Diffusion of CO_2 into the intercellular spaces inside leaves occurs mainly through stomatal pores [55]. Stomatal density and distribution on leaves differ among species. As a result, we obtained different *Ci* values even though the CO_2 concentration in the measurement chamber (400 μmol·mol^{-1}) was similar for all species. Green alder had the lowest intercellular CO_2 concentration, which was significantly different from those of the other species. The variation in intercellular CO_2 concentration allows us to understand whether the decline in net photosynthesis rate is due to stomatal limitations or to the reduction of photosynthetic activity in the leaves' cells [56].

Instantaneous carboxylation efficiency was the lowest for aspen and goat willow and the highest for green alder. This photosynthetic parameter can be considered as an estimate of Rubisco activity; generally, higher intercellular CO_2 concentration is associated with lower stomatal conductance. Water use efficiency (WUE and iWUE) was also the highest for green alder and differed significantly from those of the other species.

Regardless of the species, the highest correlation was observed between transpiration rate and stomatal conductance (Table 2). For birch only, a negative correlation was found between net assimilation rate and intercellular CO_2 concentration. The correlation coefficient between net assimilation rate and transpiration rate varied from 0.74 (birch) to 0.93 (goat willow) and was significant in all cases. Similar significant negative correlations between *Ci* and *A* have also been found for birch in a Lithuanian forest (boreal zone) [40].

3.2. Variability of Photosynthetic Parameters during the Vegetation Season

For broadleaf species, the chlorophyll content of leaves changes throughout the growing season, which induces variability in eco-physiological processes. Because of these biochemical changes in leaves, it is important to explore the variation in individual photosynthetic parameters during the vegetation season [12]. To avoid the systematic influence of diurnal variation, the gas exchange measurements were distributed during the day from 9 am to 3 pm. The net assimilation rate varied across the season for all four species, with the lowest values recorded at the end of June (Figure 2a). The lowest net assimilation rate was observed for goat willow at the end of June (7.55 μmol CO_2·m^{-2}·s^{-1}) and was significantly different from the values in other months. For green alder, the maximum value of net assimilation was measured at the end of July (18.64 μmol CO_2·m^{-2}·s^{-1}), and the minimum was recorded at the end of June (14.93 μmol CO_2·m^{-2}·s^{-1}). A similar trend in variation was found for goat willow. On the contrary, net assimilation rates for birch and aspen increased continuously during the season.

Table 2. Pearson correlation coefficients for relationships between gas exchange parameters (*A*–net photosynthetic rate, *E*–transpiration rate, *gsw*–stomatal conductance, *Ci*–intercellular CO_2 concentration).

Parameters	*E*	*gsw*	*Ci*
Green alder			
A	0.80 **	0.87 **	0.73 **
E		0.88 **	0.84 **
gsw			0.93 **
Birch			
A	0.74 **	0.77 **	−0.13
E		0.94 **	0.46 *
gsw			0.52 **
Aspen			
A	0.84 **	0.84 **	0.66 **
E		0.96 **	0.82 **
gsw			0.87 **
Goat willow			
A	0.93 **	0.94 **	0.45 *
E		0.97 **	0.66 **
gsw			0.67 **

** Significance at $p < 0.01$; * Significance at $p < 0.05$.

Studies in boreal forests indicate that, for birch, the highest rates of leaf-mass net assimilation occur under light-saturated conditions in early May after the leaves unfold, and there is minimal variation during the vegetation season [12]. For other species, the maximum leaf-area net assimilation occurs in late June and early July [31]. Studies on different *Salix* spp. have highlighted a weak correlation between biomass yield and photosynthetic rate and a positive influence of total leaf area per plant [27].

Stomatal conductance increased continuously from the beginning of the season to autumn for all species, except goat willow (Figure 2b). A similar pattern was observed for transpiration rate, but with a higher rate of increase from the beginning to the end of the season. These two parameters had a similar trend in variation because transpiration is regulated mainly by stomatal conductance [56]. The stomatal conductance of aspen and goat willow in June was significantly different from that measured in July or August. The maximum values for transpiration rate were measured in August for green alder (2.96 mmol $H_2O \cdot m^{-2} \cdot s^{-1}$), aspen (3.93 mmol $H_2O \cdot m^{-2} \cdot s^{-1}$) and birch (4.22 mmol $H_2O \cdot m^{-2} \cdot s^{-1}$) and in July for goat willow (3.35 mmol $H_2O \cdot m^{-2} \cdot s^{-1}$) (Figure 2c).

iWUE and WUE decreased during the vegetation season for all studied species, possibly linked with the lower soil water content during the measurements in July and August (see Figure 1). Water use efficiency parameters were significantly different between all three measurement timepoints only in the case of aspen. Similarly, differences between monthly values of WUE have been reported for another poplar species (*Populus angustifolia*) from North America [57]. Goat willow exhibited a different trend in variation, with slightly higher iWUE values at the end of August compared to July. Green alder had the highest values of both WUE and iWUE compared to other species.

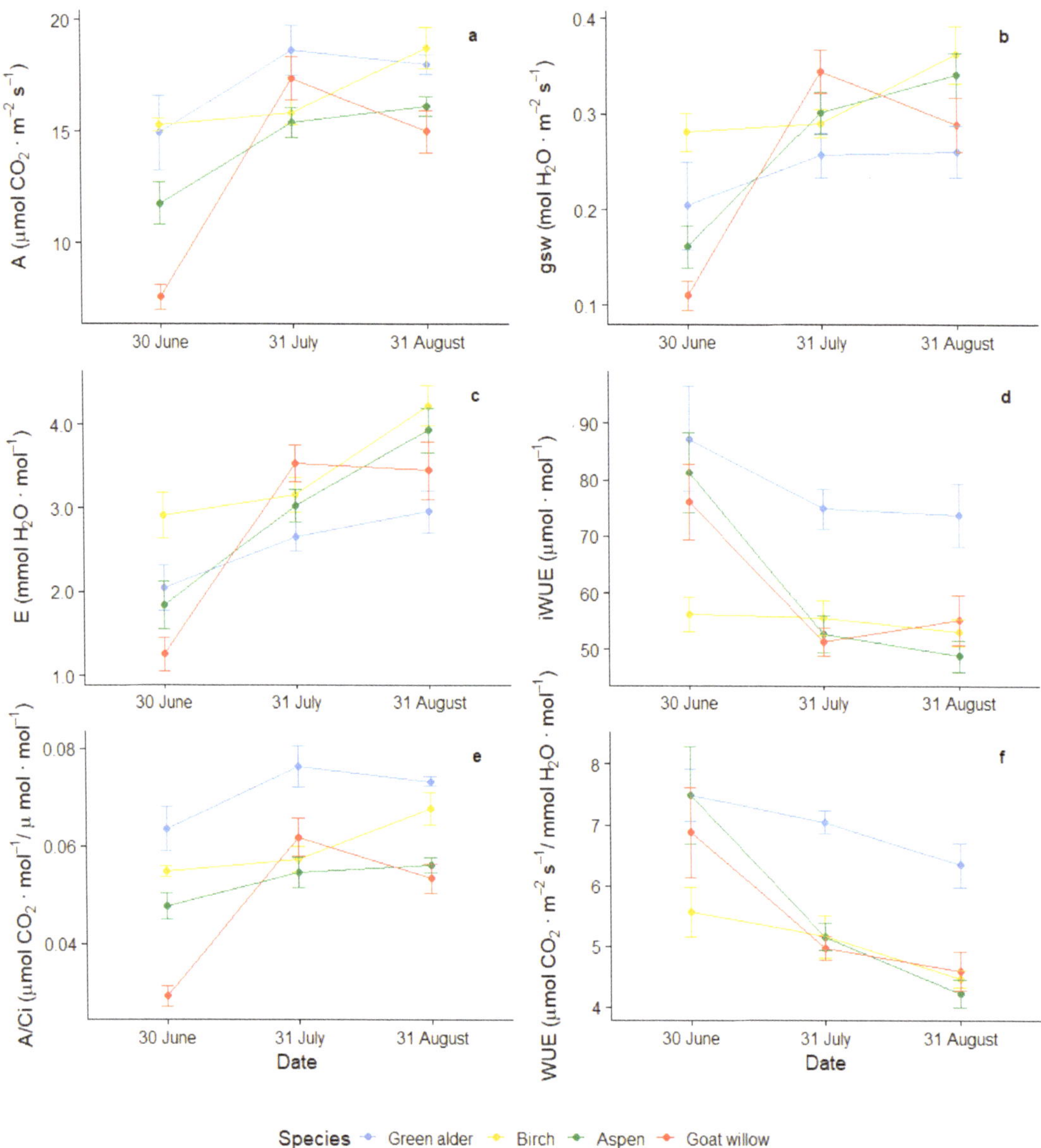

Figure 2. Variability in photosynthetic parameters during the 2019 growing season for deciduous pioneer species on a tailing dump in the Călimani Mountains. (**a**) Net assimilation rate, (**b**) stomatal conductance, (**c**) transpiration rate, (**d**) intrinsic water use efficiency, (**e**) instantaneous carboxylation efficiency, (**f**) water use efficiency (points are mean values, and whiskers represent standard errors).

Variation in iWUE and WUE during the season is a consequence of both changing environmental conditions and physiological changes in leaf structure due to ageing [58]. Both parameters reflect water use efficiency (how much carbon is fixed per unit of water loss), but WUE can also be used as a water stress indicator (drought indicator) [59]. Higher values of iWUE or WUE can be achieved through lower stomatal conductance or transpiration rate, higher assimilation capacity or a combination of both [60]. Water supply is one factor that can cause plants to have a lower stomatal conductance and higher WUE [12]. Our study was conducted in a mountain area where precipitation is not a limiting factor, but variations in WUE may still occur due to particular conditions induced by precipitation variation during vegetation season. In the second and third measurement time points, low soil water content was reported (see Figure 1), which may induce a decrease in water use efficiency parameters.

Green alder also demonstrated the highest instantaneous carboxylation efficiency compared with other deciduous pioneer species, with a maximum in late July (Figure 2e). It had a lower intercellular CO_2 concentration but maximum values for assimilation rate. One possible explanation for this is that green alder is in its optimal distribution range, is acclimatized to cold environments and uses resources more efficiently. The maximum instantaneous carboxylation efficiency occurred in July for goat willow and at the end of August for aspen and birch.

The mining dump areas are exposed to high light intensity, large temperature variability and low nitrogen availability. Low availability of soil nitrogen can be an important limiting factor of photosynthesis [19,21]. N_2-fixing plants that are capable of fixing atmospheric nitrogen, like green alder, can be more performant in terms of photosynthetic traits on degraded soils compared with other pioneer species [61]. Leaf size and structure, combined with crown size and branching patterns, play an important role in the assimilation performance of different species [62]. Field observations confirm that the leaf area and crown development of green alder are greater than for the other species.

3.3. Photosynthesis Response Curve under Increasing CO_2 Concentration

Using a variable concentration of CO_2 to understand assimilation performance is more efficient than using a constant CO_2 concentration [63]. An A/Ci curve can be constructed to understand and interpret the biochemical processes of leaf photosynthesis under various environmental conditions. The measured values of net assimilation at different CO_2 concentrations are used to estimate photosynthetic limitations and parameters of photosynthetic performance using two key parameters from the FvCB model: V_{cmax} and J_{max}.

The response curve for the relationship between assimilation and CO_2 concentration shows an initial increase followed by a relatively constant variation caused by light limitation (Figure 3). A clear difference between the three measurement time points was observed for all species, except for birch in June and July. The limitation of assimilation rate due to light availability occurred after reaching an intercellular CO_2 concentration of 500 μmol·mol^{-1}. After passing this threshold, the highest assimilation rate was measured in August for all four species. For goat willow and aspen, variation in WUE in relation to CO_2 concentration was higher in June.

Based on the FvCB model, the limitations of the photosynthetic process are characterized by two main parameters—V_{cmax} and J_{max}—and occasionally also by TPU. Estimation of these parameters is regulated by the limitations of one of three curves of the FvCB model. V_{cmax} is the maximum rate of Rubisco activity and reflects a limitation in RuBP-regeneration [48]. For green alder, the highest values of V_{cmax} were recorded in July, and the lowest in June, without significant differences during the season (Figure 4a). For birch and aspen, there were no significant differences in V_{cmax} during the season. The lowest V_{cmax} for goat willow was recorded in June, statistically different from the rest of the season, and was associated with lower assimilation levels. To derivate the seasonal mean of V_{cmax}, it is recommended to perform measurements in midsummer. Maximum values of V_{cmax} were obtained in the middle of the vegetation season (end July).

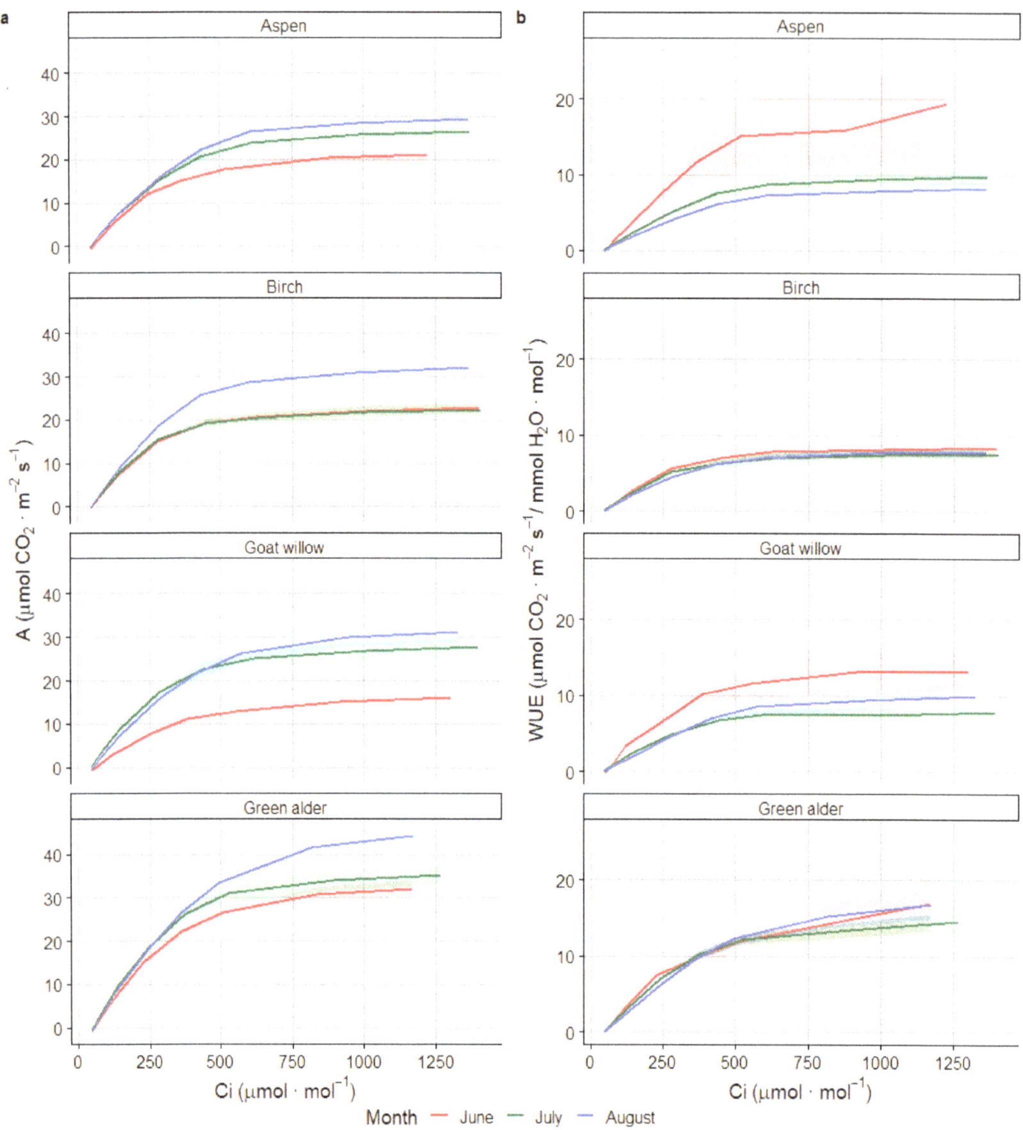

Figure 3. (**a**) Mean response curve of net assimilation relative to intercellular CO_2 concentration as modelled by the FvCB model; (**b**) water use efficiency relative to intercellular CO_2 concentration (shaded areas represent standard error).

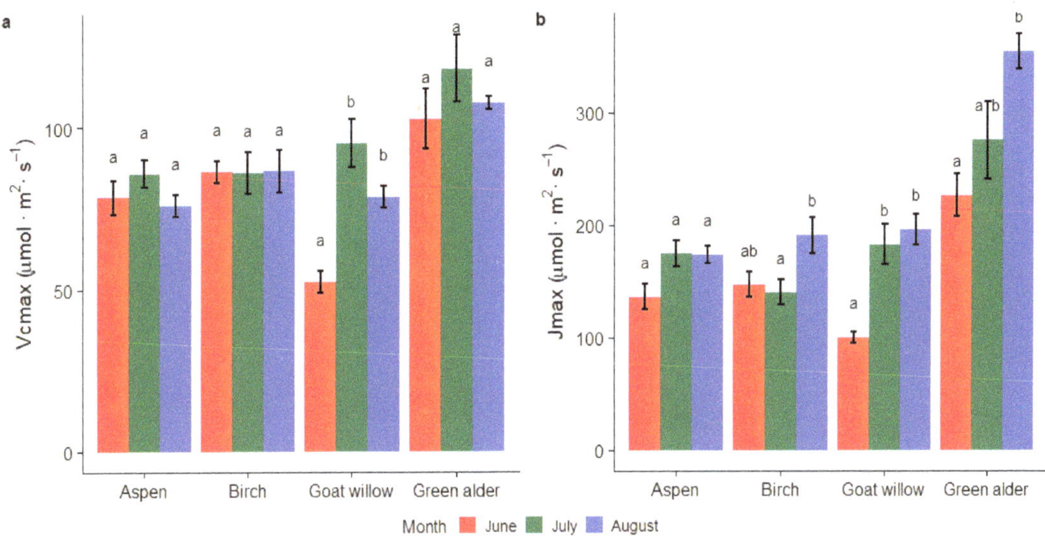

Figure 4. Mean values of FvCB model parameters for deciduous pioneer species on tailings dumps in the Călimani Mountains: (**a**) V_{cmax}—maximum rate of Rubisco activity; (**b**) J_{max}—maximum rate of electron transport for RuPB-regeneration (whiskers represent standard errors, and letters represent significant differences).

J_{max} represents the maximum rate of electron transport for RuPB-regeneration at the light intensity used in the study (1000 µmol·m^{-2}·s^{-1}) (Figure 4b) [64]. J_{max} was the lowest in June and highest in August, regardless of the species. With exception of aspen, the J_{max} differs significantly between time points during the season. V_{cmax} and J_{max} are correlated with A_{max} (the maximum value of assimilation) because these parameters all reflect limitations in photosynthetic processes. Green alder exhibited the highest values of V_{cmax} and J_{max} and also had the highest rate of photosynthesis at both ambient and saturating CO_2, while goat willow presented the lowest values for all of these variables. The ratio of J_{max} to V_{cmax} varied in the typical range (1.5 to 2.5) observed for other woody species [65].

TPU limitation occurred in less than half of the samples, as it generally requires a higher *Ci* concentration than that used in our study. This third state of photosynthesis limitation occurs when the chloroplast system's reaction is greater than the possibility of the leaf using the triose phosphate [43]. This is more likely to happen under experimental conditions than in natural situations, and this limitation state has not been taken into account in many studies [66].

Light dependence of J_{max} varied by up to 40% across the different leaves of birch [30]. Meanwhile, V_{cmax} is dependent on the temperature and increases at higher temperatures, but with limitations at very high temperatures [30]. For *Populus* species, multiple studies have documented V_{cmax} and J_{max} values lower than those reported in this study, even though the light radiance used was higher, at 200–500 µmol·m^{-2}·s^{-1} [25]. A global study that analysed more than 350 species highlighted that, for tree species, the mean values of V_{cmax} and J_{max} were 66.6 µmol·m^{-2}·s^{-1} and 114.4 µmol·m^{-2}·s^{-1}, respectively [67].

4. Conclusions

Reforestation and the stable ecological restoration of tailings dumps resulting from surface mining activities in the Călimani Mountains have been a high priority for regional and national administrations in the last two decades. Even though several solutions have been implemented (dumps stabilization, soil amendments, etc.), the complete recovery of the degraded habitats is still a challenge. Gaining a better understanding of natural colonization with pioneer woody species, such as through studying primary natural succession, can offer valuable knowledge about the species that are most adapted to these particular environmental conditions.

The most productive pioneer species in terms of photosynthetic traits was the green alder, a woody N_2-fixing shrub. It showed the highest rates of net assimilation, carboxylation efficiency and water use efficiency and can be a suitable species for reforestation based on the study conditions. During the growing season, this species' maximum photosynthetic capacity is generally observed at the end of July, with a minimum in late June.

A detailed understanding of the variability and dynamics of the photosynthetic capacity of pioneer species that naturally occur on tailings dumps is essential, offering valuable data to the process of characterizing suitable species for ecological restoration systems to heal these open wounds in the landscape.

Author Contributions: Conceptualization, I.P. and A.P.; methodology, A.P.; writing—original draft preparation, A.P.; writing—review and editing, I.P. All authors have read and agreed to the published version of the manuscript.

Funding: This research was funded by the Romanian Ministry of Research and Innovation in Core Program for forestry–BIOSERV (2019)–Project PN 19070102.

Data Availability Statement: Data is available on request formed to the corresponding author.

Acknowledgments: We thank Margareta Grudnicki for valuably suggestions and the Forestry Faculty of Suceava for supporting this study.

Conflicts of Interest: The authors declare no conflict of interest.

References

1. Shukla, P.R.; Skea, J.; Buendia, E.C.; Masson-Delmotte, V.; Pörtner, H.-O.; Roberts, D.C.; Zhai, P.; Slade, R.; Connors, S.; van Diemen, R.; et al. *IPCC, 2019: Climate Change and Land: An IPCC Special Report on Climate Change, Desertification, Land Degradation, Sustainable Land Management, Food Security, and Greenhouse Gas Fluxes in Terrestrial Ecosystems*; Intergovernmental Panel on Climate Change (IPCC): Geneva, Switzerland, 2019; pp. 1–864.
2. Liu, S.; Li, W.; Qiao, W.; Wang, Q.; Hu, Y.; Wang, Z. Effect of Natural Conditions and Mining Activities on Vegetation Variations in Arid and Semiarid Mining Regions. *Ecol. Indic.* **2019**, *103*, 331–345. [CrossRef]
3. Chauhan, S.S. Mining, Development and Environment: A Case Study of Bijolia Mining Area in Rajasthan, India. *J. Hum. Ecol.* **2010**, *31*, 65–72. [CrossRef]
4. Deng, J.; Bai, X.; Zhou, Y.; Zhu, W.; Yin, Y. Variations of Soil Microbial Communities Accompanied by Different Vegetation Restoration in an Open-Cut Iron Mining Area. *Sci. Total Environ.* **2020**, *704*, 135243. [CrossRef] [PubMed]
5. Henry, H.F.; Burken, J.G.; Maier, R.M.; Newman, L.A.; Rock, S.; Schnoor, J.L.; Suk, W.A. Phytotechnologies—Preventing Exposures, Improving Public Health. *Int. J. Phytoremediat.* **2013**, *15*, 889–899. [CrossRef] [PubMed]
6. Macdonald, S.E.; Landhäusser, S.M.; Skousen, J.; Franklin, J.; Frouz, J.; Hall, S.; Jacobs, D.F.; Quideau, S. Forest Restoration Following Surface Mining Disturbance: Challenges and Solutions. *New For.* **2015**, *46*, 703–732. [CrossRef]
7. Pietrzykowski, M. Tree Species Selection and Reaction to Mine Soil Reconstructed at Reforested Post-Mine Sites: Central and Eastern European Experiences. *Ecol. Eng. X* **2019**, *3*, 100012. [CrossRef]
8. Frouz, J.; Dvorščík, P.; Vávrová, A.; Doušová, O.; Kadochová, Š.; Matějíček, L. Development of Canopy Cover and Woody Vegetation Biomass on Reclaimed and Unreclaimed Post-Mining Sites. *Ecol. Eng.* **2015**, *84*, 233–239. [CrossRef]
9. Cenușă, E. *Instalarea Vegetației Naturale În Zone Afectate de Activități Miniere Din Parcul National Călimani*; Forestry Technical Publishing House: Voluntari, Romania, 2018.
10. Boscutti, F.; Poldini, L.; Buccheri, M. Green Alder Communities in the Alps: Phytosociological Variability and Ecological Features. *Plant Biosyst.* **2014**, *148*, 917–934. [CrossRef]
11. Mauri, A.; Caudullo, G. Alnus viridis in Europe: Distribution, Habitat, Usage and Threats. In *European Atlas of Forest Tree Species*; European Commission: Brussels, Belgium, 2016; p. 68. [CrossRef]
12. Tenkanen, A.; Keski-Saari, S.; Salojärvi, J.; Oksanen, E.; Keinänen, M.; Kontunen-Soppela, S. Differences in Growth and Gas Exchange between Southern and Northern Provenances of Silver Birch (Betula Pendula Roth) in Northern Europe. *Tree Physiol.* **2020**, *40*, 198–214. [CrossRef]
13. Beck, P.; Caudullo, G.; de Rigo, D.; Tinner, W. *Betula pendula*, *Betula pubescens* and Other Birches in Europe: Distribution, Habitat, Usage and Threats. In *European Atlas of Forest Tree Species*; European Commission: Brussels, Belgium, 2016; pp. 70–73. [CrossRef]
14. Caudullo, G.; de Rigo, D. *Populus tremula* in Europe: Distribution, Habitat, Usage and Threats. In *European Atlas of Forest Tree Species*; European Commission: Brussels, Belgium, 2016; pp. 138–139. [CrossRef]

15. Enescu, C.M.; Houston Durrant, T.; de Rigo, D.; Caudullo, G. *Salix caprea* in Europe: Distribution, Habitat, Usage and Threats. In *European Atlas of Forest Tree Species*; European Commission: Brussels, Belgium, 2016; pp. 170–171. [CrossRef]
16. Pallardy, S. Physiology of Woody Plants. *Pac. Conserv. Biol.* **1998**, *4*, 272. [CrossRef]
17. Cavaleri, M.A.; Reed, S.C.; Smith, W.K.; Wood, T.E. Urgent Need for Warming Experiments in Tropical Forests. *Glob. Chang. Biol.* **2015**, *21*, 2111–2121. [CrossRef]
18. Long, S.P.; Bernacchi, C.J. Gas Exchange Measurements, What Can They Tell Us about the Underlying Limitations to Photosynthesis? Procedures and Sources of Error. *J. Exp. Bot.* **2003**, *54*, 2393–2401. [CrossRef] [PubMed]
19. Zhao, X.; Mao, Z.; Xu, J. Gas Exchange, Chlorophyll and Growth Responses of Betula Platyphylla Seedlings to Elevated CO_2 and Nitrogen. *Int. J. Biol.* **2010**, *2*, 143–149. [CrossRef]
20. Liu, M.Z.; Jiang, G.M.; Li, Y.G.; Gao, L.M.; Niu, S.L.; Cui, H.X.; Ding, L. Gas Exchange, Photochemical Efficiency, and Leaf Water Potential in Three Salix Species. *Photosynthetica* **2003**, *41*, 393–398. [CrossRef]
21. Bojarczuk, K.; Karolewski, P.; Oleksyn, J.; Kieliszewska-Rokicka, B.; Zytkowiak, R.; Tjoelker, M.G. Effect of Polluted Soil and Fertilisation on Growth and Physiology of Silver Birch (*Betula pendula* Roth.) Seedlings. *Polish J. Environ. Stud.* **2002**, *11*, 483–492.
22. Hermle, S.; Vollenweider, P.; Günthardt-Goerg, M.S.; Mcquattie, C.J.; Matyssek, R. Leaf Responsiveness of *Populus tremula* and *Salix viminalis* to Soil Contaminated with Heavy Metals and Acidic Rainwater. *Tree Physiol.* **2007**, *27*, 1517–1531. [CrossRef] [PubMed]
23. Han, G.; Cui, B.X.; Zhang, X.X.; Li, K.R. The Effects of Petroleum-Contaminated Soil on Photosynthesis of *Amorpha fruticosa* Seedlings. *Int. J. Environ. Sci. Technol.* **2016**, *13*, 2383–2392. [CrossRef]
24. Horton, J.L.; Kolb, T.E.; Hart, S.C. Leaf Gas Exchange Characteristics Differ among Sonoran Desert Riparian Tree Species. *Tree Physiol.* **2001**, *21*, 233–241. [CrossRef]
25. Zhu, G.F.; Li, X.; Su, Y.H.; Huang, C.L. Parameterization of a Coupled CO2 and H2O Gas Exchange Model at the Leaf Scale of *Populus euphratica*. *Hydrol. Earth Syst. Sci.* **2010**, *14*, 419–431. [CrossRef]
26. Rood, S.B.; Nielsen, J.L.; Shenton, L.; Gill, K.M.; Letts, M.G. Effects of Flooding on Leaf Development, Transpiration, and Photosynthesis in Narrowleaf Cottonwood, a Willow-like Poplar. *Photosynth. Res.* **2010**, *104*, 31–39. [CrossRef]
27. Andralojc, P.J.; Bencze, S.; Madgwick, P.J.; Philippe, H.; Powers, S.J.; Shield, I.; Karp, A.; Parry, M.A.J. Photosynthesis and Growth in Diverse Willow Genotypes. *Food Energy Secur.* **2014**, *3*, 69–85. [CrossRef]
28. Talbot, R.J.; Etherington, J.R.; Bryant, J.A. Comparative Studies of Plant Growth and Distribution in Relation to Waterlogging. XII. Growth, Photosynthetic Capacity and Metal Ion Uptake in Salix Caprea and S. Cinerea Ssp. Oleifolia. *New Phytol.* **1987**, *105*, 563–574. [CrossRef]
29. Johnson, J.D.; Tognetti, R.; Paris, P. Water Relations and Gas Exchange in Poplar and Willow under Water Stress and Elevated Atmospheric CO2. *Physiol. Plant.* **2002**, *115*, 93–100. [CrossRef]
30. Aalto, T.; Juurola, E. Parametrization of a Biochemical CO2 Exchange Model for Birch (*Betula pendula* Roth.). *Boreal Environ. Res.* **2001**, *6*, 53–64.
31. Oleksyn, J.; Żytkowiak, R.; Reich, P.B.; Tjoelker, M.G.; Karolewski, P. Ontogenetic Patterns of Leaf CO 2 Exchange, Morphology and Chemistry in *Betula Pendula* Trees. *Trees Struct. Funct.* **2000**, *14*, 271–281. [CrossRef]
32. Hoshika, Y.; Watanabe, M.; Inada, N.; Mao, Q.; Koike, T. Photosynthetic Response of Early and Late Leaves of White Birch (*Betula platyphylla* Var. *Japonica*) Grown under Free-Air Ozone Exposure. *Environ. Pollut.* **2013**, *182*, 242–247. [CrossRef]
33. Côté, B.; Carlson, R.W.; Dawson, J.O. Leaf Photosynthetic Characteristics of Seedlings of Actinorhizal *Alnus* Spp. and *Elaeagnus* Spp. *Photosynth. Res.* **1988**, *16*, 211–218. [CrossRef]
34. Piper, F.I.; Cavieres, L.A. Gas Exchange of Juvenile and Mature Trees of *Alnus jorullensis* (Betulaceae) at Sites with Contrasting Humidity in the Venezuelan Andes. *Ecol. Res.* **2010**, *25*, 51–58. [CrossRef]
35. Popa, A.; Popa, I. Caracteristici ale procesului de fotosinteză la aninul verde din Munții Călimani. *Rev. Silvic. Cinegetică* **2019**, *44*, 20–23.
36. Gorea, D.; Hancu, G.; Rusu, A.; Cârje, A.; Barabás, E. Quality Assessment Indicators of Surface Waters and Soils in the Vicinity of the Former Sulfur Mine in the Călimani Mountains. *Forum Geogr.* **2014**, *XIII*, 212–218. [CrossRef]
37. Erzberger, P.; Höhn, M.; Pócs, T. Contribution to the Bryoflora of Călimani Mountains in the Eastern Carpathians, Romania, I. *Acta Biol. Plant. Agriensis* **2012**, *2*, 75–97.
38. LI-COR. *Using the LI-6800 Portable Photosynthesis System for Bluestem OSTM Version 1.3*; Li-Cor Biosciences: Lincoln, NE, USA, 2018; Volume 1.
39. Covshoff, S. *Photosynthesis Methods and Protocols*; Humana Press: Totowa, NJ, USA, 2018; Volume 1770. [CrossRef]
40. Marozas, V.; Augustaitis, A.; Pivoras, A.; Baumgarten, M.; Mozgeris, G.; Sasnauskienė, J.; Dautartė, A.; Abraitienė, J.; Bičenkienė, S.; Mordas, G.; et al. Comparative Analyses of Gas Exchange Characteristics and Chlorophyll Fluorescence of Three Dominant Tree Species during the Vegetation Season in Hemi-Boreal Zone, Lithuania. *J. Agric. Meteorol.* **2019**, *75*, 3–12. [CrossRef]
41. Terashima, I. Anatomy of Non-Uniform Leaf Photosynthesis. *Photosynth. Res.* **1992**, *31*, 195–212. [CrossRef] [PubMed]
42. Farquhar, G.D.; von Caemmerer, S.; Berry, J.A. A Biochemical Model of Photosynthetic CO_2 Assimilation in Leaves of C_3 Species. *Planta* **1980**, *149*, 78–90. [CrossRef] [PubMed]
43. Sharkey, T.D.; Bernacchi, C.J.; Farquhar, G.D.; Singsaas, E.L. Fitting Photosynthetic Carbon Dioxide Response Curves for C3 Leaves. *Plant Cell Environ.* **2007**, *30*, 1035–1040. [CrossRef] [PubMed]

44. Yin, X.; Struik, P.C.; Romero, P.; Harbinson, J.; Evers, J.B.; Van Der Putten, P.E.L.; Vos, J. Using Combined Measurements of Gas Exchange and Chlorophyll Fluorescence to Estimate Parameters of a Biochemical C3 Photosynthesis Model: A Critical Appraisal and a New Integrated Approach Applied to Leaves in a Wheat (*Triticum aestivum*) Canopy. *Plant Cell Environ.* **2009**, *32*, 448–464. [CrossRef] [PubMed]
45. Medlyn, B.E.; Dreyer, E.; Ellsworth, D.; Forstreuter, M.; Harley, P.C.; Kirschbaum, M.U.F.; Le Roux, X.; Montpied, P.; Strassemeyer, J.; Walcroft, A.; et al. Temperature Response of Parameters of a Biochemically Based Model of Photosynthesis. II. A Review of Experimental Data. *Plant Cell Environ.* **2002**, *25*, 1167–1179. [CrossRef]
46. Bernacchi, C.J.; Singsaas, E.L.; Pimentel, C.; Portis, A.R.; Long, S.P. Improved Temperature Response Functions for Models of Rubisco-Limited Photosynthesis. *Plant Cell Environ.* **2001**, *24*, 253–259. [CrossRef]
47. Duursma, R.A. Plantecophys—An R Package for Analysing and Modelling Leaf Gas Exchange Data. *PLoS ONE* **2015**, *10*, 1–13. [CrossRef] [PubMed]
48. Dubois, J.J.B.; Fiscus, E.L.; Booker, F.L.; Flowers, M.D.; Reid, C.D. Optimizing the Statistical Estimation of the Parameters of the Farquhar-von Caemmerer-Berry Model of Photosynthesis. *New Phytol.* **2007**, *176*, 402–414. [CrossRef]
49. Yin, X.; Struik, P.C. Theoretical Reconsiderations When Estimating the Mesophyll Conductance to CO2 Diffusion in Leaves of C3 Plants by Analysis of Combined Gas Exchange and Chlorophyll Fluorescence Measurements. *Plant Cell Environ.* **2009**, *32*, 1513–1524. [CrossRef] [PubMed]
50. Moualeu-Ngangue, D.P.; Chen, T.W.; Stützel, H. A New Method to Estimate Photosynthetic Parameters through Net Assimilation Rate−intercellular Space CO2 Concentration (A−Ci) Curve and Chlorophyll Fluorescence Measurements. *New Phytol.* **2017**, *213*, 1543–1554. [CrossRef] [PubMed]
51. Abdi, H.; Williams, L.J. Tukey's Honestly Significant Difference (HSD) Test. *Encycl. Res. Des.* **2010**, *3*, 1–5.
52. R Core Team. *A Language and Environment for Statistical Computing*; R Foundation for Statistical Computing: Vienna, Austria, 2019; Available online: https://www.R--project.org (accessed on 2 May 2021).
53. Maurer, S.; Matyssek, R. Nutrition and the Ozone Sensitivity of Birch (Betula Pendula) II. Carbon Balance, Water-Use Efficiency and Nutritional Status of the Whole Plant. *Trees Struct. Funct.* **1997**, *12*, 11–20. [CrossRef]
54. Xu, Z.; Hu, T.; Zhang, Y. Effects of Experimental Warming on Phenology, Growth and Gas Exchange of Treeline Birch (*Betula utilis*) Saplings, Eastern Tibetan Plateau, China. *Eur. J. For. Res.* **2012**, *131*, 811–819. [CrossRef]
55. Hanson, D.T.; Stutz, S.S.; Boyer, J.S. Why Small Fluxes Matter: The Case and Approaches for Improving Measurements of Photosynthesis and (Photo)Respiration. *J. Exp. Bot.* **2016**, *67*, 3027–3039. [CrossRef] [PubMed]
56. Collatz, G.J.; Ball, J.T.; Grivet, C.; Berry, J.A. Physiological and Environmental Regulation of Stomatal Conductance, Photosynthesis and Transpiration: A Model That Includes a Laminar Boundary Layer. *Agric. For. Meteorol.* **1991**, *54*, 107–136. [CrossRef]
57. Kaluthota, S.; Pearce, D.W.; Evans, L.M.; Letts, M.G.; Whitham, T.G.; Rood, S.B. Higher Photosynthetic Capacity from Higher Latitude: Foliar Characteristics and Gas Exchange of Southern, Central and Northern Populations of *Populus angustifolia*. *Tree Physiol.* **2015**, *35*, 936–948. [CrossRef]
58. Medrano, H.; Tomás, M.; Martorell, S.; Flexas, J.; Hernández, E.; Rosselló, J.; Pou, A.; Escalona, J.M.; Bota, J. From Leaf to Whole-Plant Water Use Efficiency (WUE) in Complex Canopies: Limitations of Leaf WUE as a Selection Target. *Crop J.* **2015**, *3*, 220–228. [CrossRef]
59. Franks, P.J.; Adams, M.A.; Amthor, J.S.; Barbour, M.M.; Berry, J.A.; Ellsworth, D.S.; Farquhar, G.D.; Ghannoum, O.; Lloyd, J.; McDowell, N.; et al. Sensitivity of Plants to Changing Atmospheric CO2 Concentration: From the Geological Past to the next Century. *New Phytol.* **2013**, *197*, 1077–1094. [CrossRef] [PubMed]
60. Alía, L.A.; Lüttschwager, D.; Ewald, D. Investigation of Gas Exchange and Biometric Parameters in Isogenic Lines of Poplar Differing in Ploidy. *Silvae Genet.* **2015**, *64*, 46–59. [CrossRef]
61. Bühlmann, T.; Hiltbrunner, E.; Körner, C. Alnus Viridis Expansion Contributes to Excess Reactive Nitrogen Release, Reduces Biodiversity and Constrains Forest Succession in the Alps. *Alp. Bot.* **2014**, *124*, 187–191. [CrossRef]
62. Koike, T. Autumn coloring, photosynthetic performance and leaf development of deciduous broad-leaved trees in relation to forest succession. *Tree Physiol.* **1990**, *7*, 21–32. [CrossRef] [PubMed]
63. Manter, D.K.; Kerrigan, J. A/Ci Curve Analysis across a Range of Woody Plant Species: Influence of Regression Analysis Parameters and Mesophyll Conductance. *J. Exp. Bot.* **2004**, *55*, 2581–2588. [CrossRef] [PubMed]
64. Qian, T.; Elings, A.; Dieleman, J.A.; Gort, G.; Marcelis, L.F.M. Estimation of Photosynthesis Parameters for a Modified Farquhar-von Caemmerer-Berry Model Using Simultaneous Estimation Method and Nonlinear Mixed Effects Model. *Environ. Exp. Bot.* **2012**, *82*, 66–73. [CrossRef]
65. Farquhar, G.D.; Von Caemmerer, S.; Berry, J.A. Models of Photosynthesis. *Plant Physiol.* **2001**, *125*, 42–45. [CrossRef] [PubMed]
66. Miao, Z.; Xu, M.; Lathrop, R.G.; Wang, Y. Comparison of the A-Cc Curve Fitting Methods in Determining Maximum Ribulose 1·5-Bisphosphate Carboxylase/Oxygenase Carboxylation Rate, Potential Light Saturated Electron Transport Rate and Leaf Dark Respiration. *Plant Cell Environ.* **2009**, *32*, 109–122. [CrossRef] [PubMed]
67. Rostamza, M.; McNickle, G.G. A Global Database of Photosynthesis Model Parameters, and Phylogenetically Controlled Analysis of Photosynthetic Responses from Every Major Terrestrial Plant Clade. *bioRxiv* **2020**. [CrossRef]

MDPI

Article

Growth Relationships in Silver Fir Stands at Their Lower-Altitude Limit in Romania

Gheorghe-Marian Tudoran [1,*], Avram Cicșa [1,2], Albert Ciceu [1,2] and Alexandru-Claudiu Dobre [1,2]

1 Department of Forest Engineering, Forest Management Planning and Terrestrial Measurements, Faculty of Silviculture and Forest Engineering, "Transilvania" University, 1, Ludwig van Beethoven Str., 500123 Brașov, Romania; cicsa_avram@yahoo.co.uk (A.C.); albert.ciceu@yahoo.ro (A.C.); dobre.alexandruclaudiu@gmail.com (A.-C.D.)

2 Department of Forest Monitoring, "Marin Drăcea" Romanian National Institute for Research and Development in Forestry, 128 Eroilor Blvd., 077190 Voluntari, Romania

* Correspondence: tudoran.george@unitbv.ro; Tel.: +40-268-418-600

Abstract: This study presents the biometric relationships among various increments that is useful in both scientific and practical terms for the silvicultural of silver fir. The increments recorded in the biometric characteristics of trees are a faithful indicator of the effect of silvicultural work measures and of environmental conditions. Knowing these increments, and the relationships among them, can contribute to adaptations in silvicultural work on these stands with the purpose of reducing risks generated by environmental factors. We carried an inventory based on tree increment cores. The sample size was determined based on both radial increment and height increment variability of the trees. The sample trees were selected in proportion to their basal area on diameter categories. Current annual height increment (CAI_h) was measured on felled trees from mean tree category. For CAI_h we generated models based on the mean tree height. Percentages of the basal area increment and of form-height increment were used to compute the current annual volume increment percentage ($PCAI_v$). For the mean tree, the CAI_h estimated through the used models had a root-mean-square error (RMSE) of 0.8749 and for the current annual volume increment (CAI_v) the RMSE value was 0.1295. In even-aged stands, the mean current volume increment tree is a hypothetical tree that may have the mean basal area of all the trees and the form-height of the stand. Conclusions: The diameter, height, and volume increments of trees are influenced by structural conditions and natural factors. The structures comprising several generations of fir mixed with beech and other deciduous trees, which have been obtained by the natural regeneration of local provenances, are stable and must become management targets. Stable structures are a condition for the sustainable management of stands.

Keywords: silver fir; current annual increment; percentage volume increment; basal area; tree diameter; tree form-height

Citation: Tudoran, G.-M.; Cicșa, A.; Ciceu, A.; Dobre, A.-C. Growth Relationships in Silver Fir Stands at Their Lower-Altitude Limit in Romania. *Forests* **2021**, *12*, 439. https://doi.org/10.3390/f12040439

Academic Editors: Alessandra De Marco and Timothy A. Martin

Received: 20 January 2021
Accepted: 3 April 2021
Published: 5 April 2021

1. Introduction

In Romania, the fir (*Abies alba* Mill.) is frequently found mixed with beech (*Fagus sylvatica* L.) at altitudes between 700 and 1200 m. Mixed of fir and beech forests are formations representative of the lower mountain zone in Romania. Of the coniferous trees, Norway spruce (*Picea abies* (L.) Karst.) descends to these altitudes sporadically, but heat and reduced precipitation become limiting factors to its lower range; the proportion of spruce increases with altitude. Fir and beech influence climatic factors differently—especially humidity, light, and heat—so that, under the shelter they provide, beech seedlings frequently become established under fir, and vice versa. Consequently, the structures of stands are greatly varied in terms of the relative proportions of the two species. The stands structures present a wide range of diameter categories and offer the most favorable conditions for promoting structure of the uneven-aged stand type, which is characteristic of the natural selection system [1–6]. Long-term experiments [7–11] revealed the effects of long-term interventions on the structure, growth, and total production of stands [12]. In order to change even-aged stand structures into structures suitable for 65 forest selection systems, silvicultural conversion treatments are applied [13–16] For even-aged beech-coniferous mixed stands in the Postăvarul Massif, the interventions had a character of transformation to uneven-aged structure.

Knowledge of the growth and relationships among trees can explain the development of stands. Interventions carried out on stands result in regulation of the ratio between the number of trees and their growth area. Change in diameter is the most dynamic biometric characteristic due to its sensitivity to silvicultural practices. A reduction in density can result in an acceleration of the diameter increment.

Although a reduction in stocking degree can generate significant growth increases, these increases may be only temporary. If a certain stocking degree is exceeded, this can lead to losses in production and exploitability. The reactions of trees to a reduction in stocking degree, through the activation of diameter increments, are particularly notable in young, vigorous trees [17]. Stand structure also dictates the growing pattern, with radial stem increments having a linear variation in even-aged stands, whereas a curvilinear variation occurs in uneven-aged stands due to a decrease in the radial increment in trees with large diameters. Therefore, the largest stem increments occur in trees in central categories. The volume increment varies depending on the correlation between height and diameter increments. These aspects—for both even- and uneven-aged stands—have been discussed in detail in the forestry literature [18–25] and were expressed in growth functions used in forest modeling [26], developed at the level of the stand, size class, or single tree [27].

When no successive inventories are performed, the current annual volume increment (CAI_v) can be determined based on the percentage current annual volume increment ($PCAI_v$). $PCAI_v$ is based on the diameter and height increments. The models that have been developed express these increments in relation to variables such as diameter, height, and age of the trees. Following the experimentation of three variants of determining Δh (measured on felled trees, estimated by dynamic height curve and by conventional height curve) the elaborated models estimate the current volume increment with reduced errors of only 4–9% [28].

Other studies relating to the physiological processes of trees highlighted that the size of aged trees [29,30] (but not necessarily their age [31]) becomes very important when considering a reduction in their height increments [32]. The correlation between radial and height increments is weak, such that the diameter increment cannot be used for estimating the height increment. The radial increment has maximum values at different points in a tree's lifespan, with fluctuations caused by silvicultural practices, site conditions, climatic factors, inter- and intraspecific competition, and niche development [33]. The correlation between severe periods of reduction in fir growth and climatic factors has also been highlighted by dendrochronological studies [34] through drought indices. This also has a significant influence on the radial increment in trees [35,36].

Diameter strongly influences the CAI_v and, consequently, a tree's current annual basal area increment (CAI_g) has an equally significant impact on the volume increment. One study on Romanian even-aged stands [19] highlighted a linear correlation relationship, with high values (r = 0.80–0.95) occurring in mid-diameter categories between the CAI_g and CAI_v.

It is important to note that all of these determinations, which are aimed at identifying the variability in stand growth and the relation between increments and tree characteristics, require the use of a large number of sample trees. To ensure a sampling error within 10%, for a coverage probability of 95%, it is necessary to take measurements from at least 40–60 trees. If measurements are taken from only 10 trees, the sampling error will be between 20 and 30% [19].

Understanding the growth behavior of the fir at the local level is essential to the adaptation of its management. The trends in annual increment are also useful for understanding tree vitality. There are differences between the current annual increments of trees and stands and the increment values determined as averages by periods (periodic annual increments). It is therefore useful to know the current increment determined for each year of increment. Furthermore, the size of the CAI_v should result from directly measurable elements that intervene in its size. If CAI_g and the CAI_h or the current annual form height increment (CAI_{fh}) participates in the CAI_v, then, in the calculation relations, the annual current values of these characteristics can be entered. The CAI_g participates with the largest share in the CAI_v, but its contribution in terms of volume increment varies during tree development. Knowing each increment's contribution to the size of the CAI_v during tree development provides information on opportunities for silvicultural work to stimulate the volume increment as well as each increment's effect on volume increment. The CAI_v can be determined by direct measurements or increment cores. When height increment is not measured, it can be determined indirectly using the correlations between height increment and other biometric characteristics of trees. The extraction of increment cores from all the inventoried trees leads to the safest results, but the procedures based on sample trees are also accessible to practice. We wished to consider all these aspects at the structure level, so as to be able to develop biometric relationships that would be useful in silvicultural practices.

The main objectives of this study were (1) develop models to estimate the $PCAI_h$ based on the mean tree height of stand, (2) assess the accuracy of the relationship (based on the $PCAI_g$ and $PCAI_{fh}$) for estimating the volume increment at the mean tree and stand level, and (3) assess the possibility of using the mean tree for determining the volume increment of the stand.

2. Materials and Methods

2.1. Materials

Table 1 shows the symbols of the variables used in this study.

Table 1. Abbreviations of variables used.

Abbreviation	Units	Description
CAI		Current annual increment
$CAI_{d,g,h,fh,v}$ or $i_{d,g,h,fh,v}$	cm, cm^2, m, m, dm^3	Current annual increment of the tree diameter, basal area, height, form-height and volume ($CAI_{d,h}$ synonymous with Δd, Δh)
$CAI_{D,G,H,FH,V}$ or $I_{D,G,H,FH,V}$	cm, m^2, m, m, m^3	Current annual increment of the stand diameter, basal area, height, form-height and volume or the current annual increment for a group of trees, at category-of-diameter level ($CAI_{G,H}$ synonymous with BAI, HI)
N	–	Number of trees per hectare or a group of trees
$\overline{i_v}$	dm^3	Arithmetic mean increment of a number of trees in one year, $\overline{i_v} = \frac{I_v}{N}$ or $\overline{CAI_v} = \frac{CAI_v}{N}$
$PCAI_{d,g,h,fh,v}$ or $p_{id}, p_{ig}, p_{ih}, p_{ifh}, p_{iv}$	%	Percentage current annual increment of the tree diameter, basal area, height, form-height, and volume ($PCAI_{g,h}$ synonymous with PBAI, PHI)
i_r	mmyr^{-1}	Annual radial increment
G	m^2	Basal area of the stand (or the basal area for a group of trees) (synonymous with BA)
V	m^3	Stand volume per hectare (standing volume) or the volume for a group of trees
dbh or d	cm	Diameter of breast height (1.3 m)
d_{max}	cm	Diameter of the thickest tree
d_{min}	cm	Diameter of the thinnest tree

Table 1. *Cont.*

Abbreviation	Units	Description
$\overline{d}$	cm	Arithmetic mean diameter
d_g	cm	Mean squared diameter or quadratic mean diameter (diameter corresponding to mean basal area of stand)
d_M	cm	Diameter of median tree
d_{gM}	cm	Diameter of median basal area tree
d_v	cm	Diameter of mean volume tree (diameter of the tree with $\overline{v}$)
d_{vM}	cm	Diameter of median volume tree
g	cm^2	Basal area of tree (at 1.3 m) (synonymous with ba)
$\overline{g}$	cm^2	Mean basal area of all the trees of the stand, $\overline{g} = \frac{G}{N}$. Basal area corresponding to *mean basal area tree(MBAT)* or the basal area of the tree with d_g (i.e., g_g). For stand with even-aged structures, the tree with mean basal area is considered the *mean tree* of the stand
h_g	m	Height of the tree with d_g (height of the mean basal area tree)
v	m^3	Volume of tree
$\overline{v}$or v_v	m^3	Mean volume of all the trees of the stand, $\overline{v} = \frac{V}{N}$. Volume corresponding to *mean volume tree (MVT)* (i.e., v_v)
f	–	Form factor (determined by the ratio v/gh). For mean basal area tree is f_g
h_g/d_g	–	Slenderness of the mean basal area tree
fh	–	Form-height of the tree, determined by the ratio v/g. For the tree with $\overline{i_v}$, it is determined by the ratio $\overline{v}/\overline{g}$.
FH	–	Form height of the stand (determined by the ratio V/G or $\overline{v}/\overline{g}$)
SD	m^2	Stocking degrees determined by the ratio $G_{\text{observed}}/G_{\text{normal (yield tables)}}$

Study area: The research was carried out in stands of fir with beech situated in the Valea Cetății watershed in the Postăvarul Massif. The stands are situated at altitudes ranging between 700 and 950 m, on slopes with inclinations of between 20 and 35° and various slope aspects. The area is characterized by multiannual average temperatures of 7–8.3 °C and annual precipitation of around 780 mm. The soils are deep and intermediate-depth eutricambosoils, with a groundcover of *Galium odoratum–Cardamine bulbifera* type. Our study was limited to stands in which fir constituted at least 30% of stands composition (to ensure the sample representativeness). From the stands, we selected a key surface area comprising 10 stands covering 131.8 ha (please see the Supplementary material).

Field measurements: The stands had average ages of 100–130 years, with disseminated elements that may have reached 140–150 years old. The stands basal area are ranging between 35 and 45 m^{2}ha^{-1}. In each stand, observations were made and measurements performed on the site conditions, herbaceous flora and seedlings. To determine the structure of the stands, experimental areas of 0.25–1.0 ha were studied, in which all the trees with diameters greater than 3 cm were inventoried. For each inventoried tree, several biometric characteristics were measured (diameter, height, pruning height, diameter of the crown). In this study, we exemplify the main biometric characteristics for one stand (45°37′05″ N, 25°35′39″ E) representative of a sample area of 1 ha (100 × 100 m) (Table 2). Of 133 inventoried mature fir trees, we extracted increment cores from 64 trees, with two diametrically opposed cores taken from each, following a direction parallel to the contour line (from 45 trees), and four cores from each, in two perpendicular directions, from trees where the inclination of the ground also permitted the extraction of cores from downstream (from 19 trees, i.e., 166 cores). The trees from which the cores were extracted were selected in report to their basal area on diameter categories. Therefore, for large diameter categories (due to the smaller number of tree) samples were extracted also form outside the 1.0 ha plot. Core samples were extracted from 14 diameter categories (characterized by amplitude that varied between 2 and 8 cm). Thus, the obtained radial growth and the curve of radial increments address the trees growth trend corresponding to all diameter categories. The size of the sample ensured an error of ±8% under the conditions of a probability of 95%, which was established following an examination of the variability of radial increments. In seven felled sample trees with dimensions close to the mean basal area tree, we determined height increments by measuring the stem internode distance from the top to the base. Towards the inferior part of trees (i.e., the first 3 m of stem) where the position of the verticil was not suitable, we used increment cores extracted to the pith (to establish the trees age at different heights).

Table 2. Biometric characteristics (100 × 100 m).

Species and Generations		d_g (cm)	d_{gM} (cm)	h_g (m)	G (m^2 ha^{-1})	V (m^3 ha^{-1})	SD	d_{min} (cm)	d_{max} (cm)	% Number of Trees by Category of Diameter/ % Volume by Category of Diameter				
										4–12 (cm)	16–24 (cm)	28–36 (cm)	40–48 (cm)	≥52 (cm)
Fir	1	45.0	47.8	31.6	21.11	299.1	0.35	3.1	78.5	7/–	2/–	23/13	46/49	22/38
	2	7.6	11.7	5.8	0.05									
Beech	1	33.3	34.6	28.5	12.70	177.7	0.36	4.0	67.0	51/2	17/10	25/54	6/27	1/7
	2	9.6	13.8	10.1	1.57									
Spruce		40.5	46.9	31.1	2.19	30.5	0.03	23.9	57.8	–/–	5/2	54/34	23/30	18/34
Total		–		–	37.62	507.3	0.74	–	–	37/1	12/4	26/30	18/39	7/26

Seedlings: composition: fir (55%), beech (35%), spruce (8%), sycamore, and Norway maple (2%)—coverage: 40% of the stand area.

In what concern the stands included in the key area, we present general data in the Supplementary Table S1.

2.2. Methods

Calculating the growth of trees: The increment cores (extracted to the pith) of trees with different diameters were measured. The measurement of the diameter increment in trees of different ages enabled us to determine their development in relation to the diameter and basal area. Annual height increments, as determined in the sample trees, were expressed as a percentage in relation to the height at the end of the growing period. These data yielded models that expressed the variation in annual increments in relation to the heights the trees reached during their lifetimes. We used the ratio of the heights of mean trees to reconstitute the heights from the beginning of the growing period (0) to the end (n), h_{g0}/h_{gn}.

The diameter increment was determined on the basis of radial increments (i_r), according to the relation $d_0 = d_n - 2i_r$, and the radial increments that we used in the calculations came from the equation for the radial increments curve. Radial increments were measured using a digital positiometer.

The CAI_g (or i_g) still resulted from the difference between the two moments (at the end and beginning of the chosen period), on the basis of a diameter increment, according to the relation:

$$i_g = \pi(d_n i_r k - {i_r}^2 k^2) \tag{1}$$

We used a coefficient for the bark (k) of 1.054, which we determined experimentally as being specific to the stands studied. Bark thickness was measured in 142 fir trees.

CAI_h were measured in the felled sample trees within the category of mean basal area tree.

The increment percentages of each biometric characteristic (d, g, h, fh, or v) were analyzed at the level of the sample trees from the category of the mean basal area tree (felled). To calculate the percentages of annual increments of the biometric characteristics at the level of average tree (diameter, basal area, height, volume), we applied Pressler's formula [22].

The CAI_v was determined at the level of the mean tree of the stand and at the level categories of diameter as well as the entire stand. At the category-of-diameter and stand levels, the calculation was similar to that used for an individual tree. To obtain CAI_v (or i_v), we performed a single inventory based on increment cores.

Based on the diameter and height increments, the diameter and height of the trees could be determined at the beginning of the growth year. The CAI_v was determined by subtracting the volume at the end (n) from that of the beginning (0) of the chosen period; in the case of the tree, this was

$$i_v = v_n - v_0. \tag{2}$$

The volume of the trees was determined based on diameter and height by regression equation used for this species in Romania [23]:

$$\log v = a_0 + a_1 \log d + a_2 \log^2 d + a_3 \log h + a_4 \log^2 h \tag{3}$$

In the Equation (3): $a_0 = -4.46414$; $a_1 = 2.19479$; $a_2 = -0.12498$; $a_3 = 1.04645$; $a_4 = -0.016848$.

The relationships between the volume increments at the tree level (i_v) and stand level (I_V) were developed by introducing the basal areas (g and G, respectively) and form heights (fh) from the beginning and end of the chosen period. For the tree, we reached the known relation $i_v = g_n h_n f_n - g_0 h_0 f_0$. We expressed the form-height from the beginning of the chosen period (fh_0), based on the form-height increment (CAI_{fh} or i_{fh}). We also introduced the CAI_g (or i_g) into the relation of the CAI_v, respectively $i_g = g_n - g_0$ and we reached the known relationships: [18] $i_v = i_g fh_n + g_0 i_{fh}$ and [19,23]:

$$i_v = g_n i_{fh} + i_g fh_n - i_g i_{fh} \tag{4}$$

The CAI_v values we obtained by applying the Equations (2) and (4) were considered as reference values. With the CAI_v determined by the two relations, we compared the CAI_v obtained by applying the simplified relationship-based $PCAI_v$.

The characteristics of the mean trees (of the basal area and of the volume) resulted, indirectly, from the following calculations: d_g of the mean tree from the mean basal area (determined by the ratio G/N), the volume of the mean volume tree from V/N, and the form-height (FH) of the stand from V/G.

Finally, we present the value of the CAIs for fir mean tree and stand. They have been determined for a single year of growth, based on the values of the variables measured at the level of year 2017, which followed a period in which the health state of fir in the study area had continuously deteriorated.

3. Results

3.1. Volume Increment of the Mean Tree

The current increment percentages of the mean tree: The CAI_v is the result of diameter increment, height, and changes in the stem shape of the trees. In the $PCAI_v$, the $PCAI_g$ of the mean basal area tree can be used. The $PCAI_g$ has a relatively significant influence on the $PCAI_v$.

At ages between 110 and 130 years, the CAI_h models indicated a height increment (CAI_h) ranging between 0.8 and 0.14 m. Increments of 0.14 m were recorded in structurally closed stands situated on the lower parts of slopes. The increments decreased toward the higher parts of slopes and toward stand densities less than 0.8. Models assist in predicting annual or periodic height increments in trees of various ages in relation to their heights. Such a model is represented by the equation:

$$PCAI_h = 0.147097(h - 1.21892)^{-0.42450}(40.20507 - h)^{1.18566}, \tag{5}$$

which characterizes the experimental distribution in Figure 1b, or

$$PCAI_h = 0.020125(h - 1.181557)^{-0.359495}(40.158443 - h)^{1.750728} \quad (6)$$

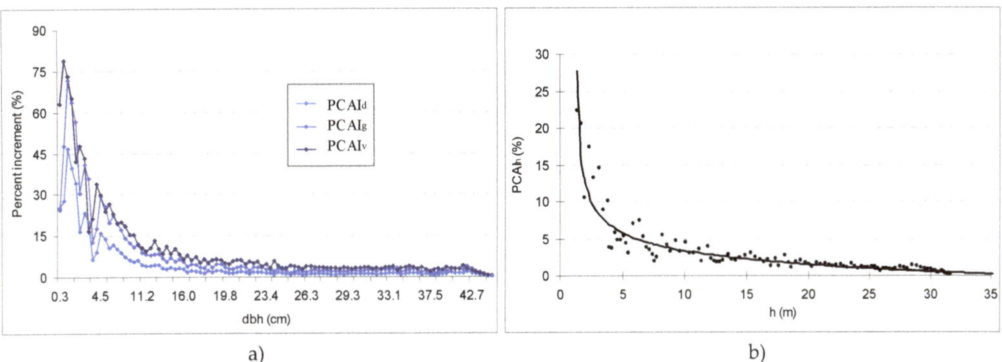

Figure 1. Percentages of mean fir tree increment: (**a**) percentage of diameter annual increments, basal area and volume (determined by Pressler's formula); (**b**) percentage height increment of the mean tree by Equation (5) (in relation to the height of the mean tree at the end of the chosen period).

In Equations (5) and (6), the CAI_h of the trees is expressed as a percentage in relation to their height at the end of the chosen period. The models become applicable if the height of the tree is known, based on a single inventory. In turn, the variation in height in relation to age can be explained through the development function. The height can also be determined by means of the height curve, which expresses the variation of the height in relation to the diameter of the trees. Equations (5) and (6) explained 84–86% of the variation units in the CAI_h. The height and diameter significantly influence the percentage annual height increment ($PCAI_hs$) and explain its tendency (R^2 = 0.91–0.93).

The percentages of the increment of each biometric characteristic ($d, h, g,$ or v) diminish as a tree advances in age (Figure 1).

Relationships between the current increment percentages: Throughout a tree's development, the percentage increments $PCAI_g$ and $PCAI_h$ changes from one year to the next and from one development period to the next (Table 3). Together, $PCAI_g$ and $PCAI_h$ determine the tree's volume increment and ensure its tendency, as expressed by the functions of growth and development.

Table 3. Annual mean percentages of mean fir growth.

Tree Height (h)	$PCAI_d$	$PCAI_h$	$PCAI_g$	$PCAI_v$	$PCAI_{fh}$
	(a) Felled sample trees				
0.1–0.5 h (15–55 years old)	6.10	4.10	12.20	16.50	4.30
0.51–0.8 h (56–85 years old)	1.21	1.61	2.42	3.84	1.42
0.81–1.0 h (86–115 years old)	1.17	0.80	2.35	2.92	0.57
Total height (15–115 years old)	3.23	2.40	6.45	8.80	2.35
	(b) Mean tree (in last year of growth, 2017)				
Total	0.32	0.23	0.64	0.80	0.16

Results obtained from the analysis of the stem of the felled sample trees.

Fir trees presented the highest percentages of growth during the period between 15 and 55 years. The greatest contribution of the $PCAI_h$ in the $PCAI_v$ (42%) is made during the period between 56 and 85 years. After this interval, the contribution of $PCAI_h$ in $PCAI_v$ climbs as high as 20%. Between 85 and 115 years, $PCAI_v$ was 2.92% on average and this is largely (80%) attributable to the $PCAI_g$.

The annual percentages of the increments established at the sample-tree level prove known relationships:

$$PCAI_v = PCAI_g + PCAI_{fh} \quad (7)$$

and:

$$PCAI_g = 2PCAI_d \quad (8)$$

The differences between the experimental values of the percentage of height increment and those estimated by the models diminished as the trees grew in height (Figure 2a). The average deviations of the experimental values of %ih, in comparison to the values estimated by the equation (BIAS), were +0.001483 and the average of the squares of the deviations RMSE was 0.87490. Using Equation (7) to calculate the $PCAI_v$ yielded an average deviation in the experimental values of $PCAI_v$, compared to the values estimated by the relation, of −0.027425, with an average of the squares of deviations of 0.129521 (Figure 2b and Table 4). When CAI_{fh} was related to the height of the trees at the end of period for the calculation of the $PCAI_{fh}$, the BIAS was 0.019223 and the RMSE was 0.050536. Furthermore, the values of the deviations decreased for volume—just as they did for height—as the volume of the trees increased (Figure 2b). This can be explained by the reduced weight of the percentage $PCAI_h$ in relation to volume as the trees advanced in age. The accumulation in volume, then, is largely due to annual basal area increments.

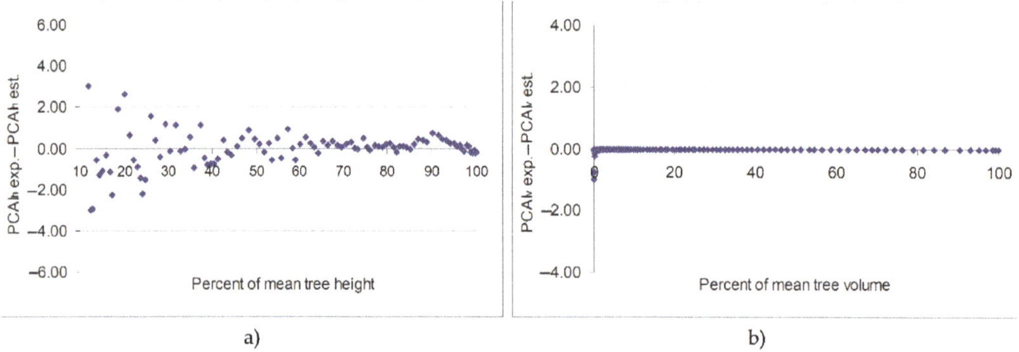

Figure 2. Errors between the experimental values and those predicted by models: (**a**) between the experimental values ($PCAI_h$) and those calculated based on percentages predicted by Equation (5); (**b**) between the experimental values ($PCAI_v$) and those calculated using Equation (7). The differences calculated were expressed in relation to the height (**a**) and volume (**b**) of the mean tree.

Table 4. Proportion of the $PCAI_{fh}$ in the $PCAI_v$ equation (Equation (7)).

Tree Height, h = 31.6 m	$PCAI_v = PCAI_g + k_iPCAI_{fh}$ (Equation (7))					
	k_i (Computed)	BIAS	RMSE	k_i (Adopted)	BIAS	RMSE
3.1–6 m (0.1–0.2 h)	0.9616	−0.00346	0.20752	1.0	−2.55778	0.39196
6.1–9 m (0.21–0.3 h)	0.9960	0.00646	0.00832	1.0	−0.14937	0.02038
9.1–12 m (0.31–0.4 h)	0.9990	−0.00744	0.00231	1.0	−0.03445	0.00510
12.1–15 m (0.41–0.5 h)	0.9994	0.00045	0.00031	1.0	−0.01185	0.00123
15.1–20 m (0.51–0.6 h)	0.9996	0.00089	0.00029	1.0	−0.00946	0.00081
20.1–25 m (0.61–0.8 h)	0.9998	0.00018	0.00004	1.0	−0.00383	0.00024
25.1–32 m (0.81–1.0 h)	0.9998	0.000217	0.00005	1.0	−0.00321	0.00013
Total	–	0.00097	0.10642	–	−0.02742	0.12952

Results obtained from the analysis of the stem of the felled sample trees.

During tree development, the contribution of each biometric characteristic to the volume differs. Therefore, it is natural that, at different moments in tree development, the percentages of the biometric characteristics would have variable proportions relative to the $PCAI_v$. For Equation (7) the participation of the respective percentages in the $PCAI_v$ was expressed by the coefficient k_i (i = 0.1 to 1.0 h). For Equation (7) we noted the stability of the k_i, which decreased throughout the development of the trees (Table 4). Thus, this relationship can be recommended for determining the current volume increment.

Current annual increment of the mean tree: The CAI_v of the mean tree resulted from the application of Equations (2) and (4) adapted to the level of the individual trees, as well as Equation (7). Through the Equations (2) and (4), the same CAI_v value resulted: 17,937 dm^3.

$PCAI_v$-based relationship (expressed by Equation (7)):

$$CAI_v = 0.01(PCAI_g + PCAI_{fh})v_v \quad (9)$$

also led to a value close to the size of 17,937 dm^3, the difference being +0.4%.

3.2. Stand Growth

The CAI_V of the stand was produced by applying Equations (2), (4), and (7) to the category-of-diameter and entire-stand levels. The CAI_h was determined by the regression Equation (5).

At the category-of-diameter and stand level, the calculation was similar to that used for an individual tree. At the level of the entire stand, we introduced into the calculations the values of the biometric characteristics from the level of the entire stand. By applying Equations (2) and (4) similar value for CAI_V as those obtained (i.e., 2.38 m^3 yr^{-1}). Of the simplified Equation (7), based on the percentage-of-form-height increment ($PCAI_{fh}$), presented the greatest stability. This can be applied to determine the volume increment of individual trees (Figure 3a) and the stand (Figure 3b).

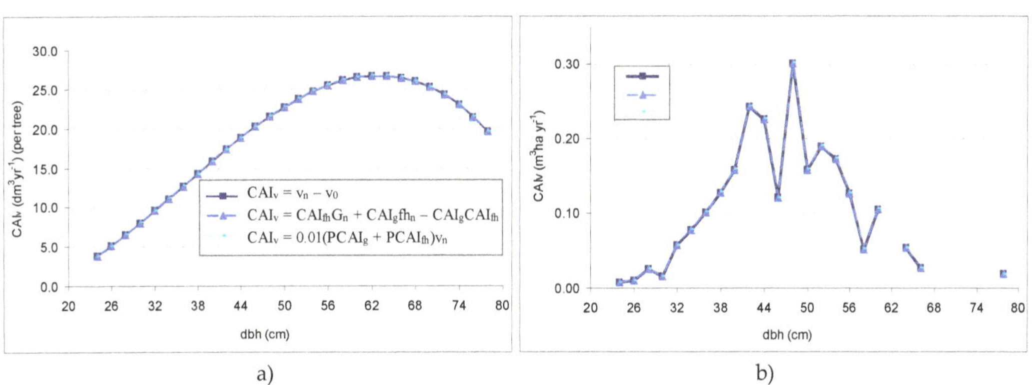

Figure 3. Current annual volume increment for individual trees (**a**) and stand (**b**).

By applying the simplified Equation (7), the volume difference compared to Equations (2) and (4) was +0.4%.

The Mean Tree of the Stand

Given the elements that impact volume increment, more attention should be paid to the basal area of the stand and the CAI_G than to the form-height of the stand (FH).

When applying Equations (4) and (7), the form-height can be deduced from the volume of the mean volume tree or the stand volume. Using only the two trees (i.e., mean basal area tree or mean volume tree) in the calculation of volume increment leads to +8.6 and

+8.9% errors (Table 5). The tree with the mean current annual volume increment ($\overline{i_v}$) is a hypothetical tree that may have the $\overline{g}$ and FH.

Table 5. Mean fir volume increment.

Mean Tree	d (cm)	h (m)	v (m^3)	f	CAI_v (dm^3)	Difference Compared to $\overline{i_v}$ = 17.937 dm^3	
						m^3	%
Median tree	43.42 (d_M)	31.3	2.100	0.453	18.425	+0.488	+2.7
Arithmetic mean tree	43.97 ($\overline{d}$)	31.4	2.156	0.452	18.825	+0.888	+5.0
Mean volume tree	44.89 (d_v)	31.6	2.249	0.450	19.475	+1.538	+8.6
Mean basal area tree	44.96 (d_g)	31.6	2.257	0.450	19.530	+1.593	+8.9
Median volume tree	47.79 (d_{vM})	32.1	2.556	0.444	21.417	+3.480	+19.4
Median basal area tree	47.83 (d_{gM})	32.1	2.561	0.444	21.443	+3.506	+19.5

Furthermore, upon analyzing the results from Table 5, it results that, for the structural conditions of the studied stands (i.e., even-aged structures), the diameter of the real tree with represent about 95% of the d_g. It results that the growth determination based on measurements of trees from the categories of the two trees (i.e., mean basal area tree and mean volume tree) can lead to values of current volume increment close to the values of the tree with $\overline{i_v}$. Thus, the field works can be simplified.

In Figure 3a, it is clear that the trees in the superior-diameter categories, due to their large exchange surfaces, also recorded the greatest volume increments. It was found that the CAI_v per tree was influenced by weight of trees by category of diameter, and that the measure of the CAI_v of a stand depended on the respective size of the growing stock on the distribution of tree volume by diameter category.

In Table 5, the CAI_vs of the mean trees were determined using the Equation (4). For comparative purposes, we used the average value of the CAI_v per tree: 17.937 dm^3.

4. Discussion

4.1. Relationships between Tree Growth and Biometric Characteristics

Due to fluctuations in the radial increments, the maximum diameter increment occurred at different moments in the lifetimes of the trees. The trees with higher radial growths also have large volume growth. Remarkable volume growths were found in the diameter categories with the greatest volumes. However, these trees, in the conditions of the lower-altitude limit, do not have the largest radial increments. For older stands, the radial growth-diameter relationship can be expressed by a second-degree parabola. Expressing radial growth or diameter growth by a line or a logarithmic equation may overestimate the radial growths of older trees. Because stands are suited to uneven-aged structures, the study of increments is of interest, especially for establishing target diameters after which the stand structures may be modeled.

The maximum height increment was influenced by the position of the trees, their vitality and their stationary conditions. Just as with the diameter increment, the height increments of the studied stands, expressed in Equations (5) and (6), presented reduced values. The height increment values measured in the sample trees, expressed as a percentage in relation to the height, diameter or age, indicated the same decreasing tendency, with height and age explaining a variation in percentage of the height growth of 85%. Other studies [28,31] show that the age of trees does not statistically explain the reduced height gain, but the size of the trees significantly influences it.

Throughout the life of a tree, the rhythm of the diameter increment differs from that of the height increment. At the sample-tree level, the $PCAI_v$ is particularly influenced by $PCAI_g$, with the correlation being very strong (R^2 = 0.99). The $PCAI_h$ also influences the volume increment (R^2 = 0.86). In the $PCAI_v$ of the trees, $PCAI_h$ has a low share, around 20% (Table 3). Therefore, the CAI_v of trees is achieved mostly (approximately 80%) based on the $PCAI_g$.

Trees distribute their maximum radial increments at different times in their lives, depending on the structural conditions. The measurements performed on the felled sample-trees, from the category of the mean tree, show that the height increment contributed the greatest weight to the volume increment, among trees aged 55–65 years, when the trees grew 0.5–0.6 of the mean tree height at the end of the growing year, and 0.2 of their volume. At this point, the diameter increment had the lowest proportion of volume increment, but after this moment, the proportion of the diameter increment began to increase, while the proportion of height increment decreased. Thus, after 85 years, a progressively accentuated thickening of the stem occurs. In this period, trees grow 22% of their height increment and 16% of their form-height.

The form factor also decreased as the trees grew in diameter, from a value of 0.505 (for trees with d = 24 cm and h = 26 m) to 0.392 (for trees with d = 78 cm and h = 35.1 m). Fluctuations in the form factor depended on two variables: tree diameter and height. For short periods of time, the form factor gave reduced variations or remained constant [37], so that its value did not influence the measurement of the CAI_v, which was determined by successive inventories. For the example presented, the value of the form factor of the stand remained almost unchanged (from 0.4486 in 2016 to 0.4482 in 2017).

The $PCAI_h$ determined on the basis of experimental data produced values of height increment close to those reported in the existing literature [38] for fir stands in Romania [39].

4.2. Precision in Estimating the Current Volume Increment Using Simplified Relationship

Efforts have been made to establish simplified relationships for determining the CAI_v. In the literature, formulae based on the $PCAI_v$ have been proposed by Dvoreţki, Tiurin, Anucin, Pressler, Breymann, Schneider, and Prodan (see Giurgiu, 1979). We have used only the equation for the $PCAI_v$: $p_{iv} = p_{ig} + p_{ih} + p_{if}$ [22]. The precision of the different formulae varies. The formulae of Tretiacov and Dvoreţki [19] lead to mean squared errors when determining the $PCAI_V$, ranging between ±9 and ±13%, but the determination errors can reach up to 20%, especially for short observation periods of 3–5 years.

Simplified relationships are based on strong correlations between the percentages of mean-tree growth. It is known that, throughout a tree's lifetime, the percentages of biometric-characteristic increments vary from one year to the next, but, on the whole, they preserve a decreasing tendency. Elements such as CAI_g, CAI_{fh}, basal area, and form-height at the end of the growth period decisively influence the volume increment [18].

The precision of the CAI_v determinations based on successive inventory (Equation (2)) was categorically influenced by errors in the respective volumes $e_{Vn\%}$ and $e_{V0\%}$. These volumes were determined using Equation (3). Errors in the two volumes depended on the errors with which the variables introduced in their calculus were determined. For a probability of 95%, the interval of errors reaches of 8–10% [37]. Under our study conditions, the tree diameters were determined based on circumference, the heights were measured in all 133 samples, and the form factor was found using the volume of the trees, so we can consider that the error in volume increment determined by successive inventories reached a maximum of 10%.

According to Krenn's equation [18], in the case of successive inventories, the greater the volume of the stand and its determination error, the less the growth in the volume of the stand, and the greater the volume increment error. For periods of 10 years, the error in CAI_v, obtained by applying the equation based on form-height and basal area (Equation (4)) from the beginning of the observation period, was ±8%, decreasing to ±2.4% when the period was extended to 40 years [18].

In the case of simplified relationship (Equation (7)), the error in CAI_v (e_{CAIv}) depends on the error involved in determining the volume (e_v), the volume to which $PCAI_v$ is applied, and the error in the $PCAI_v$ (e_{PCAIv}). Thus, the error in volume increment could be written:

$$e_{PCAIv}{}^2 = e_v{}^2 + e_{PCAIv}{}^2 \quad (10)$$

The error involved in determining the volume of the trees and the stand (e_v) can be considered to be ±10%. In determining the error in the $PCAI_v$, the error in $PCAI_g$ (e_{PCAIg}) and the error in the $PCAI_{fh}$ (e_{PCAIfh}) intervene. By analyzing radial increments from 64 trees, the error e_{PCAIg} was found to represent ±8%. Given that e_{PCAIfh} represents around 25% of e_{PCAIg}, we can consider that e_{PCAIph} represents 2%. By introducing these values into each relationship (10), errors in volume increment fall within ±13% (in ~90% of the cases). These errors also include possible errors in determining the volume of the trees using Equation (3). Only the error in determining the $PCAI_v$ (e_{PCAIv}), obtained by applying the relation (7), is ± 8%.

4.3. Practical Utility of Volume Increment Relationship

The relationships used to determine the CAI_v can also be applied to cases where a single inventory is carried out at the end of the chosen period. In such cases, to determine CAI_g, the extraction of increment cores is necessary. If the CAI_h cannot be measured, it can be determined with the help of Equations (5) and (6). Because the form factor fluctuates throughout the lifetime of trees, it is best that the form-height be determined from the volume of the stand. However, the use of the volume, both at the end and beginning of the chosen period, leads to diminishing errors in determining volume increments.

At the level of the mean tree, using simplified Equation (7) to has led to values of the CAI_v (17.937 dm^3) that are close to those obtained using Equations (2) and (4) (Table 6). Annual increments in height of 0.08 m were obtained by other studies carried out on fir located at its southernmost distribution limit, but at tree heights of 42.5 m [32].

Table 6. Methods used to determine the current annual volume increment.

Level of Inventory		Method		
		Successive Inventory (SI) (2)	Relationship Based on Basal Area and Form-Height Increments (4)	Simplified Relationship Based on $PCAI_v$ (8)
CAI_v of mean tree of stand, dm^3		17.937	17.9367	18.009
Difference compared to SI	dm^3	–	0	+0.073
	%	–	0	+0.4
CAI_V of stand, m^3		2.386	2.386	2.395
Difference compared to SI	m^3	–	0	+0.009
	%	–	0	+0.4

Biometric characteristics of stand: 115 years old, volume 299.1 m^3yr^{-1}, SD stand 0.74, SD fir 0.35, N 133 trees.

At the level of the stand, the $PCAI_v$ of the mean tree (0.80 from Table 3), obtained using Equation (7), applied to the volume of the fir stand, produced a measure of growth of 2.39 m^3yr^{-1}, which is 0.009 m^3 greater than the (+0.4%) current growth of the tree (Table 6).

Limits to the simplified relationship: The simplified relationship, Equation (7), introduced the entire measure of $PCAI_{fh}$ into the calculation, together with the $PCAI_g$. This made it applicable to any structural conditions, both at the level of individual trees, category of diameter and the entire stand.

The adaptation of silvicultural work in fir stands: This study showed that, at their lower-altitude limit, old, even-aged stands have lower growth. Future structures of fir and beech mixed stands will need to be adapted to ensure a better representation of younger generations in the stands [5,6]. These generations are capable of ensuring the continuity of the protection functions attributed to such stands as well as their stability. To create such structures, silvicultural interventions in the stands need not delay placing installed seedlings in sunlight or extracting trees in the central categories that hinder the stand's development. For stands in transformation, the necessity of installing and developing new generations dictates the timing and intensity of such interventions. By the age of around 80 years, silvicultural works should aim to stimulate the growth of trees in height—and after that age, growth in diameter. The deciduous species that naturally regenerate and are encountered in mixed stands, such as sycamore maple and mountain elm, need to be given the chance to contribute to the composition of these stands. The creation of diversified vertical structures must be a condition of sustainable silvicultural work on such stands.

Thus, stands will be able to maintain their vitality and their capacity for regeneration at any moment in their existence. The natural regeneration of mixed stands also creates the premise of being able to capitalize on local provenances that are much better adapted to the prevailing environmental conditions [40]. Such characteristics should constitute targets for silvicultural work on the future structures of fir and beech mixed forests.

5. Conclusions

By applying silvicultural works, the stand structure and, implicitly, the relationships between the trees change, against a background of site conditions. These influences are recorded by the trees and are displayed in their growth. The CAI_v is the result of diameter increment, height, and changes in the stem shape of the trees. The percentages of the increment of each biometric characteristic (*d, h, g, or v*) diminish as a tree advances in age. The height increment of the mean tree can be determined by the models expressed by Equations (5) and (6). The differences between the experimental values of the percentage of height increment and those estimated by the models diminished as the trees grew in height. The average deviations of the experimental values of $PCAI_h$, in comparison to the values estimated by the equation, was 0.001483, and the RMSE was 0.87490. The CAI_v of the tree was produced by applying relationship of the $PCAI_v$. Using Equation (7) to calculate the $PCAI_v$ yielded an RMSE in the experimental values of $PCAI_v$, compared to the values estimated by the relation, of 0.129521. When CAI_{fh} was related to the height of the trees at the end of period for the calculation of the $PCAI_{fh}$ the RMSE was 0.050536. At the level of the mean tree and stand, the simplified equation (Equation (7)) has led to values of the CAI_v, and respectively CAI_V, that are close (0.4%) to those obtained using relationship based on basal area and form-height increments and relationship based on successive inventory.

The CAI_V of a stand can be determined on the basis of the CAI_v of the mean current volume increment tree. The tree with the average value of the CAI_v of all trees is influenced by the structure of the stand, and its biometric characteristics are difficult to determine because they are specific to each structure. The characteristics of such a tree can be linked to the characteristics of the mean basal area tree and of the mean volume tree, which are easy to determine by means of a single inventory at the end of the chosen period. In even-aged stands, the mean current volume increment tree is a hypothetical tree that may have the mean basal area and the form-height of the stand. Such research should be continued on several felled sample trees in order to further clarify this statement in different structural conditions.

This study on the trees in the area shows that the fir trees maintain their ability to record active height increments until around the age of 80 (with a maximum between 50 and 60 years). After this age, the trees continue to accumulate significant increases in diameter. The reduction of stand density through silvicultural work must be correlated with these moments, which may prevent volume growth losses and increase final production.

Supplementary Materials: The following are available online at https://www.mdpi.com/article/10.3390/f12040439/s1, Table S1 presents the general characteristics of the 10 research areas located in the stands included in the study. The sample plots of 0.5 ha have a circular shape and are allocated five to a survey (four in the direction of the cardinal points and one in the center). The sample plots of 1 ha have a regular shape (100 × 100 m).

Author Contributions: Conceptualization, G.-M.T.; Data curation, G.-M.T. and A.C. (Avram Cicsa); Formal analysis, G.-M.T.; Investigation, G.-M.T., A.C. (Avram Cicsa), A.C. (Albert Ciceu), and A.-C.D.; Methodology, G.-M.T.; Visualization, G.-M.T.; Writing—original draft, G.-M.T.; Writing—review & editing, G.-M.T., A.-C.D., and A.C. (Avram Cicsa). All authors have read and agreed to the published version of the manuscript.

Funding: This research received no external funding.

Acknowledgments: We are deeply indebted to N. Rucăreanu who, through his silvicultural management work, supported studies on transformation cuttings and initiated forestry biometry research in the forests of this watershed, starting in 1960. We also owe thanks to the Romanian Ministry of Waters and Forests, which, since its endorsement of silvicultural management in these forests, has expressed its approval of maintaining experimental areas with a view to monitoring the structure, measurement and growth of the stands.

Conflicts of Interest: The authors declare no conflict of interest.

References

1. Biolley, E.H. *L'amênagement des Fôrets par la Méthode Expérimentale et Spécialment la Méthode du Contrôle*; Editeurs Neuchatel: Paris, France, 1921; pp. 23–91.
2. Schaeffer, A.; Gazin, A.; d'Alverny, A. *Sapinières. La Jardinage par Contenanace (Méthode du Controle par les Courbes)*; Les Presses Universitaires de France: Paris, France, 1930; pp. 47–100.
3. Meyer, H.A. *Forest Management*; Penn. Valley Publ., Inc. State College, Penn.: Philadelphia, PA, USA, 1953; pp. 12–21, 327–333.
4. Schütz, J.P.H. *Sylviculture 2. La Gestion des Forêts Irrégulières et Méelangées*; Presses Polytechniques et universitaire romandes: Laussanne, Switzerland, 1997; pp. 112–178.
5. Cavlovic, J.; Bozic, M.; Boncina, A. Stand structure of an unevenaged fir-beech forest with an irregular diameter structure: Modeling the development of the Belevine forest, Croatia. *Eur. J. For. Res.* **2006**. [CrossRef]
6. Danescu, A.; Albrecht, A.T.; Bauhus, J.; Kohnle, U. Geocentric alternatives to site index for modeling tree increment in uneven-aged mixed stands. *For. Ecol. Manag.* **2017**. [CrossRef]
7. Boncina, A.; Gaspersic, F.; Diaci, J. Long-term changes in tree species composition in the Dinaric mountain forests of Slovenia. *For. Chron.* **2003**. [CrossRef]
8. Diaci, J.; Firm, D. Long-term dynamics of a mixed conifer stand in Slovenia managed with a farmer selection system. *For. Ecol. Manag.* **2011**. [CrossRef]
9. Dănescu, A.; Kohnle, U.; Bauhus, J.; Weiskittel, A.; Albrecht, A.T. Long-term development of natural regeneration in irregular, mixed stands of silver fir and Norway spruce. *For. Ecol. Manag.* **2018**. [CrossRef]
10. Pretzsch, H. Stand density and growth of Norway spruce (*Picea abies* [L.] Karst.) and European beech (*Fagus sylvatica* [L.]). Evidence from long-term experimental plots. *Eur. J. For. Res.* **2005**. [CrossRef]
11. Pretzsch, H. Density and growth of forest stands revisited. Effect of the temporal scale of observation, site quality, and thinning. *For. Ecol. Manag.* **2020**. [CrossRef]
12. Zeide, B. Thinning and growth: A full turnaround. *J. For.* **2001**. [CrossRef]
13. Hanewinkel, M. Comparative economic investigations of even-aged and uneven-aged silvicultural systems: A critical analysis of different methods. *Forestry* **2002**. [CrossRef]
14. Hanewinkel, M.; Pretzsch, H. Modelling the conversion from even-aged to uneven-aged stands of Norway spruce (*Picea abies* L. Karst.) with a distance-dependent growth simulator. *For. Ecol. Manag.* **2000**. [CrossRef]
15. Nyland, R.D. Even-to uneven-aged: The challenges of conversion. *For. Ecol. Manag.* **2003**. [CrossRef]
16. O'Hara, K.L. *Multiaged Silviculture: Managing for Complex Forest Stand Structures*; Oxford University Press: Oxford, UK, 2014; pp. 115–116. [CrossRef]
17. Plauborg, K.U. Analysis of radial growth responses to changes in stand density for four tree species. *For. Ecol. Manag.* **2004**. [CrossRef]
18. Assmann, E. *The Principles of Forest Yield Study*; Pergamon: New York, NY, USA, 1970; pp. 151–152.
19. Giurgiu, V. *Dendrometrie şi auxologie forestieră*; Editura Ceres: Bucureşti, Romania, 1979; pp. 126–568.
20. Avery, T.E.; Burkhart, H.E. *Forest Measurement*; McGraw-Hill: New York, NY, USA, 1994; pp. 303–349.
21. Leahu, I. *Dendrometrie*; Editura Didacticăşi Pedagogică: Bucureşti, Romania, 1994; pp. 275–350.
22. Prodan, M.; Peters, R.; Cox, F.; Real, P. *Mensura Forestal*; Instituto Interamericano de Cooperación para la Agricultura: San José, CA, USA, 1997; pp. 386–387, 408–416, 431–435.
23. Giurgiu, V.; Drăghiciu, D. *Metode şi Tabele Dendrometrice*; Editura Ceres: Bucureşti, Romania, 2004; pp. 44–110.
24. Laar, A.; Akça, A. *Forest Mensuration*; Springer: Dordrecht, The Netherlands, 2007.
25. Pretzsch, H. *Forest Dynamics, Growth and Yield*; Springer: Berlin, Germany, 2009; pp. 381–420. [CrossRef]
26. Burkhart, H.E.; Tomé, M. *Modeling Forest Trees and Stands*; Springer: New York, NY, USA, 2012; pp. 5–129, 239–258.
27. Vanclay, J.K. *Modelling Forest Growth and Yield: Applications to Mixed Tropical Forests*; CAB International: Wallingford, UK, 1994; p. 312.
28. Marziliano, P.A.; Menguzzato, G.; Scuderi, A.; Corona, P. Simplified methods to inventory the current annual increment of forest standing volume. *IForest* **2012**, 276–282. [CrossRef]
29. Koch, G.W.; Sillett, S.C.; Jennings, G.M.; Davis, S.D. The limits to tree height. *Nature* **2004**. [CrossRef] [PubMed]
30. Bond, B. Age-related changes in photosynthesis of woody plants. *Trends Plant Sci.* **2000**, *5*, 349–353. [CrossRef]
31. Marziliano, P.A.; Tognetti, R.; Lombardi, F. Is tree age or tree size reducing height increment in *Abiesalba*Mill.at its southernmost distribution limit? *Ann. For. Sci.* **2019**. [CrossRef]
32. Bond, B.J.; Czarnomski, N.M.; Cooper, C.; Day, M.E.; Greenwood, M.S. Developmental decline in height growth in Douglas-fir. *Tree Physiol.* **2007**. [CrossRef] [PubMed]
33. Parviainen, J. Virgin and natural forests in the temperate zone of Europe. *For. Snow Landsc. Res.* **2005**, *79*, 9–18.

34. Becker, M. The role of climate on present and past vitality of silver fir forests in the Vosges mountains of northeastern France. *Can. J. For. Res.* **1989**. [CrossRef]
35. Toromani, E.; Sanxharu, M.; Pasho, E. Growth responses to climate and drought in silver fir (*Abies alba*) along an altitudinal gradient in southern Kosovo. *Can. J. For. Res.* **2011**. [CrossRef]
36. Prodan, M. *Forest Biometrics (Transl. by S.H. Gardiner)*; Pergamon Press: Oxford, MS, USA, 1968; pp. 341–394.
37. Giurgiu, V.; Decei, I.; Drăghiciu, D. *Modele Matematico-Auxologice şi Tabele de Producţie Pentru Arborete*; Editura Ceres: Bucureşti, Romania, 2004; pp. 47–143.
38. Husch, B. *Forest Mensuration and Statistics*; The Roland Press Company: New York, NY, USA, 1963; p. 240.
39. Giurgiu, V.; Decei, I.; Armăşescu, S. *Biometria arborilor şi arboretelor din România*; Editura Ceres: Bucureşti, Romania, 1972; pp. 69–71, 116–122.
40. Tudoran, G.M.; Zotta, M. Adapting the planning and management of Norway spruce forests in mountain areas of Romania to environmental conditions including climate change. *Sci. Total Environ.* **2020**. [CrossRef] [PubMed]

Article

Impact of Industrial Pollution on Radial Growth of Conifers in a Former Mining Area in the Eastern Carpathians (Northern Romania)

Cristian Gheorghe Sidor [1], Radu Vlad [1], Ionel Popa [1,2], Anca Semeniuc [1], Ecaterina Apostol [3] and Ovidiu Badea [3,4,*]

1 "Marin Drăcea" National Institute for Research and Development in Forestry, Câmpulung Moldovenesc Station, 73 bis Calea Bucovinei, 725100 Câmpulung Moldovenesc, Romania; cristi.sidor@yahoo.com (C.G.S.); vlad.radu2@gmail.com (R.V.); popaicas@gmail.com (I.P.); semeniuc.anca@yahoo.ro (A.S.)
2 Center of Mountain Economy-INCE-CE-MONT Vatra Dornei, 49 Petreni Street, 725700 Vatra Dornei, Romania
3 "Marin Drăcea" National Institute for Research and Development in Forestry, 128 Eroilor Blvd., 077190 Voluntari, Romania; cathyches@yahoo.com
4 Department of Forest Engineering, Forest Management Planning and Terrestrial Measurements, Faculty of Silviculture and Forest Engineering, "Transilvania" University, 1 Ludwig van Beethoven Str., 500123 Braşov, Romania
* Correspondence: obadea@icas.ro; Tel.: +40-744-650-981

Abstract: The research aims to evaluate the impact of local industrial pollution on radial growth in affected Norway spruce (*Picea abies* (L.) Karst.) and silver fir (*Abies alba* Mill.) stands in the Tarniţa study area in Suceava. For northeastern Romania, the Tarniţa mining operation constituted a hotspot of industrial pollution. The primary processing of non-ferrous ores containing heavy metals in the form of complex sulfides was the main cause of pollution in the Tarniţa region from 1968 to 1990. Air pollution of Tarniţa induced substantial tree growth reduction from 1978 to 1990, causing a decline in tree health and vitality. Growth decline in stands located over 6 km from the pollution source was weaker or absent. Spruce trees were much less affected by the phenomenon of local pollution than fir trees. We analyzed the dynamics of resilience indices and average radial growth indices and found that the period in which the trees suffered the most from local pollution was between 1978 and 1984. Growth recovery of the intensively polluted stand was observed after the 1990s when the environmental condition improved because of a significant reduction in air pollution.

Keywords: air pollution; increment cores; Norway spruce; radial growth series; silver fir

Citation: Sidor, C.G.; Vlad, R.; Popa, I.; Semeniuc, A.; Apostol, E.; Badea, O. Impact of Industrial Pollution on Radial Growth of Conifers in a Former Mining Area in the Eastern Carpathians (Northern Romania). *Forests* **2021**, *12*, 640. https://doi.org/10.3390/f12050640

Academic Editor: Riccardo Marzuoli

Received: 7 April 2021
Accepted: 13 May 2021
Published: 19 May 2021

1. Introduction

Climate change and air pollution represent the main drivers of global change, significantly impacting forest health and sustainable development [1]. Accelerated industrial development after World War II increased pollutant inputs in many parts of the globe, and Central and Eastern Europe were significantly affected [2]. The changes in political regimes in Eastern Europe and national and world environmental policy after the 1900s allowed for a significant reduction in air pollution in most regions. Nonetheless, investigating the environmental impact of air pollution resulting from anthropogenic industrial activities remains critically important [3].

Air pollution negatively affects forest ecosystems and soil quality worldwide. Industrial emissions from the Ivano-Frankivsk and Chernivtsi regions (Ukraine) have led to high concentrations of heavy metals in forest soils, high levels of tree crown defoliation and ecosystem changes such as biodiversity decline or reduced productivity [4]. Soil acidification in the region has risen progressively due to the increased content of heavy metals [5], which indirectly influences forest ecosystem vegetation [6]. In Germany, the areas near recently halted mining operations were investigated to determine the uptake of heavy metals by forest trees in heavily contaminated ecosystems. The researchers also described the levels of damage caused by heavy metal toxicity [7]. A similar study in the Czech Republic analyzed the health status of Norway spruce (*Picea abies*) forests according to the phenology and radial growth of trees in relation to air pollutants, especially NO_2 and SO_2 [8]. The distribution and accumulation of heavy metals in forest soils in the Rozocze National Park (SE Poland) was found to be related to anthropogenic pollution through local and background emission sources [9]. In the Carpathian Mountains, the results of long-term monitoring activities indicated that the combined effects of O_3, SO_2 and NO_2 could negatively affect forest stands and highlighted the association between air pollution levels and tree growth [10]. Furthermore, elevated levels of N and S deposition at the levels found in the Carpathian Mountains may have negatively affected forest health status and biodiversity, including visible leaf injury, losses in stand growth and productivity and higher sensitivity to biotic and abiotic stressors [11,12]. The accumulation of heavy metals with accompanying S and associated soil and foliar nutrient imbalances and reduced soil water holding capacity can restrict the recolonization of plant communities in the forest ecosystem [13]. In the recent past, some industrial activities in Romania (Copșa Mică, Zlatna, Baia Mare) were known to create regional hotspots of pollution, negatively affecting forest vegetation [14–17]. Toxicity thresholds for the forest environment in Romania were highlighted based on air quality analysis [18].

Trees are sensitive to environmental factors, and any changes in growing conditions are reflected in tree ring parameters. The reduction in tree growth is generally associated with unfavorable climate conditions and an increase in specific ecosystem competition. Furthermore, air pollution can be associated with narrow growth rings for several decades.

The evolution of forest ecosystems affected by past pollution in highly industrialized areas and damage dynamics assessments offer crucial knowledge needed to develop management strategies for the conservation and improvement of the environment. Thus, to accurately assess how severely trees or forests have been affected by pollutants, it is critical to study the issue in well-defined ecological areas. For the northeastern part of Romania, the Tarnița mining exploitation constituted a hotspot of industrial pollution. The primary processing activity of non-ferrous ores containing heavy metals in complex sulfide forms was the main cause of pollution in the Tarnița region. Tarnița mining exploitation began in 1968, and the amount of production increased until 1990 through the exploitation of new deposits. With the political regime change in Romania, concomitant with the economic recession after the fall of communism, there was a sharp decline in mining activity, followed by a cessation in 1998 [18].

Considering this specific pollution history, the aim of this research was to evaluate the impact of local past industrial pollution on radial growth in affected Norway spruce (*Picea abies*) and silver fir (*Abies alba*) stands in the Tarnița study area, Suceava (northern Romania).

2. Materials and Methods

A network of experimental plots was established in five representative yield management units in the Tarnița region, Suceava county, within the Stulpicani Forest District (FD) (Table 1). This network of experimental plots included 30 plots (15 for silver fir and 15 for Norway spruce) from which radial increment cores were collected (40 trees for each species per plot). In order to highlight the level of pollution intensity on the silver fir and Norway spruce stands in the Tarnița area, the 30 plots were located spatially at different distances from the main source of pollution. Taking into account previous research [19,20], in which several stands located at different distances from the source of pollution were analyzed and it was concluded that the 2 km length was the approximate distance to and from which the stands were affected by pollution to varying degrees of intensity, we tested this hypothesis and we considered in this study that intensively polluted stands are located at a maximum distance of 2 km from the main source of pollution, moderately polluted stands are located between 2 and 6 km and largely unpolluted stands are located at a distance greater than

6 km. Thus, based on the results obtained in these studies, in order to test our hypothesis, we considered these distances as limits between different pollution intensities.

Table 1. The main characteristics of the experimental plots network from Tarnița region.

Forest Management Unit (u.a.)	Yield Management Unit (UP)	Distance from the Polluting Source (km)	Area (ha)	Age	Composition *	Exposure	Slope (Centesimal Degrees)	Altitude (m)	Canopy Cover	Yield Class	Standing Volume (m^3/ha^{-1})
73C	V Tarnița	6.0	4.90	95	9MO1BR	NE	28	1120	0.8	2	585
62A	V Tarnița	3.1	6.34	105	4MO3BR2FA1PAM	NE	25	985	0.7	2	416
39A	V Tarnița	0.5	4.13	85	8MO2FA	NE	16	860	0.4	2	304
111E	V Tarnița	1.2	4.67	105	4MO4BR 2FA	SE	26	980	0.6	3	396
118A	V Tarnița	2.5	18.45	140	4MO4FA2BR	S	26	1125	0.5	3	314
18F	V Tarnița	3.2	19.54	140	5MO3FA2BR	NW	36	1125	0.6	3	370
14C	V Tarnița	1.6	6.87	95	5BR3FA2MO	S	26	900	0.7	3	402
126I	V Tarnița	1.2	5.2	100	6MO2BR 2FA	SE	16	875	0.6	2	346
5A	VI Botoșana	4.5	29.84	95	7MO2BR1FA	N	27	905	0.7	1	572
17B	VI Botoșana	3.1	10.66	75	8MO2BR	N	18	1000	0.8	2	589
45A	VI Botoșana	6.9	8.57	115	6MO2BR2FA	N	18	775	0.7	2	601
61A	VI Botoșana	6.0	16.22	110	4MO3FA2BR1PAM	SE	22	1000	0.7	2	482
43B	IV Porcăreț	6.2	24.2	95	6MO3BR1FA	NE	24	990	0.7	2	513
22A	II Negrileasa	9.7	28.78	95	5MO4BR1FA	N	22	850	0.7	2	560
4B	VIII Gemenea	12.0	18.10	110	8MO2BR	NW	12	825	0.6	2	482
101C	VIII Gemenea	12.1	13.03	125	8BR2MO	SW	33	740	0.6	2	474

* Degrees of participation (in tenths) of the species in the mix forest stand; Norway spruce (MO), silver fir (BR), European beech (FA), sycamore maple (PAM).

Concerning the characteristics of the studied stands, the trees included in the forest stands of the network within the Tarnița study area were between 75 and 157 years old. The tree stand composition was generally a mixture of Norway spruce (the predominant species), silver fir and European beech (*Fagus sylvatica* L.). Plot altitude varied between 750 and 1150 m. The studied stands had a canopy cover between 0.6 and 0.8 and were mostly of higher productivity (classified in the 2nd relative yield class). The volume per hectare was between 346 and 601 m^3.

For the classification of the radial growth series in relation to the distance from the polluting source, both the distance from the source and the predominant direction of airmasses (NE–SW) were taken into account. Thus, there were six series of growth (three silver fir and three Norway spruce) located in the intensively polluted area, 14 series in the moderately polluted area (seven silver fir and seven Norway spruce) and 10 series in the largely unpolluted area (five silver fir and five Norway spruce). In the direction of the main valley, the stands were intensively polluted up to distances of 6–7 km (Figure 1).

The following specific statistical parameters were calculated, both for radial growth series and tree ring index series [21,22]: the period covered by each series with a mini-mum replication of 10 individual series, sample depth, mean tree ring width, average sensitivity (average percentage change of annual ring width relative to the next annual ring [21]) and average Rbar (correlation coefficient between all individual series).

The degree of reduction and recovery of growth due to the influence of local industrial pollution was determined through the resilience indices, presented as a 5-year moving average. Tree resilience is its post-disturbance ability to reach the level of radial growth it experienced before the disturbance, calculated as the ratio between the pre- and post-disturbance growth [23]. The tree resilience calculations were performed for each growth series analyzed, revealing the capacity of trees to grow after the disturbing events that caused reduced radial growth during certain periods.

Radial growth indices of intensively and moderately polluted trees were compared to the radial growth indices of the unpolluted trees, considered reference values. The calculations and analyses were performed for the period common to all analyzed series from 1951 to 2018. The highlighting and quantification of the growth changes of the stands in the areas affected by the industrial pollution were performed using software such as CooRecorder 7.4 [24], CDendro 7.6 [24], TsapWin [25], COFECHA [26,27] and R studio [28].

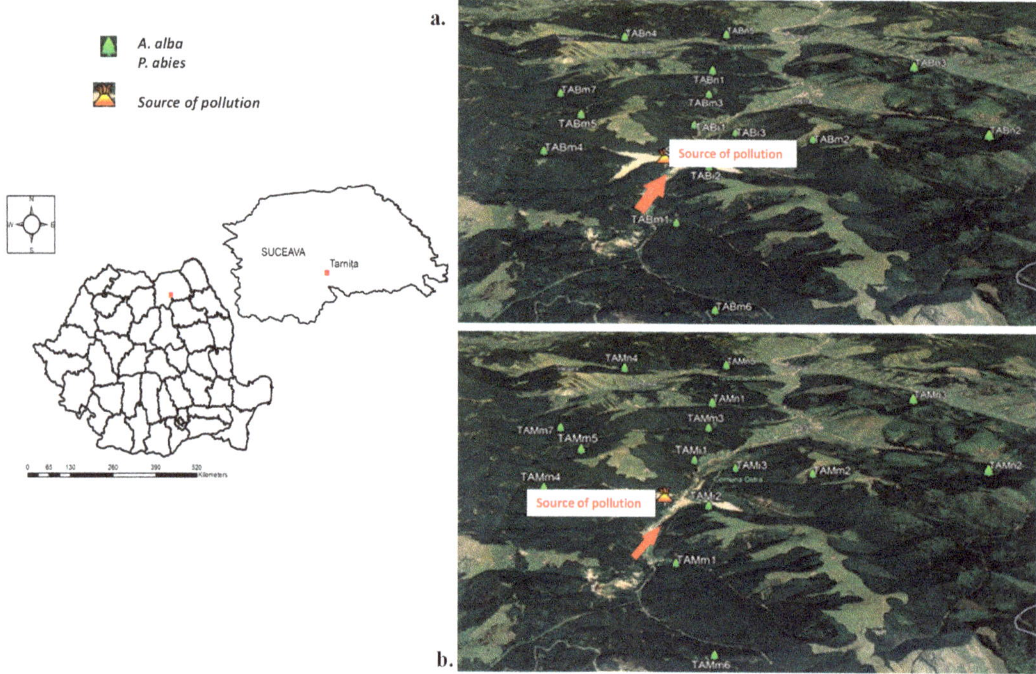

Figure 1. Study location: (**a**) Silver fir from Tarnița region; (**b**) Norway spruce from Tarnița region (the red arrow indicates the source of pollution and the predominant direction of airmasses in the area is NE–SW, according to the valley orientation).

3. Results

3.1. The Statistical Parameters of the Series of Average Radial Growth

The main statistical parameters for all average radial growth series studied are shown in Table 2. The length of the analyzed silver fir series from the Tarnița area varied between 75 and 157 years, with an average ring width value between 1.704 and 3.346 mm. The mean values of sensitivity varied from 0.196 to 0.253, and the mean Rbar values of the residual series were between 0.219 and 0.411. The Norway spruce growth series lengths varied between 72 and 143 years, similar to silver fir. The average value of the tree ring width varied between 1.812 and 3.447 mm. The Norway spruce Rbar values were lower than the silver fir, varying between 0.194 and 0.364 (Table 2).

3.2. Analysis of Growth Changes of Trees

The average radial growth of the silver fir in the Tarnița area showed a similar dynamic regardless of the plot distance from the pollution source (Figure 2A). The exception was the period 1978 to 1990, during which the radial increments of intensively polluted silver fir were significantly lower than those from unpolluted areas. The average radial growth values of silver fir in moderately polluted areas during this period were intermediate between intensively polluted and unpolluted plots. According to average radial growth indices, the Norway spruce trees were the most affected by local pollution from 1978 to 1984 (Figure 2B). The silver fir trees in the Tarnița area demonstrated average radial growth dynamics (Figure 2A) that indicated they were more affected by the pollution than the Norway spruce.

Table 2. Statistical parameters of the average radial growth series for silver fir and Norway spruce in the Tarnița area.

Serial Code	No. of Cores	Overlapping Period > 10 Cores	Average Radial Growth	Average Sensitivity	Average Rbar	Location (FD/UP/u.a.)
TABi1	43	1943–2018	3.205	0.231	0.411	Stulpicani/V/126I
TABi2	40	1925–2018	2.881	0.236	0.311	Stulpicani/V/39A
TABi3	40	1861–2018	1.704	0.253	0.374	Stulpicani/V/14C
TABm1	41	1922–2018	3.104	0.216	0.219	Stulpicani/V/62A
TABm2	40	1930–2018	2.912	0.198	0.291	Stulpicani/V/18F
TABm3	41	1921–2018	2.686	0.211	0.280	Stulpicani/VI/5A
TABm4	43	1879–2018	2.409	0.209	0.234	Stulpicani/V/118B
TABm5	40	1927–2018	2.729	0.185	0.237	Stulpicani/VI/17B
TABm6	42	1896–2018	2.041	0.193	0.322	Stulpicani/V/73C
TABm7	41	1895–2018	2.578	0.226	0.275	Stulpicani/VI/61A
TABn1	40	1912–2018	2.882	0.195	0.256	Stulpicani/VI/45A
TABn2	40	1893–2018	2.301	0.203	0.343	Stulpicani/IV/43B
TABn3	40	1905–2018	2.632	0.228	0.319	Stulpicani/II/22A
TABn4	40	1919–2018	2.765	0.196	0.238	Stulpicani/VIII/4B
TABn5	40	1911–2018	3.346	0.216	0.390	Stulpicani/VIII/101A
TAMi1	43	1946–2018	3.447	0.221	0.313	Stulpicani/V/126I
TAMi2	40	1935–2018	2.993	0.210	0.304	Stulpicani/V/39A
TAMi3	41	1907–2018	1.872	0.297	0.314	Stulpicani/V/14C
TAMm1	40	1920–2018	2.738	0.228	0.209	Stulpicani/V/62A
TAMm2	40	1926–2018	2.473	0.198	0.397	Stulpicani/V/18F
TAMm3	40	1918–2018	2.463	0.217	0.306	Stulpicani/VI/5A
TAMm4	42	1891–2018	2.303	0.204	0.304	Stulpicani/V/118B
TAMm5	41	1938–2018	3.108	0.177	0.194	Stulpicani/VI/17B
TAMm6	42	1891–2018	1.812	0.193	0.311	Stulpicani/V/73C
TAMm7	40	1916–2018	2.680	0.213	0.365	Stulpicani/VI/61A
TAMn1	40	1909–2018	2.657	0.224	0.322	Stulpicani/VI/45A
TAMn2	41	1893–2018	2.201	0.210	0.248	Stulpicani/IV/43B
TAMn3	40	1901–2018	2.357	0.270	0.364	Stulpicani/II/22A
TAMn4	40	1915–2018	2.075	0.219	0.282	Stulpicani/VIII/4B
TAMn5	42	1875–2018	2.361	0.228	0.237	Stulpicani/VIII/101C

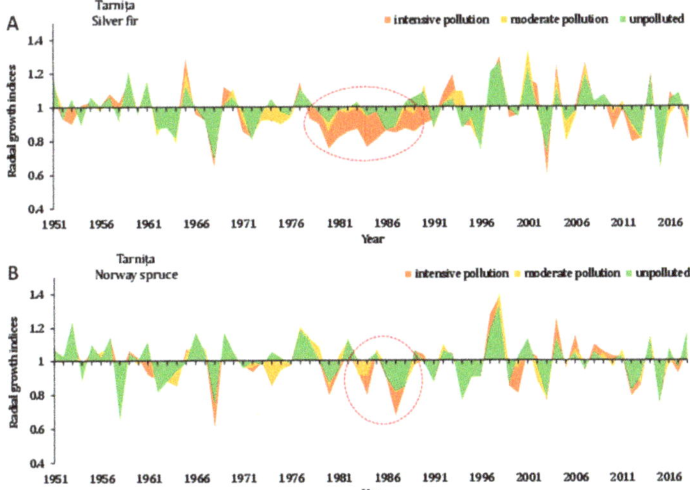

Figure 2. The average series of radial growth indices developed for each of the 3 categories of stands studied (the dotted circle represents the time interval in which the trees were most affected by air pollution); (A) silver fir; (B) Norway spruce.

From 1978 to 1990 (Figure 3A), only those resilience indices corresponding to the analyzed trees in the intensively polluted area had negative values. The analysis of the resilience indices (Figure 3B) revealed that the Norway spruce trees were also affected by the local pollution from 1978 to 1990, but to a much lesser extent than the silver fir.

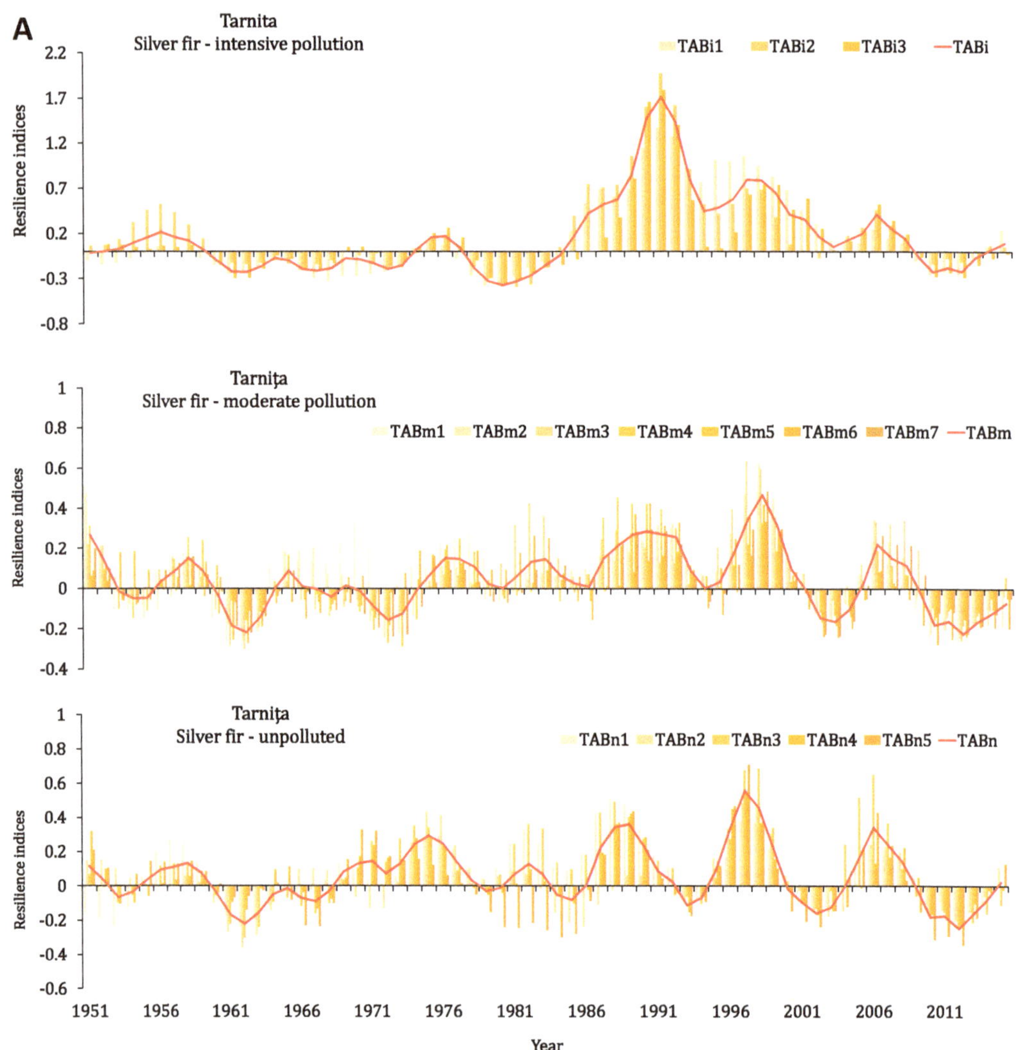

Figure 3. *Cont.*

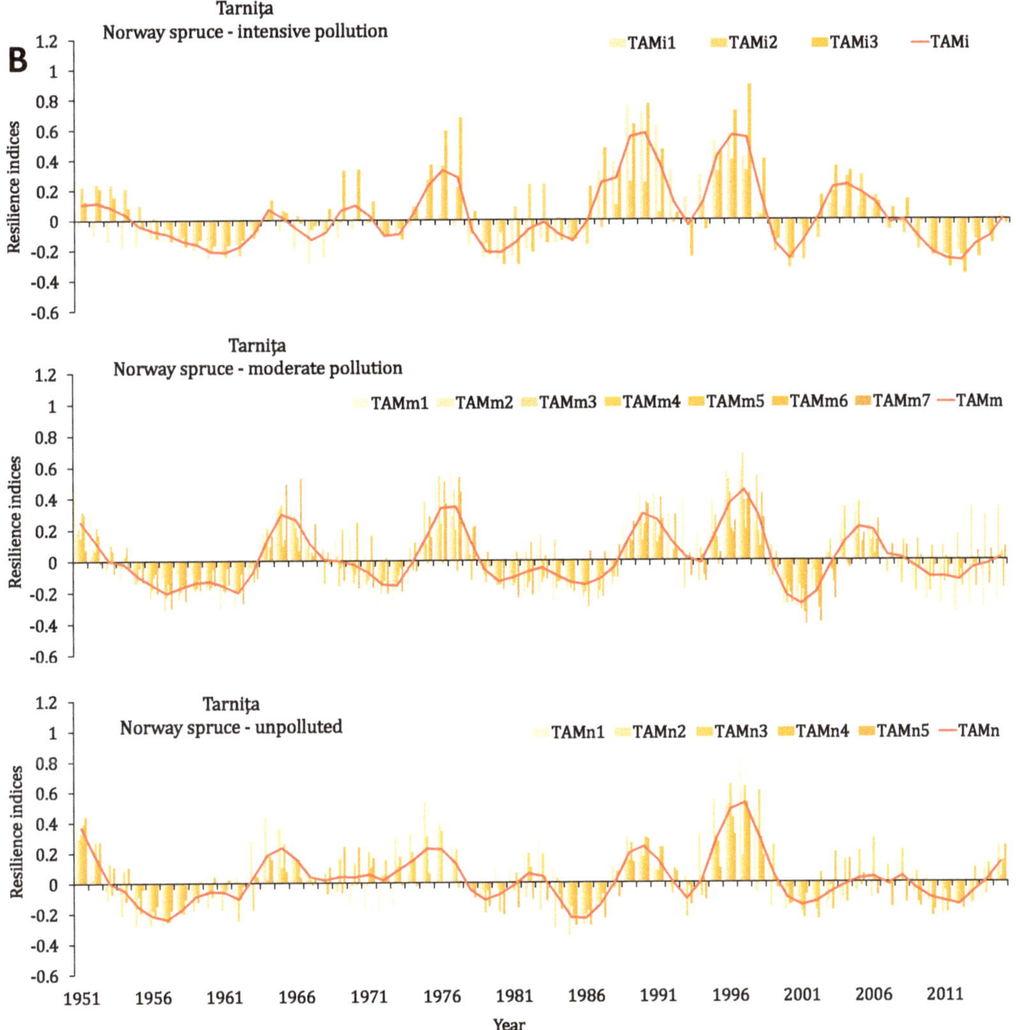

Figure 3. Resilience indices of the average radial growth series of silver fir (**A**) and Norway spruce (**B**).

For intensely polluted silver fir trees, after the cessation of the polluting activity, the resilience index values were significantly higher than those of trees in unpolluted areas (Figure 4A). From 1978 to 1990, for the silver fir trees from the intensively polluted area, the resilience indices (Figure 3A) and average radial growth indices for Norway spruce (Figure 4B) were much lower than those of the silver fir trees in the unpolluted areas.

The quantification of growth losses for silver fir reflects reductions of up to almost 20% (in 1984 and 1989) in heavily polluted areas (Figure 5A). The growth losses of trees located in moderately polluted areas were not as significant (up to 5–7% relative to normal). The average losses throughout the highlighted period for heavily polluted silver fir were approximately 14%. Compared to the silver fir tree, the Norway spruce suffered much smaller diameter growth losses (Figure 5B). The average loss of diameter growth of the intensively polluted Norway spruce during the entire period of pollution exposure was 5%, and the loss was only 2% for the moderately polluted.

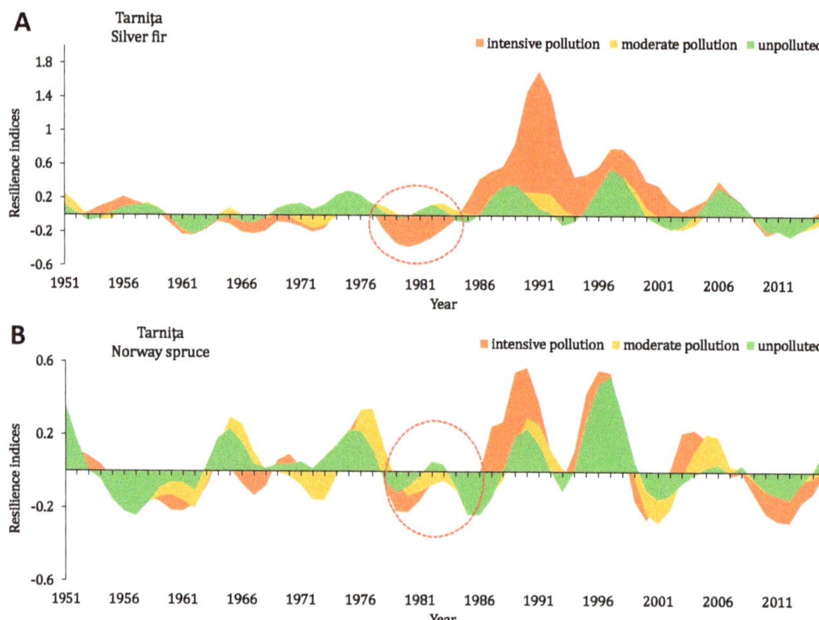

Figure 4. Average resilience indices for each of the 3 categories of silver fir (**A**) and Norway spruce (**B**) stands in the Tarnița area (the dotted circle represents the time interval in which the trees were most affected by the influence of pollution).

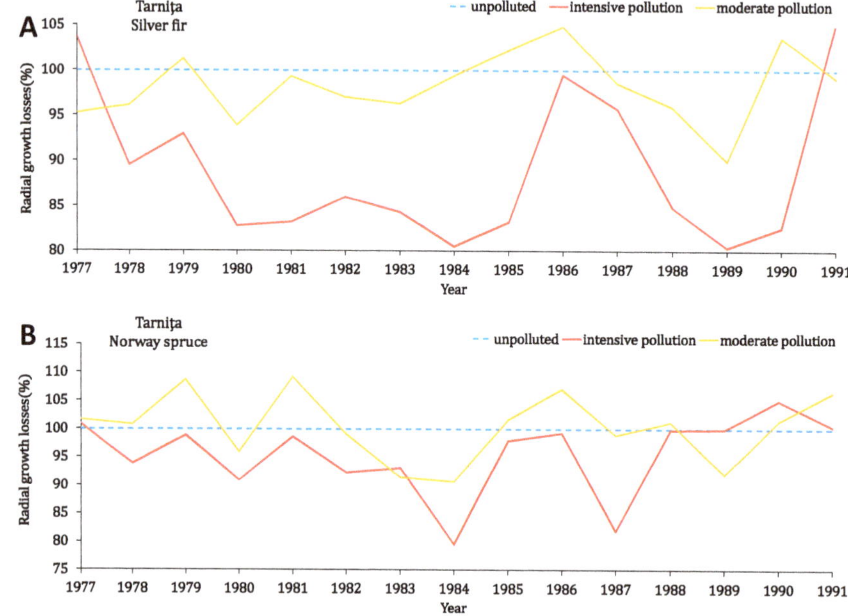

Figure 5. Radial growth losses recorded by the trees ((**A**) silver fir; (**B**) Norway spruce) in the Tarnița area affected by moderate and intensive air pollution.

At the stand level, the most affected trees by the local industrial pollution were the silver fir trees, while Norway spruce trees were less affected.

4. Discussion

As in the recent study developed in the same area [18], whose results confirmed that the frequency of growth events is determined by the distance from the sources of pollution, our results indicate that pollutant emissions near the local pollution zone significantly impacted the growth and development of coniferous trees in the Tarnița region, Suceava. In the studied area, the negative effect of pollution on the radial growth of coniferous trees (silver fir and Norway spruce) was greatest from 1978 to 1990. During this period, silver fir trees in the intensively polluted area experienced radial growth losses of up to almost 20% in 1984 and 1989. The growth losses of trees located in moderately polluted areas were not as significant, up to 5–7% compared to normal. The average loss throughout the highlighted period for heavily polluted silver fir was approximately 14%. In the case of Norway spruce, from 1978 to 1990, the trees were much less negatively affected by local pollution than the silver fir tree. The period in which the Norway spruce trees were most affected by local pollution was between 1978 and 1984, followed, except for 1987, by a period in which the trees did not experience as much growth reduction. Compared to the silver fir, the Norway spruce in this area showed much smaller radial growth losses. The average loss in radial growth of the intensively polluted and moderately polluted Norway spruce during the entire period of local pollution influence was 5 and 2%, respectively. After the 1990s, we observed a significant improvement in radial growth linked with the reduction in air pollution due to the closing of the mine. These results confirm those obtained in previous studies in this area [29]. Similarly, growth losses were registered as an effect of the long period of excessive drought [30]. The impact of air pollution on tree growth revealed by our results is slightly underestimated, because in the analysis were included only the trees that survived until the present. The growth decrease would likely be more evident in trees that did not survive the period of high industrial activity [18].

The effects of air pollution on forests are observed mainly as a direct impact on tree health, by crown damages and abnormal defoliation, favorable to losing tree vitality and in some cases, even death [31]. Coniferous trees are more sensitive to the effects of air pollution and acid rain than broad-leafed trees because of their greater capacity to intercept water from precipitation [32]. Similarly, in other regions of Europe with high air pollution, the silver fir is more pollution-sensitive than spruce [2,33].

Mihaljevič et al. [34] analyzed the annual tree rings relative to the mining period and a potential source of contamination, a high concentration of cobalt (Co) that corresponded to maximum mining production. Hojdová et al. [35] assessed contaminated soils and vegetation surrounding mining areas. The authors found a strong correlation between Hg concentration in beech and mining metal production and no correlation with spruce trees located closer to the source of pollution. Numerous studies on this topic have proposed that emissions of heavy metals cause imbalances in the forest ecosystem and beyond. Shparyk et al. [4] showed that the highest levels of defoliation of trees were close to sources of industrial emissions. Biochemical investigations performed on leaves and phloem in tree trunks revealed decreased assimilative pigments in trees in those areas affected by pollution [17]. Changes in the processes of photosynthesis, respiration and transpiration occur differently in intensity both at the individual and species levels [36,37]. Barium mining activities can affect the quality of sediments and soil water through pollution with Fe, Hg and Pb, indicating an unacceptable risk to human health and the environment. After assessing the degree of contamination and the risks due to barite mining, the authors showed that the average concentrations of Fe, Hg and Pb were above allowable levels [38].

The radial growth of trees is certainly influenced by the variation and capacity of mining production. In our study, differences found in the dynamics of radial growth indices between the two species analyzed were determined by the sensitivity and reaction of each species to the imbalance of the ecophysiological process. The Norway spruce had much smaller diameter growth losses than the silver fir.

5. Conclusions

Air pollution from Tarnița mining operations induced strong growth reduction from 1978 to 1990, undoubtedly related to a decline in tree health and vitality due to airborne pollutants. The growth decline in stands further away (over 6 km) from the pollution source was weaker or absent, and the tree ring width variability was related to climate variation. Growth recovery of the intensively polluted stand was observed after the 1990s when the environmental condition improved because of a significant reduction in air pollution.

Of the two species analyzed (silver fir and Norway spruce), the silver fir demonstrated a higher sensitivity to local pollutants. Analyzing the dynamics of resilience indices and average radial growth indices in an integrated and comparative way allowed us to determine that the period in which the spruce trees suffered the most from the effect of local pollution was from 1978 to 1984, followed by, except for 1987, a period in which the trees experienced less growth reduction.

Author Contributions: Conceptualization, C.G.S.; methodology, C.G.S., R.V. and A.S.; software, C.G.S.; validation, C.G.S., O.B., I.P. and R.V.; formal analysis, C.G.S.; investigation, C.G.S.; resources, C.G.S.; data curation, C.G.S., R.V. and A.S.; writing—original draft preparation, C.G.S.; writing—review and editing, C.G.S., E.A., O.B. and I.P.; visualization, C.G.S., E.A., O.B. and I.P.; supervision, C.G.S., E.A., O.B. and I.P.; project administration, C.G.S. and O.B.; funding acquisition, C.G.S. and O.B. All authors have read and agreed to the published version of the manuscript.

Funding: This study was funded by the Romanian Ministry of Research and Innovation, within the Nucleu National Programme, Project-PN-19070104/Contract no. 12N/2019.

Acknowledgments: We would like to thank the Stulpicani Forest District for permission to conduct field research and the research team within the project.

Conflicts of Interest: The authors declare no conflict of interest.

References

1. Elling, W.; Dittmar, C.; Pfaffelmoser, K.; Rötzer, T. Dendroecological assessment of the complex causes of decline and recovery of the growth of silver fir (*Abies alba* Mill.) in Southern Germany. *For. Ecol. Manag.* **2009**, *257*, 1175–1187. [CrossRef]
2. Vallero, D. *Fundamentals of Air Pollution*, 5th ed.; Elsevier Academic Press: Oxford, UK, 2014; p. 996.
3. Szaro, R.C.; Oszlanyi, J.; Godzik, B.; Bytnerowicz, A. (Eds.) *Effects of Air Pollution on Forest Health and Biodiversity in Forests of the Carpathian Mountains*; NATO Science Series 345; IOS Press: Amsterdam, The Netherlands, 2002; pp. 23–96.
4. Shparyk, Y.S.; Parpan, V.I. Heavy metal pollution and forest health in the Ukrainian Carpathians. *Environ. Pollut.* **2004**, *130*, 55–63. [CrossRef] [PubMed]
5. Staszewski, T.; Łukasik, W.; Kubiesa, P. Contamination of Polish national parks with heavy metals. *Environ. Monit. Assess.* **2012**, *184*, 4597–4608. [CrossRef] [PubMed]
6. Dharani, N.; Onyari, J.M.; Maina, D.M.; Mavuti, K.M. The distribution of Cu and Pb levels in soils and Acacia xanthophloea Benth. From Lake Nakuru national park Kenya. *Environ. Contam. Tox.* **2007**, *79*, 172–177. [CrossRef] [PubMed]
7. Truby, P. Impact of heavy metals on forest trees from mining areas. In Proceedings of the Sudbury Mining and the Environment Conference, Sudbury, ON, Canada, 25–28 May 2003.
8. Král, J.; Vacek, S.; Vacek, Z.; Putalová, T.; Bulušek, D.; Štefančík, I. Structure, development and health status of spruce forests affected by air pollution in the western Krkonoše Mts. in 1979–2014. *Lesn. Cas. For. J.* **2015**, *61*, 175–187. [CrossRef]
9. Mazurek, M.; Kowalska, J.; Gąsiorek, M.; Zadrożny, P.; Józefowska, A.; Zaleski, T.; Kępka, W.; Tymczuk, M.; Orłowska, K. Assessment of heavy metals contamination in surface layers of Roztocze National Park forest soils (SE Poland) by indices of pollution. *Chemosphere* **2017**, *168*, 839–850. [CrossRef]
10. Muzikaa, R.M.; Guyettea, R.P.; Zielonkab, T.; Liebholdc, A.M. The influence of O3, NO2 and SO2 on growth of *Picea abies* and *Fagus sylvatica* in the Carpathian Mountains. *Environ. Pollut.* **2004**, *130*, 65–71. [CrossRef] [PubMed]
11. Bytnerowicz, A.; Badea, O.; Popescu, F.; Musselman, R.; Tanase, M.; Barbu, I.; Vasile, C. Air pollution, precipitation chemistry and forest health in the Retezat Mountains, Southern Carpathians, Romania. *Environ. Pollut.* **2005**, *137*, 546–567. [CrossRef] [PubMed]
12. Silaghi, D.; Badea, O.; Iacoban, C.; Neagu, Ș.; Leca, Ș. Air pollutants concentrations (O3, NO2 and NH3) registered in selected forest ecosystems (core plots) in the Romanian Intensive Monitoring Network (Level II). *Rev. Pădurilor* **2011**, *126*, 85–92.
13. Salemaa, M.; Vanha-Majamaa, I.; Derome, J. Understorey vegetation along a heavy-metal pollution gradient in SW Finland. *Environ. Pollut.* **2001**, *112*, 339–350. [CrossRef]

14. Leșan, M. Poluarea pădurilor din vecinătatea municipiului Baia Mare și consecințele ei asupra acumulării de masă lemnoasă. *Rev. Pădurilor* **2003**, *118*, 12–14.
15. Stancu, P.T. Studii Geochimice şi Transformări Mineralogice Rezultate în Urma Proceselor Secundare de Exploatare Minieră Şi tehnologii de Remediere a Zonelor Poluate cu Metale Grele şi/sau Rare în zona Zlatna (Geochemical Studies and Mineralogical Transformations Resulting from Secondary Mining Processes and Technologies for Remediation of Areas Polluted with Heavy and/or Rare Metals in the Zlatna Area). Ph.D. Thesis, Bucharest University, Faculty of Geology and Geophysics, Bucharest, Romania, 2013.
16. Ianculescu, M.; Budu, E.C. Indicatori biochimici de evaluare a rezistenței speciilor forestiere la poluarea industrială cu compuși ai sulfului în acțiune sinergică cu metalele grele în zona Copșa Mică (Biochemical indicators for assessing the resistance of forest species to industrial pollution with sulfur compounds in synergistic action with heavy metals in the Copșa Mică area). *An. ICAS* **2007**, *50*, 99–119.
17. Vasile, D.; Bolea, V. Compararea calităţii aerului din Parcul Naţional Piatra Craiului cu alte Parcuri Naturale şi Naţionale din ţară (Air quality of Parcul Naţional Piatra Craiului compared with another Naturals and Nationals Parks of our country). *Rev. Silvic. Cinegetică* **2009**, *25*, 31–34.
18. Flocea, N.M. Cercetări Auxologice în Păduri de Molid din Nordul Țării Aflate în zone Afectate de Poluare (Auxological Research in the Northern Spruce Forests of Romania Located in Areas Affected by Pollution). Ph.D. Thesis, Ștefan cel Mare University, Suceava, Romania, 2013.
19. Jacoban, C.; Risca, I.M.; Roibu, C.; Ciorna, E.T.; Necula, R.; Ilieva, D.; Sandu, I.; Drochioiu, G. Tarnita Polluted Area: Accumulation of Heavy Metals and Nutrients from the Soil by Woody Species. *Rev. Chim.* **2019**, *70*, 753–758. [CrossRef]
20. Ianculescu, M.; Ionescu, M.; Lucaci, D.; Neagu, S.; Măcărescu, C.M. Dynamic of pollutants concentration in forests stands from Copșa Mică industrial area. *Ann. For. Res.* **2009**, *52*, 207–226.
21. Fritts, H.C. *Tree Rings and Climate*; Elsevier Academic Press: London, NY, USA, 1976; 567p.
22. Schweingruber, F.H. *Tree Rings and Environment-Dendrochronology*; Haupt: Bern, Switzerland, 1996; 609p.
23. Tilman, D.; Downing, J.A. Biodiversity and stability in grasslands. *Nature* **1994**, *367*, 363–365. [CrossRef]
24. Cybis Elektronik: CDendro and CooRecorder. 2016. Available online: https://www.cybis.se/forfun/dendro/index.htm (accessed on 22 November 2020).
25. Rinntech. *Tsap User Reference*; Triwerk GmbH & Co. KG Hardtstr: Heidelberg, Germany, 2005; 110p.
26. Holmes, R.L. Computer-assisted quality control in tree-ring dating and measurement. *Tree Ring Bull.* **1983**, *43*, 69–75.
27. Cook, E.R.; Kairiukstis, L.A. (Eds.) *Methods of Dendrochronology. Applications in the Environmental Sciences*; Kluwer Academic Publishers: Dordrecht, The Netherlands, 1990; 394p.
28. R Core Team. *R: A Language and Environment for Statistical Computing*; R Foundation for Statistical Computing: Vienna, Austria, 2017; Available online: https://www.R-project.org (accessed on 20 October 2020).
29. Popa, I.; Barbu, I. Evaluarea gradului de vătămare a ecosistemelor forestiere din zona Tarnița prin tehnici GIS de analiză spațială. *Rev. Pădurilor* **2001**, *6*, 8–11.
30. Schwarz, J.; Skiadaresis, G.; Kohler, M.; Kunz, J.; Schnabel, F.; Vitali, V.; Bauhus, J. Quantifying Growth Responses of Trees to Drought–a Critique of Commonly Used Resilience Indices and Recommendations for Future Studies. *Curr. For. Rep.* **2020**, *6*, 185–200. [CrossRef]
31. Chojnacka-Ożga, L.; Ożga, W. Impact of air pollution on Scots pine stands growing in Poland on the basis of dendrochronological analysis. *Environ. Sci. Proc.* **2021**, *3*, 77. [CrossRef]
32. Godek, M.; Sobik, M.; Błaś, M.; Polkowska, Z.; Owczarek, P.; Bokwa, A. Tree rings as an indicator of atmospheric pollutant deposition to subalpine spruce forests in the Sudetes (Southern Poland). *Atmos. Res.* **2015**, *151*, 259–268. [CrossRef]
33. Hauck, M.; Zimmermann, J.; Jacob, M.; Dulamsuren, C.; Bade, C.; Ahrends, B.; Leuschner, C. Rapid recovery of stem increment in Norway spruce at reduced SO_2 levels in the Harz Mountains, Germany. *Environ. Pollut.* **2012**, *164*, 132–141. [CrossRef] [PubMed]
34. Mihaljevič, M.; Ettler, V.; Šebek, O.; Sracek, O.; Kříbek, B.; Kyncl, T.; Majer, V.; Veselovský, F. Lead Isotopic and Metallic Pollution Record in Tree Rings from the Copperbelt Mining–Smelting Area, Zambia. *Water Air Soil Pollut.* **2011**, *216*, 657–668. [CrossRef]
35. Hojdová, M.; Navrátil, T.; Rohovec, J.; Žák, K.; Vaněk, A.; Chrastný, V.; Bače, R.; Svoboda, M. Changes in Mercury Deposition in a Mining and Smelting Region as Recorded in Tree Rings. *Water Air Soil Pollut.* **2011**, *216*, 73–82. [CrossRef]
36. Ianculescu, M.; Tisesscu, A. Efectele poluării industriale pe bază de compuși ai sulfului în acțiune sinergică cu metalele grele asupra creșterii arboretelor din zona Copșa Mică și evaluarea pagubelor produse (The effects of industrial pollution based on sulfur compounds in synergistic action with heavy metals on the growth of stands in the Copșa Mică area and the assessment of the damages caused). *Rev. Pădurilor* **1989**, *104*, 170–175.
37. Ianculescu, M.; Bândiu, C.; Budu, C.E. Modificări ale principalelor procese-ecofiziologice la arborii forestieri ca urmare a influenței poluării în zona Copșa Mică (Changes of the main ecophysiological processes in forest trees due to the influence of pollution in the Copșa Mică area). *Rev. Pădurilor* **1989**, *104*, 64–68.
38. Adamu, C.I.; Nganje, T.N.; Aniekan, E. Heavy metal contamination and health risk assessment associated with abandoned barite mines in Cross River State, southeastern Nigeria. *Environ. Nanotechnol. Monit. Manag.* **2015**, *3*, 10–21. [CrossRef]

MDPI

Article

Diagnostic Assessment and Restoration Plan for Damaged Forest around the Seokpo Zinc Smelter, Central Eastern Korea

A Reum Kim [1], Bong Soon Lim [1], Jaewon Seol [1], Chi Hong Lim [2], Young Han You [3], Wan Sup Lee [4] and Chang Seok Lee [1,*]

[1] Division of Chemistry and Bio-Environmental Sciences, Seoul Women's University, Seoul 01797, Korea; dkfma@swu.ac.kr (A.R.K.); bs6238@swu.ac.kr (B.S.L.); seol_jaewon@swu.ac.kr (J.S.)
[2] Division of Ecological Survey Research, National Institute of Ecology, Seoul 33657, Korea; sync03@nie.re.kr
[3] Department of Biology, Kongju National University, Kongju 32588, Korea; youeco21@kongju.ac.kr
[4] Samseong Landscape Co., Ltd., Andong 36665, Korea; kjc3700@hanmail.net
* Correspondence: leecs@swu.ac.kr; Tel.: +82-2-970-5666

Abstract: *Research Highlights*: This study was carried out to diagnose the forest ecosystem damaged by air pollution and to then develop a restoration plan to be used in the future. The restoration plan was prepared by combining the diagnostic assessment for the damaged forest ecosystem and the reference information obtained from the conservation reserve with an intact forest ecosystem. The restoration plan includes the method for the amelioration of the acidified soil and the plant species to be introduced for restoration of the damaged vegetation depending on the degree of damage. *Background and Objectives*: The forest ecosystem around the Seokpo smelter was so severely damaged that denuded lands without any vegetation appear, and landslides continue. Therefore, restoration actions are urgently required to prevent more land degradation. This study aims to prepare the restoration plan. *Materials and Methods*: The diagnostic evaluation was carried out through satellite image analysis and field surveys for vegetation damage and soil acidification. The reference information was obtained from the intact natural forest ecosystem. *Results*: Vegetation damage was severe near the pollution source and showed a reducing trend as it moved away. The more severe the vegetation damage, the more acidic the soil was, and thereby the exchangeable cation content and vegetation damage were significantly correlated. The restoration plan was prepared by proposing a soil amelioration method and the plants to be introduced. The soil amelioration method focuses on ameliorating acidified soil and supplementing insufficient nutrients. The plants to be introduced for restoring the damaged forest ecosystem were prepared by compiling the reference information, the plants tolerant to the polluted environment, and the early successional species. The restoration plan proposed the *Pinus densiflora*, *Quercus mongolica*, and *Cornus controversa–Juglans mandshurica* communities as the reference conditions for the ridge, slope, and valley, respectively, by reflecting the topographic condition. *Conclusions*: The result of a diagnostic assessment showed that ecological restoration is required urgently as vegetation damage and soil acidification are very severe. The restoration plan was prepared by compiling the results of these diagnostic assessments and reference information collected from intact natural forests. The restoration plan was prepared in the two directions of soil amelioration and vegetation restoration.

Keywords: air pollution; diagnostic assessment; forest ecosystem; reference information; restoration plan

Citation: Kim, R.A.; Lim, B.S.; Seol, J.; Lim, C.H.; You, Y.H.; Lee, W.S.; Lee, C.S. Diagnostic Assessment and Restoration Plan for Damaged Forest around the Seokpo Zinc Smelter, Central Eastern Korea. *Forests* **2021**, *12*, 663. https://doi.org/10.3390/f12060663

Academic Editors: Ovidiu Badea, Alessandra De Marco, Pierre Sicard and Mihai A. Tanase

Received: 16 April 2021
Accepted: 21 May 2021
Published: 24 May 2021

1. Introduction

Most developed countries have decreased anthropogenic air pollution emissions by implementing abatement polices [1,2]. Korea has also practiced such a policy, and thus environmental conditions around the industrial complexes of large scale have improved greatly [3,4].

However, factories of a small scale that are far away from the public's attention, such as the Seokpo smelter where this study was carried out, still emit air pollutants and thereby cause vegetation damage and acidify soil. The forest ecosystem around the Seokpo zinc smelter was severely destroyed and therefore landslides sometimes occur. Industrial activities have resulted in the enormous emissions of air pollutants for about 40 years since the 1970s when the smelter was constructed in Seokpo in central eastern Korea. The pollutants have continued to affect the surrounding forests and other ecosystems. Forest vegetation has become sparse and poor as trees have withered, undergrowth has disappeared, and bare ground has appeared throughout the wide area.

Pollutants discharged beyond the limits of the buffering capacity of an ecosystem prevent it from maintaining its normal structure and function. Excessive land and energy use and the ecological imbalance that it brings appear to be major factors that threaten environmental stability on local as well as global levels [4–8]. The vegetation decline and subsequent soil erosion and landslides observed in the vicinity of the Seokpo zinc smelter correspond to such an example. In fact, global environmental problems such as climate change are also due to this functional imbalance between the pollution source and the sink [9–11].

If the population grows and the land and energy use continue to intensify, such ecological imbalance is likely to increase even more in the future [7,12,13]. Indeed, industrialized and urbanized areas have been expanding steadily, and the real size of degraded vegetation, such as grassland and shrubland, has increased proportionally to such land transformation in industrial areas of Korea [4,7,8,14,15]. Moreover, vegetation decline induces the structural simplification and functional weakening of plant communities, consequently leading to negative effects on ecosystem service, which provides invaluable benefits to us [12,13]. In this respect, the restoration of degraded ecosystems is urgently required to prevent the spread of such additive pollution damage [4,7,13,16].

Ecological restoration is aimed at recovering the sound natural conditions before destruction. Ecological restoration is an ecological technology that heals the damaged nature by imitating the system and function of the integrated nature, thus providing habitats for various creatures and seeking to secure the future environment of humankind [17–20]. Ecological restoration has been considered as improving ecological productivity in degraded lands, conserving biological diversity, and mitigating lost or damaged ecosystems [19–28]. Human aids are often required to restore the damaged ecosystems and prevent further damage [16,19,20,29], and provisions of extra propagules and site amendment may initiate recovery processes [30].

In order to heal the damaged nature, we must first check what problems the target has, such as how much it is degraded or what is the cause of the damage. In other words, a diagnostic assessment of the restoration target should be made [19,20].

All restoration projects are with targets to reduce the negative ecological impacts of the past and to restore the natural potential of the restoration target as much as possible. Ecological restoration means copying nature by studying a system of the integrate nature. There are several planning steps, based on the results of diagnostic assessments, to find the deficits and the targets for planning to improve the degraded ecosystem. First of all, we have to prepare such measures by obtaining diverse ecological information on an area to be restored because specific restoration efforts have to be applied in the field [31–33].

The restoration of an ecosystem damaged by environmental pollution can be achieved either through improvement of the environment polluted or by establishing plants tolerant to the pollutants [7,8,13,16,29,34–36]. Species tolerant to environmental pollution can persist through growth and reproduction or even expand their distribution range in the polluted environment [5–8,37,38].

The Seokpo smelter is uniquely located on a small mountain village and is thus far away from the public's attention. Therefore, little is known about the situation and, moreover, no academic research has been carried out. However, due to the effects of air pollutants emitted over a long period of time and the soil pollution resulting from them, the forest damage began to become visible, and it has led to landslides and damage to surrounding rivers in good condition, causing worry in recent years. There is, therefore, a growing demand for restoration. The major pollution source of the Seokpo smelters, which refines primarily processed ore rather than raw ore, is sulfur dioxide generated from the combustion of fossil fuels used as an energy source. The damage state was similar to that of Ulsan and Yeocheon industrial complexes, major industrial complexes in Korea [4,7,8,12,36].

Restoration is an ecological technology that ameliorates degraded nature by imitating integrated and healthy nature. Restoration is achieved through a series of procedures, such as a survey of the existing conditions, a statement of the goals and objectives, the designation and description of a reference, the preparation of a master plan, the establishment of a restoration plan, restoration practices, monitoring, adaptive management, and evaluation [39–42]. Such ecological restoration is common as a means to solve such problems in developed countries, which correctly recognize that the environmental problems at a global level, such as climate change, are due to the functional imbalance between the artificial environment as an environmental stress source and the natural environment as its sink. However, in most developing countries, including Korea, most restoration projects have neglected such procedures and thus have not met the restoration goals, in spite of great expense and labor [43–50]. A series of procedures are required to achieve successful ecological restoration. However, these procedures usually tend to be ignored in most restoration projects implemented in Korea. Diagnostic evaluation is generally omitted. Even if a diagnostic evaluation is made, there are very few cases in which the level and method of restorative treatment are determined based on the results, and most restoration projects progress only by active methods without any relation to the degree of damage [46,49]. Therefore, cost and energy are wasted, and the effect is very little [46,47,50]. In most restoration projects, the reference information is not used, and restoration is performed based on the subjective decisions of the project manager. Thus, restoration projects are conducted without any model or goals. Consequently, exotic species, which should be excluded thoroughly in a restoration project, are introduced frequently, and the spatial distribution range for plant species is barely considered [46,47,50]. Therefore, most restoration projects remain at the level of past afforestation or classical landscaping.

This study was attempted to implement ecological restoration of the advanced level, which is beyond this low level of restoration. This study conducted a diagnostic evaluation for the damaged forest from air and soil pollution, collected reference information from intact forest without any damage, and prepared a restoration plan by combining such information. In order to ecologically restore the area, the following situations need to be considered: first, since pollutants continue to be emitted, tolerant plants that can withstand such pollution should be selected and introduced. Second, soils contaminated by the effects of pollutants discharged for a long time should be improved. Third, many species have already disappeared due to environmental pollution, and thus we need to introduce species that have disappeared based on the ecological information obtained from the reference ecosystem. Finally, since the environment has been severely damaged to the point where landslides occur, measures should also be taken to prevent landslides when establishing the planting bed. This study aims to clarify the extent and the type of damaged forest and, furthermore, to recommend a restoration plan suitable for the ecological condition of the target site as well as the damage degree based on the principle of restoration ecology. In order to arrive at these goals, we assessed vegetation damage based on satellite image interpretation and field checks. Vegetation damage was also assessed based on species composition and species diversity. Furthermore, we also diagnosed the damaged state of soil based on its physic-chemical properties. We recommended the restoration plan by synthesizing the results of the diagnostic assessment and the reference information, including pollution tolerant species and early successional species considering the environmental condition of this area.

2. Materials and Methods

2.1. Study Site

This study was carried out in the forest ecosystem around the Seokpo smelter, located on the central eastern Korea (Figure 1). The Seokpo smelter has been in operation since 1970. The smelter produces zinc, cadmium sulfate, copper sulfate, and manganese sulfate and has emitted many air pollutants around it [51]. This area is surrounded by mountainous areas with steep slopes and has topographical features that make it difficult for air pollutants to spread (Figure 2) [4,6,16]. In addition, the temperature inversion, which occurs as the cold air on the mountain descends along the surrounding mountain slopes and is trapped at the bottom of the basin after sunset, makes it more difficult to spread air pollutants, causing an increase in pollution damage in this narrow valley. During this temperature inversion time, dense smoke often settles in low-lying areas and becomes trapped due to temperature inversions—when a layer within the lower atmosphere acts as a lid and prevents vertical mixing of the air. Steep canyon walls act as a horizontal barrier, concentrating the smoke within the deepest parts of the canyon and increasing the strength of the inversion [52,53].

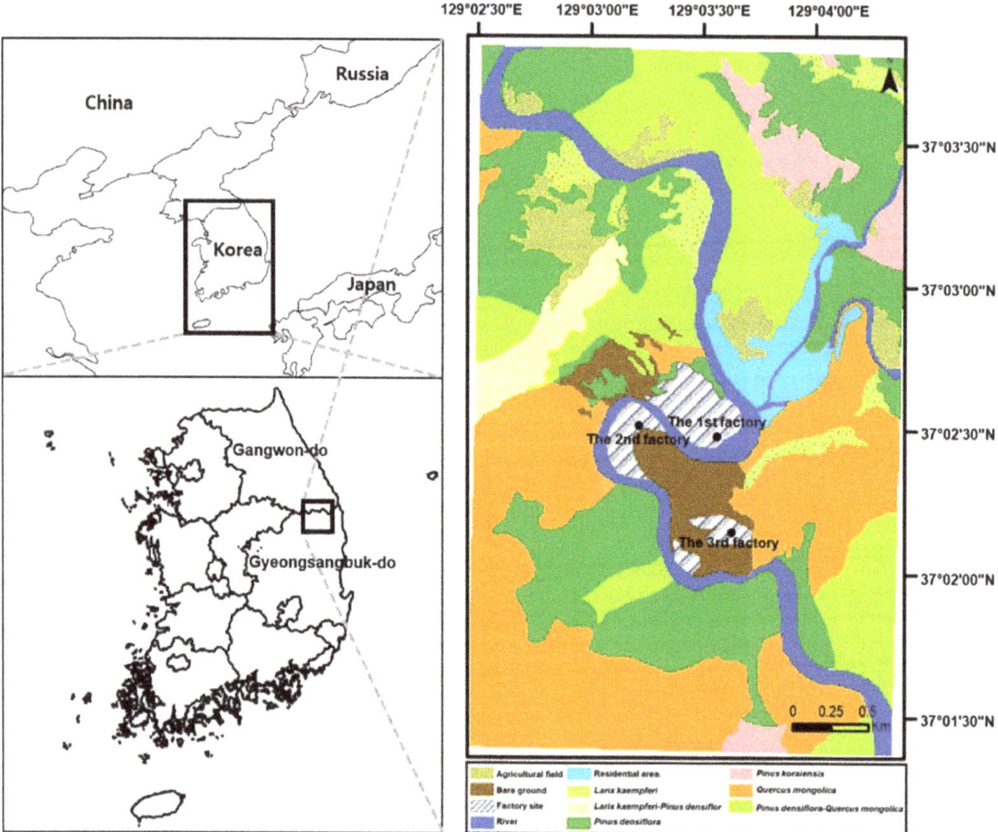

Figure 1. A map showing the study area, the Seokpo zinc smelter, which is located in central eastern Korea. Seokpo zinc smelter is composed of three factories. A colored map shows vegetation and land use types established around the smelter. Dots and the parts expressed with oblique lines around them indicate factories and factory sites.

As is shown in a vegetation map in Figure 1, the vegetation of this area is dominated by the *Quercus mongolica* community. However, the *Pinus densiflora* community is established on the slopes of mountainous areas with steep slopes or mountain ridges and peaks with shallow soil depth due to edaphic characteristics. There is also a mixed forest in which two species forming a community together. On the other hand, there are plantations where *Larix kaempferi* and *P. koraiensis* are introduced artificially in some areas, and there are places where mixed forests are formed where natural vegetation is mixed with planted species. Meanwhile, agricultural and residential areas are established in the lowlands.

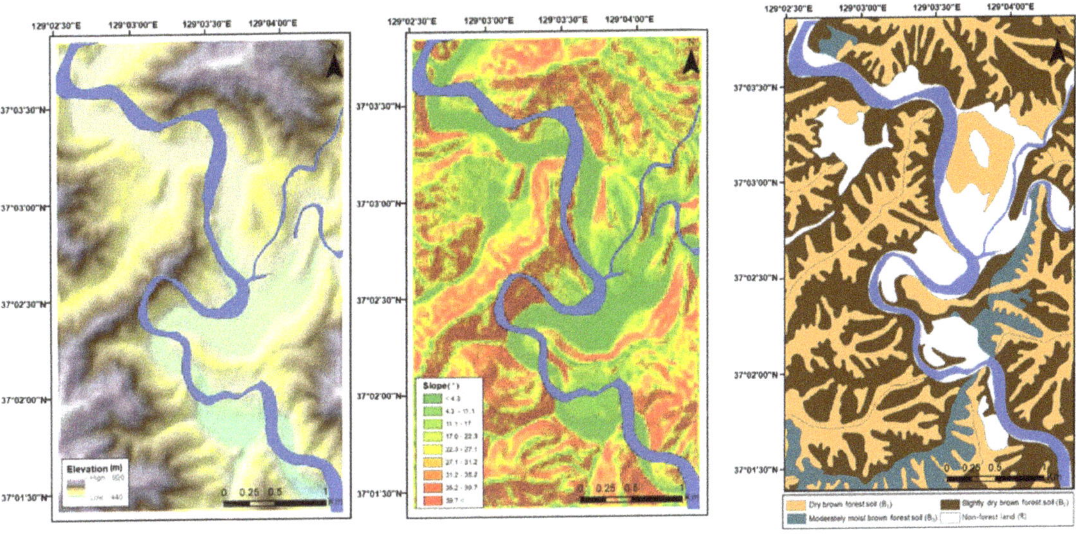

Figure 2. Maps showing the spatial distribution of elevation (m), slope (degree), and soil type in the study area.

The climate of Bonghwa is continental, with warm and moist summers and cold and dry winters. The mean annual temperature is 9.9 °C and the high and low mean temperatures are recorded as 28.6 °C and −10.3 °C in August and January, respectively. The mean annual precipitation is 1217.9 mm; about 60% falls in the rainy season from June to August and, including the typhoon season of September, about 70% is concentrated in both periods [54].

The elevation of the study area ranges from 400 to 900 m above sea level. The slope degree is as steep as more than 20° in most of the mountainous land except the valley. The parent rock of the study area consists mostly of granite, and in the flat land beside rivers and streams consists of alluvium. Soil in this area is composed of dry (B1), slightly dry (B2), and moderately moist brown forest soil (B3), which were developed on granite bedrock [55] (Figure 1).

The reference forest, used for comparison, was designated as the Korean red pine (*Pinus densiflora* Siebold & Zucc.) and oak (*Quercus mongolica* Fisch. Ex Ledeb., and *Q. variabilis* Blum communities), which are the representative late successional vegetation types in Korea, and *Cornus controversa* Hemsl. ex Prain, which represents the valley forest [56]. The reference forests were selected in the Uljin genetic resource reserve, which are about 15 km from this study area and therefore retain a healthy vegetation. The reference forests were selected as the forests that are from 50 to 100 years old, which is not an old growth forest but a stable forest. The number of plots chosen for the survey for the reference forests was 10, 10, and 10 for the *P. densiflora*, *Q. mongolica*, and *Cornus controversa* communities, respectively.

2.2. Methods

A vegetation map was made based on image interpretation and field checks. Aerial photo images (1:5000 scale) were used to identify the vegetation types and boundaries, which appear as a homogeneous patch. These vegetation types were confirmed by field checks. Vegetation types were overlapped onto topographical maps at a 1:5000 scale. Patches smaller than 1 mm on the map were excluded from this study because of the uncertainty of their sizes and shapes [57]. The final map was constructed with ArcGIS program (ver. 10.0, ESRI, Redlands, CA, USA) [58].

To determine the vitality of vegetation in the study area, Landsat images taken on 2 October 2018 were downloaded to analyze the normalized differential vegetation index (NDVI). Vegetation damage based on vitality was assessed through supervising analysis on the satellite image [59]. This study applied a supervised classification-maximum likelihood algorithm to classify the vegetation damage state around the Seokpo zinc smelter using Landsat images in the ArcGis10.1 program. The maximum likelihood algorithm is the most common method in remote sensing image data analysis [60], which is mainly controlled by selecting the pixels that are representative of the desired classes [61]. Using the signature file creation tool, vegetation damage was classified into five classes of very severe, severe, moderate, light, and none. The damage degrees classified were verified through field checks as follows.

Visible damage was investigated by recording the degree of necrosis that appeared on the leaf surface of plants appearing in the process of the vegetation survey. The damage degree was classified into five groups based on the percentage of injury shown on the leaf surface: very severe (V, more than 75% of total leaf area damaged), severe (IV 50–75% damaged), moderate (III, 25–50% damaged), light (II, less than 25% damaged), and none (I, 0%) [7].

The vegetation structure damage was assessed by the deformation of vegetation stratification based on [62]. More than 50% of the land, which is covered with barren ground was assessed as 'very severe'. Grassland or barren ground without any woody plants was assessed as 'severe'. Vegetation that had lost some stratification was assessed as 'moderate'. Vegetation with visible damage only to integrate structure, with all strata composed of canopy, understory, shrub, and herb layers shown without any loss of stratification was assessed as 'light'. A map expressing vegetation damage was prepared by applying the GIS program (ver. 10.0, ESRI, Redlands, CA, USA).

The vegetation survey was carried out from May to September in 2018 and 2019. The vegetation survey was carried out by recording the Braun–Blanquet's cover class of plant species appearing in quadrats of 2 m × 2 m, 5 m × 5 m, and 20 m × 20 m size in grassland, shrubland, and forest, respectively, installed randomly [63]. The vegetation survey was carried out in 99 plots (28, 28, 1, 5, 5, 19, and 13 plots for *Pinus densiflora* community, *Quercus mongolica* community, *Q. variabilis* community, valley forest, shrubland, grassland, and cut slope, respectively) from May to September in 2018 and 2019.

Soil samples were collected with a sampling spade in June–August 2019 from the top 10 cm after removing the litter at five random points in each plot, after which they were pooled, air dried at room temperature, and sieved through 2 mm mesh. A total of 18, 18, 1, 3, 3, 12, and 9 soil samples were collected from the *Pinus densiflora* community, the *Quercus mongolica* community, the *Q. variabilis* community, valley forest, shrubland, grassland, and cut slope, respectively. Soil properties were diagnosed for pH and Ca^{2+}, Mg^{2+}, and Al^{3+} content. Soil pH was measured with a bench top probe after mixing the soil with distilled water (1:5 ratio, w/v) and filtering the extract (Whatman No. 44 paper). Organic matter (OM) concentration was estimated by loss of dry mass on ignition at 400 °C. Total nitrogen was measured with the micro-Kjeldahl method [64]. Available P was extracted in 1-N ammonium fluoride (pH = 7.0) and exchangeable Ca^{2+}, Mg^{2+}, and Al^{3+} contents were extracted with 1N ammonium acetate (pH = 7.0 for Ca and Mg and pH = 4.0 for Al) and measured by ICP (inductively coupled plasma atomic emission spectrometry; Shimadzu ICPQ-1000) [65]. The results of the analysis on the physic-chemical properties of soil were reinforced by the simple kriging model. Maps expressing the physic-chemical properties of the soil were prepared by applying the GIS program (Version 10.1). The soil properties (pH, OM, N, P, Ca, Mg and Al) of sites showing different degrees of damage to vegetation were compared with one-way analysis of variance (ANOVA) and Tukey's honestly significant difference (HSD) test at $\alpha = 0.05$ [66].

The restoration plan was prepared by recommending a soil amelioration method and the plant species to be introduced depending on the degree of damage to the vegetation and soil. Dolomite and organic fertilizer were recommended for soil amelioration. The dolomite requirement was calculated by applying the following equation: dolomite requirement (t/ha) = (target pH–current pH) × soil texture factor. We decided to set the target pH as 5.5, based on the normal pH of the natural forest soil, and soil texture factor as 3, reflecting the soil texture of this area [67]. Dolomite raises the soil pH and increases available Ca^{2+} and Mg^{2+} due to the following chemical reactions in the soil solution [68]:

$$\text{Initial chemical reaction: } Ca\bullet\, Mg(CO^3)_2 + 2H^+ \rightarrow 2HCO_3{}^- + Ca^{2+} + Mg^{2+}, \quad (1)$$

$$\text{Second reaction: } 2HCO_3{}^- + 2H^+ \rightarrow 2CO^2 + 2H_2O, \quad (2)$$

$$\text{Net reaction: } Ca\bullet\, Mg(CO^3)_2 + 4H^+ \rightarrow Ca^{2+} + Mg^{2+} + 2CO^2 + 2H_2O, \quad (3)$$

The amount of organic fertilizer applied was determined to be half the level of dolomite, referring to previous study results [8]. The chemical characteristics of the organic fertilizer are given in Appendix A, Table A1.

The introduction of vegetation for restoration took the form reinforcing the lost part in the vegetation stratification. Therefore, we planned to introduce the disappeared species compared to the species composition of the natural reference site. Furthermore, we added tolerant and early successional species in our restoration plan, considering the environmental condition of this area, where air and soil pollution continues and vegetation is so severely damaged that bare ground can appear and severe soil erosion occurs. The plant species to be introduced were selected by applying indicator species analysis. Indicator species analysis was carried out using the function 'vegan', 'indicspecies' of the R statistical package (version 4.0.2). In addition, we reinforced the plant species to be introduced by referring to national vegetation information [69] and existing research data conducted on the reference site of this study [70], considering that this study was conducted in a limited place.

3. Results

3.1. Vegetation Damage

The spatial distribution of NDVI showed that its value was lower near the factory and tended to increase as it moved away from it (Figure 3). The value was also related to the topographic condition; thus, it was low in the ridge and high in the valley (Figure 3).

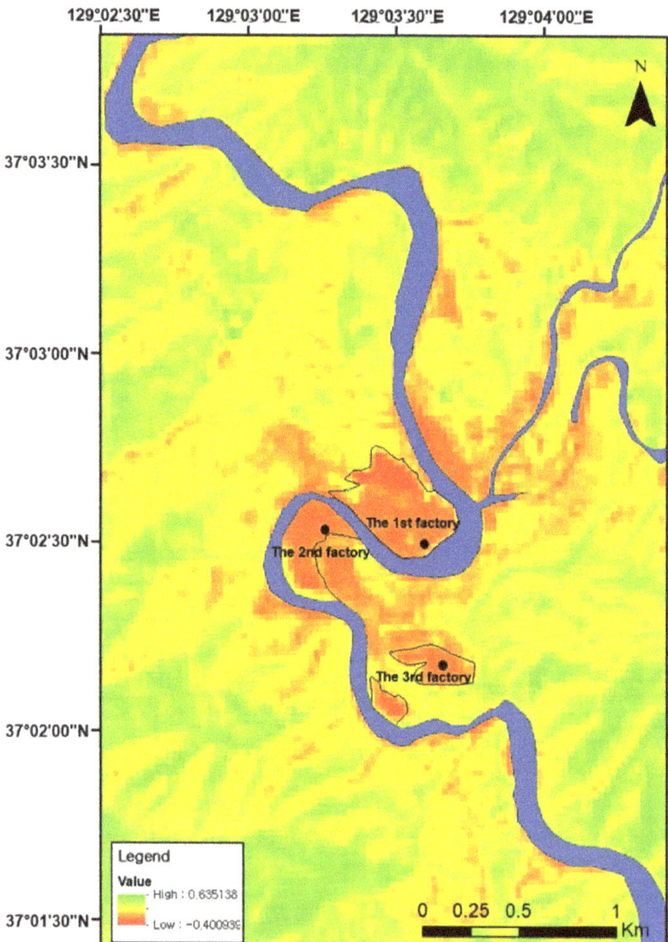

Figure 3. Spatial distribution of NDVI in the study area.

Vegetation damage identified from the satellite image interpretation depended on the distance from the smelter and topography. The damage appeared was severer in sites closer to the smelter and decreased farther away (Figure 4). The degree of damage was also dominated by topographic conditions, and therefore damage was restricted within the first ridge from the pollution source, little damage thus appearing on the opposite slope or beyond the first ridge from the smelter (Figure 4).

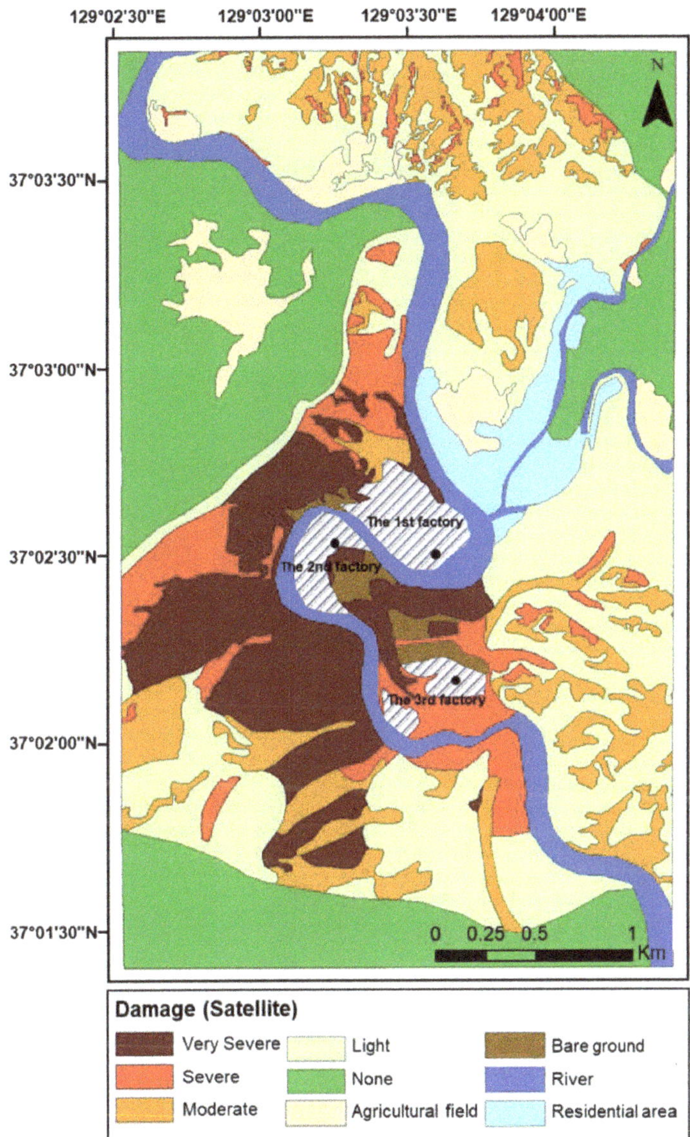

Figure 4. Spatial distribution of vegetation damage based on satellite image interpretation in the study area.

Damage based on vegetation stratification showed a trend similar to the abovementioned results. Vegetation in the site where the damage was light showed the integrate structure with all strata composed of canopy, understory, shrub, and herb layers. However, vegetation structure became simplified with the increase of the damage, and thus grassland or barren ground appeared in sites where the damage was the severest (Figure 5).

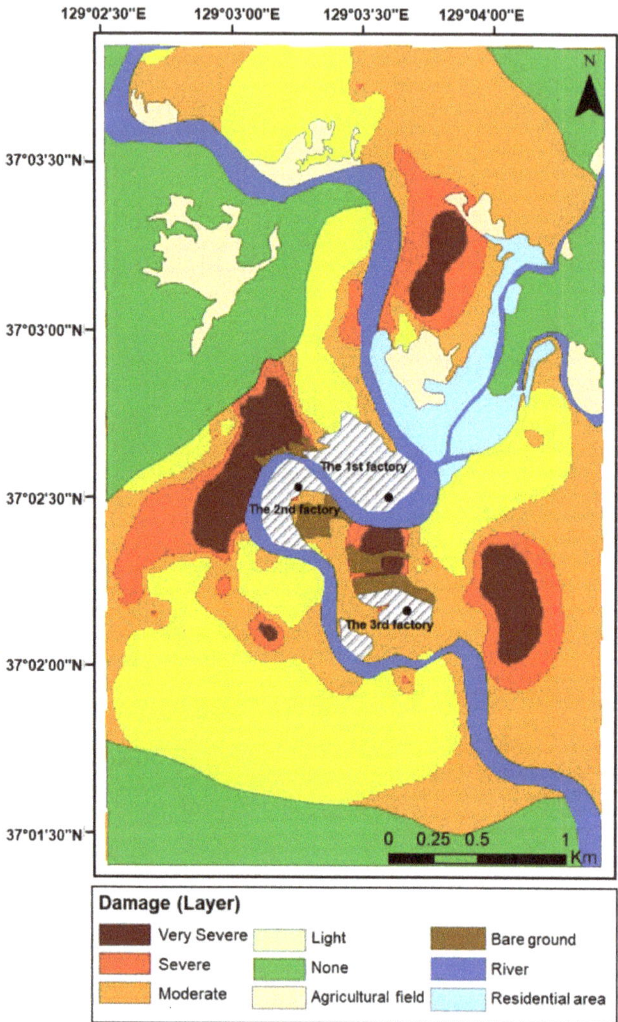

Figure 5. Spatial distribution of vegetation damage based on vegetation stratification in the study area.

In a map where vegetation damage by damage class is shown (Figure 4), very severely damaged vegetation appeared in areas located in the northwestern direction of the first factory, the western and southern directions of the second and third factory, and surrounded by the first, second, and third factories. Severely damaged vegetation appeared in the areas farther than the very severely damaged vegetation from the three factories in all four directions of east, west, south, and north. Moderately damaged vegetation appeared in the areas located in the eastern and western directions, with the third factory at the center. Lightly damaged vegetation appeared in the areas far from them within the first ridge from the factories.

3.2. Soil Degradation

Spatial distribution of the physic-chemical properties of the soil reflected a trend of vegetation damage. Soil pH was usually low compared with the unpolluted area but was lower in sites close to pollution sources and became higher farther away (Figure 6).The

Ca^{2+} and Mg^{2+} content showed trends similar to that of the pH, whereas Al^{3+} content represented a reverse trend (Figure 6).

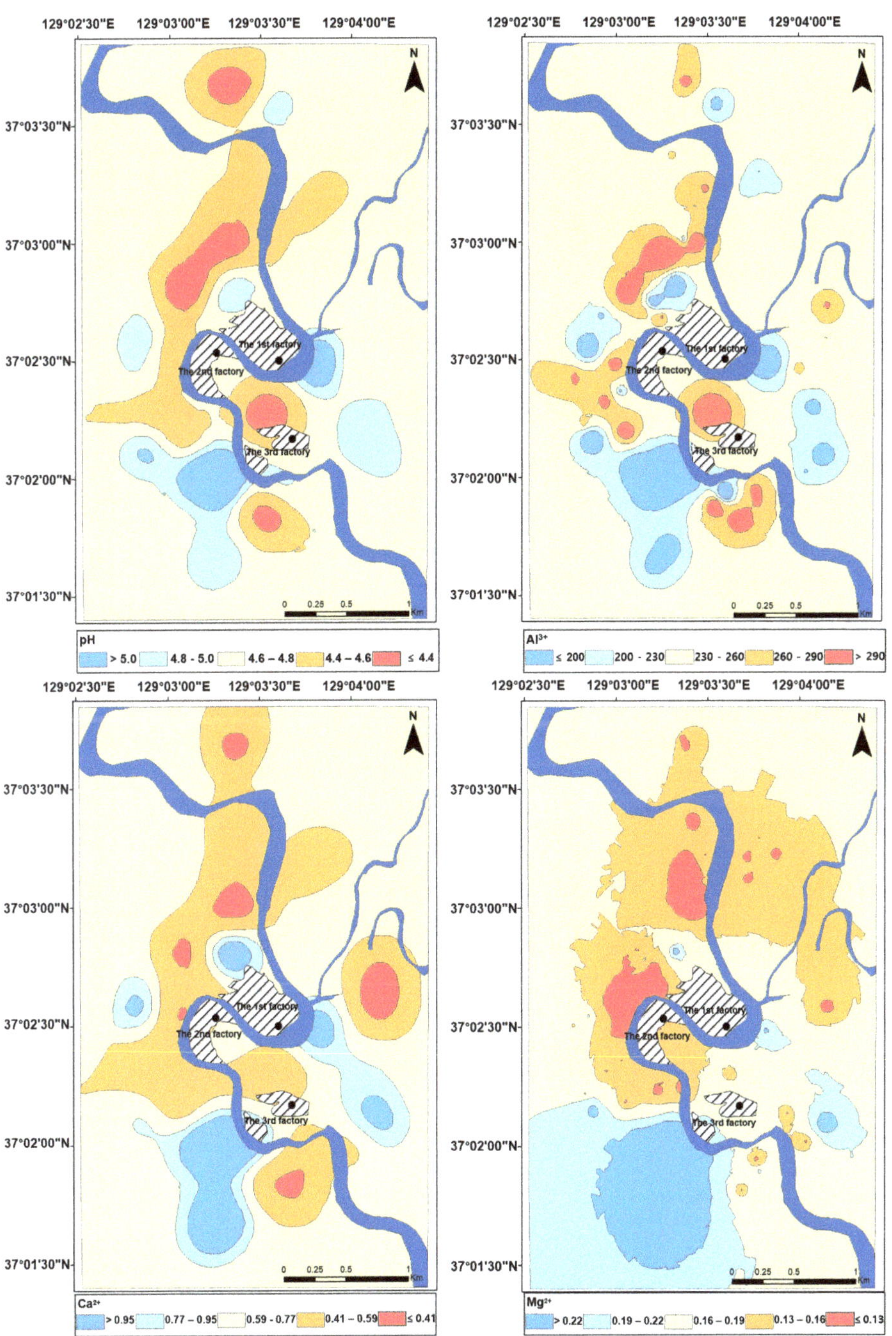

Figure 6. Spatial distribution of soil pH and Ca^{2+}, Mg^{2+}, and Al^{3+} contents of soil in the study area.

Soil pH tended to be relatively low in the northern and western directions of the three factories, the area surrounded by those factories, and the southern direction of the third factory, whereas it was relatively high in the southwestern and northeastern directions of the third factory (Figure 6).

The Ca^{2+} and Mg^{2+} content showed trends similar to that of the pH, while Al^{3+} content represented a reverse trend (Figure 6).

The physic-chemical properties of the soil were compared with those of the reference site and among the degrees of damage to the vegetation (Figure 7). The pH and Ca^{2+}, Mg^{2+}, and available phosphorus contents were lower than those in the reference site, whereas the total nitrogen content was vice versa. However, organic matter and Al^{3+} content did not show any significant difference between both sites. On the other hand, pH and Ca^{2+} and Mg^{2+} contents showed significant difference among the degrees of damage to the vegetation, but the other factors did not show any significant differences among the degrees of damage.

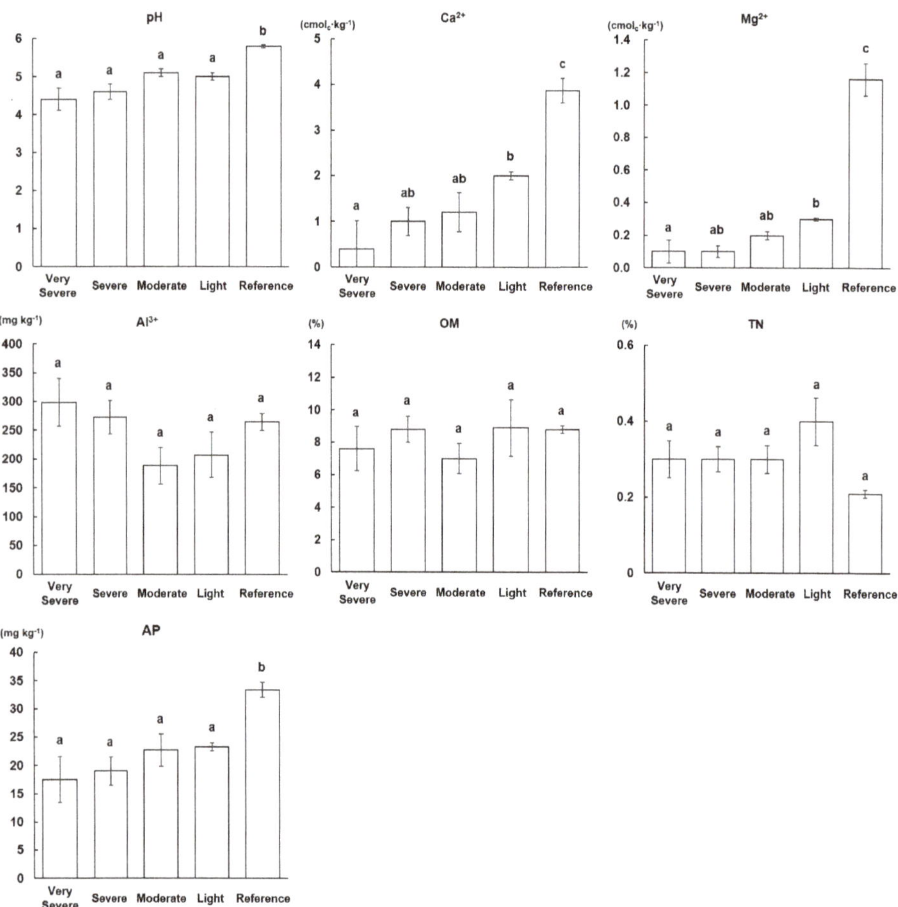

Figure 7. A comparison of the physic-chemical properties of the soil among damage degrees of vegetation and with that of the reference site. Very severe, severe, moderate, and light indicate damage degree and reference indicates the unpolluted site selected for comparison. OM: organic matter; TN: Total nitrogen; AP: available phosphorus. Each bar was expressed with mean and standard error of mean. Tukey's honestly significant difference (HSD) test was conducted on each of the parameters that show a statistically significant difference among the four types of damage degrees at $\alpha = 0.05$; the means with the same alphabetical character (in superscript), for each parameter, were not different from each other.

3.3. Species Composition

As the result of stand ordination, arrangement of stands reflected vegetation damage (Figure 8). The reference stands were arranged on the left on the AXIS I and very severely or severally damaged stands on the right, and moderately and lightly damaged stands were arranged between both groups. Moderately and lightly damaged stands tended to be arranged depending on the topographical position as *P. densiflora* stands, *Q. Mongolica* stands, and stands established in the valley were arranged in the mentioned order as moves from bottom upward on the AXIS II. Meanwhile, cut slope stands were arranged in the left part on the AXIS I like the reference stands but separated from them on the AXIS II.

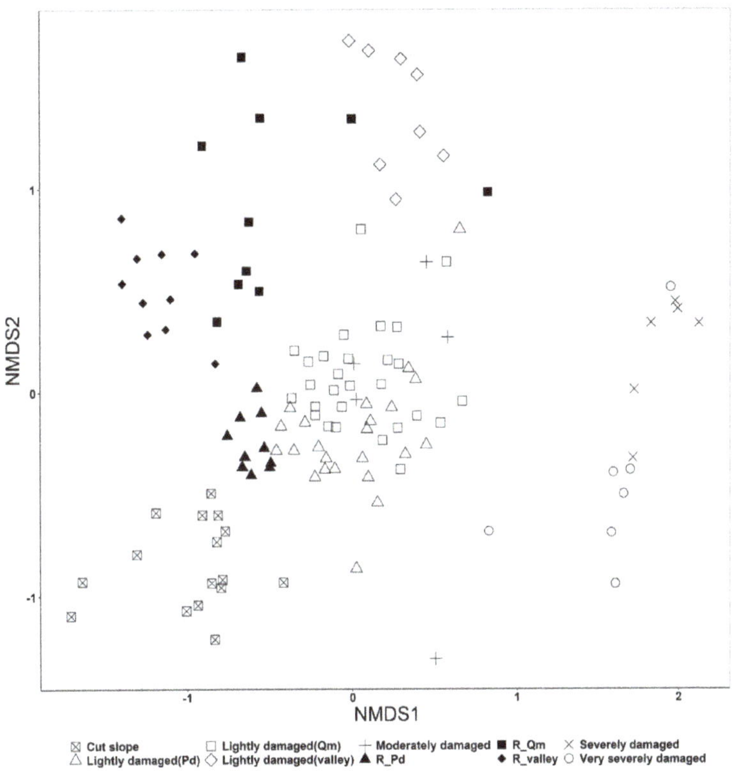

Figure 8. Ordination of vegetation stands established around the Seokpo smelter and on the natural reference forest, Uljin Forest Genetic Resources Conservation Reserve, central eastern Korea. Legends expressed as damage degree represent the damaged stands established around the Seokpo smelter. R_Pd: *Pinus densiflora* stands established in the reference site; R_Qm: *Quercus mongolica* stands established in the reference site; R_Valley: stands established in the reference site; Cut slope: stands established on the incised slope along the forest road ($p = 0.001$, stress = 0.1702836).

3.4. Species Diversity

As a result of comparing the species diversity by the species rank–dominance curve, the species diversity of the damaged sites was lower than that in the reference sites (Figure 9). In the damaged sites, species diversity tended to decrease in proportion to the damage degree (Figure 8). The species diversity was also dominated by the topographic condition and thus high in the vegetation established in the valley and low in the pine forest on the ridge, and the species diversity of broad-leaved forests on the mountain slope was located between both vegetation types (Figure 9).

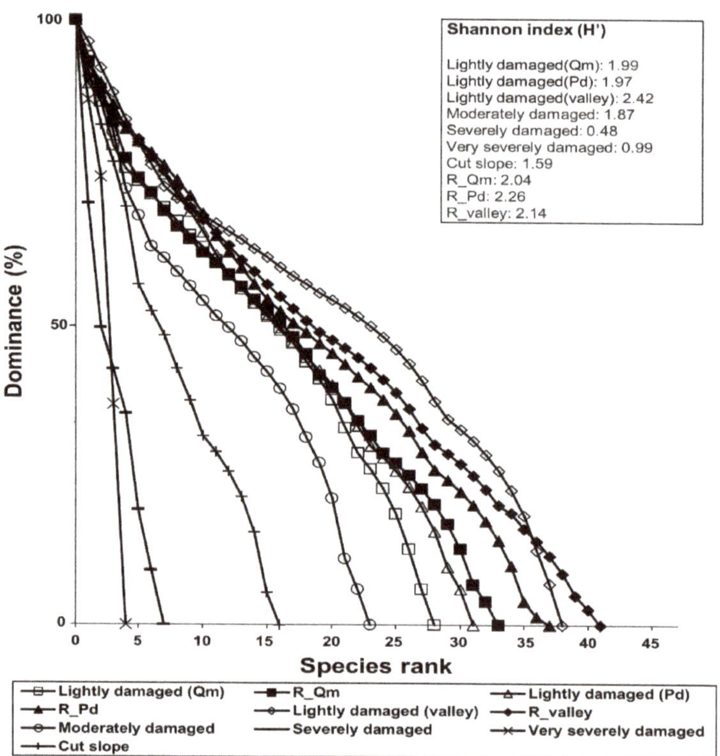

Figure 9. Species rank–dominance curves of vegetation stands established around the Seokpo smelter and on the natural reference forest, Uljin Forest Genetic Resources Conservation Reserve, southeastern Korea. Legends are the same as those in Figure 8.

3.5. Selection of Plant Species for Vegetation Restoration

We selected disappeared, tolerant, and early successional species by applying the indicator species analysis (Tables 1–4). First, we selected species to be introduced for vegetation restoration by comparing all vegetation data between polluted and natural reference sites and cut slope. We selected species which appear in the natural reference site but do not appear in the polluted site as the disappeared species. The disappeared species mean species that should be introduced for vegetation restoration. We selected species which showed the reverse trend to the disappeared species as the tolerant species. The species which appear characteristically on the cut slope were selected as the early successional species. Ecological restoration should copy the environment by studying a system of the integrate nature. However, considering the environmental condition of this area where air and soil pollution continues and vegetation is so severely damaged that bare ground can appear and severe soil erosion occurs, we added the tolerant and early successional species in our restoration plan. In addition, we added the plant species to be introduced by referring to the existing research data conducted around this study area [69,70] to enhance the stability of the restoration plan.

Table 1. The result of indicator species analysis for selecting the disappeared species, tolerant species to the polluted environment, and early successional species based on data collected in all study sites. Cut slope: early successional species; Severe, Moderate, Light: tolerant species in severely, moderately, and lightly damaged sites, respectively; Valley: tolerant species in valley; Reference(Pd), Reference(Qm), Reference(VA): disappeared species in the natural reference sites of *Pinus densiflora* forest, *Quercus mongolica* forest, and valley forest, respectively.

Species Name	Site Type	Stat	*p*-Value
Betula schmidtii	Cut slope	0.675	0.001 ***
Calamagrostis arundinacea	Cut slope	0.787	0.001 ***
Lindera obtusiloba	Moderate	0.802	0.001***
Tripterygium regelii	Moderate	0.436	0.009 **
Rhododendron schlippenbachii	Light	0.463	0.006 **
Actinidia arguta	Valley	0.525	0.002 **
Aralia elata	Valley	0.482	0.012 *
Deutzia parviflora	Valley	0.502	0.002 **
Fraxinus rhynchophylla	Valley	0.414	0.026 *
Schisandra chinensis	Valley	0.396	0.037 *
Athyrium yokoscense	Very severe	0.904	0.001 ***
Miscanthus sinensis var. *purpurascens*	Severe	0.952	0.001 ***
Betula chinensis	Reference(Pd)	0.496	0.004 **
Disporum smilacinum	Reference(Pd)	0.585	0.001 ***
Fraxinus sieboldiana	Reference(Pd)	0.677	0.001 ***
Aster scaber	Reference(Qm)	0.513	0.007 **
Athyrium vidalii	Reference(Qm)	0.449	0.005 **
Callicarpa japonica	Reference(Qm)	0.470	0.005 **
Carex siderosticta	Reference(Qm)	0.440	0.007 **
Hydrangea serrata f. *acuminata*	Reference(Qm)	0.425	0.016 *
Potentilla freyniana	Reference(Qm)	0.564	0.002 **
Styrax obassis	Reference(Qm)	0.417	0.030 *
Carex humilis var. *nana*	Reference(VA)	0.441	0.006 **
Cornus controversa	Reference(VA)	0.874	0.001 ***
Juglans mandshurica	Reference(VA)	0.402	0.050 *

*** significant at 0.1% level, ** significant at 1% level, * significant at 5% level.

Table 2. The result of indicator species analysis for selecting the disappeared species and tolerant species to the polluted environment based on data collected in both polluted and natural *Pinus densiflora* forests. Polluted: tolerant species in the polluted site of *Pinus densiflora* forest; Reference: disappeared species in the natural reference site of *Pinus densiflora* forest.

Species Name	Site Type	Stat	*p*-Value
Athyrium yokoscense	Polluted	0.315	0.039 *
Lindera obtusiloba	Polluted	0.447	0.004 **
Miscanthus sinensis var. *purpurascens*	Polluted	0.333	0.040 *
Quercus mongolica	Polluted	0.446	0.011 *
Rhododendron schlippenbachii	Polluted	0.461	0.024 *
Atractylodes ovata	Reference	0.418	0.033 *
Betula chinensis	Reference	0.391	0.034 *
Carex humilis var. *nana*	Reference	0.704	0.001 ***
Dendranthema zawadskii var. *latilobum*	Reference	0.419	0.031 *
Disporum smilacinum	Reference	0.568	0.001 ***
Fraxinus sieboldiana	Reference	0.533	0.002 ***
Lespedeza bicolor	Reference	0.886	0.001 ***
Rhododendron micranthum	Reference	0.634	0.001 ***
Rhododendron mucronulatum	Reference	0.422	0.031 *

*** significant at 0.1% level, ** significant at 1% level, * significant at 5% level.

Table 3. The result of indicator species analysis for selecting the disappeared species and tolerant species to the polluted environment based on data collected in both polluted and natural *Quercus mongolica* forests. Polluted: tolerant species in the polluted site of *Quercus mongolica* forest; Reference: disappeared species in the natural reference site of *Quercus mongolica* forest.

Species Name	Site Type	Stat	p-Value
Athyrium yokoscense	Polluted	0.310	0.034 *
Fraxinus sieboldiana	Polluted	0.419	0.008 **
Lindera obtusiloba	Polluted	0.514	0.001 ***
Rhododendron schlippenbachii	Polluted	0.498	0.004 **
Acer pseudosieboldianum	Reference	0.397	0.006 **
Ainsliaea acerifolia	Reference	0.543	0.001 ***
Artemisia keiskeana	Reference	0.449	0.002 **
Aster scaber	Reference	0.407	0.018 *
Athyrium vidalii	Reference	0.351	0.015 *
Atractylodes ovata	Reference	0.448	0.001 ***
Betula schmidtii	Reference	0.380	0.012 *
Callicarpa japonica	Reference	0.419	0.015 *
Carex humilis var. *nana*	Reference	0.419	0.016 *
Carex siderosticta	Reference	0.341	0.002 **
Hydrangea serrata f. *acuminata*	Reference	0.360	0.015 *
Lespedeza bicolor	Reference	0.306	0.027 *
Polystichum tripteron	Reference	0.391	0.017 *
Potentilla freyniana	Reference	0.645	0.001 ***
Rubus crataegifolius	Reference	0.419	0.015 *

*** significant at 0.1% level, ** significant at 1% level, * significant at 5% level.

Table 4. The result of indicator species analysis for selecting the disappeared species and tolerant species to the polluted environment based on data collected in both polluted and natural valley forests. Pol-luted: tolerant species in the polluted site of valley forest; Reference: disappeared species in the natural reference site of valley forest.

Species Name	Site Type	Stat	p-Value
Actinidia arguta	Polluted	0.423	0.028 *
Athyrium yokoscense	Polluted	0.548	0.019 *
Lindera obtusiloba	Polluted	0.655	0.001 ***
Miscanthus sinensis var. *purpurascens*	Polluted	0.554	0.028 *
Quercus mongolica	Polluted	0.432	0.038 *
Cornus controversa	Reference	0.759	0.001 ***
Rhododendron mucronulatum	Reference	0.496	0.048 *

*** significant at 0.1% level, * significant at 5% level.

3.6. Zonning and Design for Restorative Treatment

The spatial range for which the restoration was required was restricted within the first ridge from the pollution source, considering the impact extent of the air pollution. We divided the district for restoration into four zones of very severely, severely, moderately, and lightly damaged zones based on the damage degree (Table 5) and three zones of ridge, slope, and valley based on the topographic conditions (Table 6) [13].

Table 5. Level and method of restoration recommended based on a diagnostic evaluation of the forest ecosystem around Seokpo zinc smelter, central eastern Korea.

Damaged Degree	Vegetation Status	Soil pH	Restoration Method
Very severe	Grassland with low coverage or bare ground	4.4	Soil amelioration: dolomite 4.5 ton/ha + organic fertilizer 2.3 ton/ha Introduction of plants forming all layers of vegetation
Severe	Canopy layer disappeared and shrub and herb layers are poor	4.6	Soil amelioration: dolomite 3.0 ton/ha + organic fertilizer 1.5 ton/ha Introduction of plants forming canopy tree, shrub, and herb layers
Moderate	All vegetation strata exist but coverage is poor	5.1	Soil amelioration: dolomite 1.5 ton/ha + organic fertilizer 0.8 ton/ha Introduction of plants forming shrub and herb layers
Light	Development of undergrowth is poor	5.0	Soil amelioration: dolomite 1.0 ton/ha + organic fertilizer 0.5 ton/ha Vegetation restoration is left to passive restoration

Table 6. Species to be introduced for restoration by layer of vegetation in each topographic condition. This species information was prepared by incorporating disappeared species compared with the natural reference site, tolerant species to the polluted environment, and early successional species. In very severely and severely damaged zones, plant species forming all layers of vegetation stratification including canopy tree, understory tree, shrub, and herb layers are recommended for restoration. Plant species forming shrub and herb layers are recommended in moderately damaged zone. Passive restoration is recommended in lightly damaged zone. Vegetation type, such as Korean red pine forest established as an edaphic climax type on the upper slope, usually lacks understory tree layer due to topographic condition, which is dry and infertile.

Vegetation Stratum	Ridge	Slope	Valley
Canopy tree layer	*Betula schmidtii* *Betula chinensis* * *Pinus densiflora* *Quercus variabilis* etc.	*Betula davurica* * *Betula schmidtii* * *Quercus aliena* * *Quercus mongolica* *Quercus variabilis* etc.	*Acer pictum* subsp. *Mono* * *Cornus controversa* * *Fraxinus rhynchophylla* *Juglans mandshurica* *Quercus mongolica* ** etc.
Understory tree layer	*Lindera obtusiloba* **	*Acer pseudosieboldianum* * *Fraxinus rhynchophylla* * *Lindera obtusiloba* ** etc.	*Lindera obtusiloba* ** *Magnolia sieboldii* * etc.
Shrub layer	*Fraxinus sieboldiana* * *Lespedeza bicolor* * *Lespedeza cyrtobotrya* ** *Rhododendron micranthum* ** *Rhododendron mucronulatum* * *Rhododendron schlippenbachii* ** *Toxicodendron trichocarpum* ** *Vaccinium hirtum* var. *koreanum* ** *Weigela florida* * etc.	*Callicarpa japonica* * *Clerodendrum trichotomum* * *Fraxinus sieboldiana* ** *Lespedeza bicolor* * *Lespedeza maximowiczii* * *Lindera glauca* * *Rhododendron mucronulatum* ** *Rhododendron schlippenbachii* ** *Rubus crataegifolius* * *Toxicodendron trichocarpum* ** *Symplocos sawafutagi* * *Vaccinium hirtum* var. *koreanum* ** etc.	*Alangium platanifolium* var. *trilobum* * *Cimicifuga simplex* *Corylus heterophylla* *Rhododendron mucronulatum* ** *Styrax obassia* * *Weigela subsessilis* * etc.

Table 6. *Cont.*

Vegetation Stratum	Ridge	Slope	Valley
Herb layer	*Arundinella hirta* *** *Athyrium yokoscense* ** *Carex humilis* var. *nana* * *Carex siderosticta* * *Dendranthema zawadskii* var. *latilobum* * *Disporum smilacinum* * *Melampyrum roseum* * *Miscanthus sinensis* var. *purpurascens* *** *Pteridium aquilinum* var. *latiusculum* * *Spodiopogon sibiricus* ** etc.	*Ainsliaea acerifolia* * *Artemisia keiskeana* * *Aster scaber* * *Athyrium vidalii* * *Athyrium yokoscense* ** *Atractylodes ovata* * *Calamagrostis arundinacea* *** *Carex humilis* var. *nana* * *Carex siderosticta* * *Disporum smilacinum* * *Hydrangea serrata* f. *acuminata* *Polystichum tripteron* * *Potentilla freyniana* * etc.	*Actinidia arguta* * *Angelica decursiva* *Athyrium yokoscense* ** *Carex humilis* var. *nana* * *Cimicifuga dahurica* * *Corydalis speciose* * *Dryopteris crassirhizoma* *Isodon excisus* * *Miscanthus sinensis* var. *purpurascens* ** *Persicaria filiformis* *Polystichum tripteron* *Scutellaria dependens* * etc.

*, **, and *** indicate disappeared species, tolerant species, and pioneer species, respectively.

The restorative treatment was determined by reflecting the damaged level of vegetation and soil acidification. In very severely damaged zones, dolomite of 4.5 ton/ha and organic fertilizer of 2.25 ton/ha were recommended for soil amelioration. Plant species forming all layers of vegetation stratification, including canopy tree, understory tree, shrub, and herb layers, were recommended for vegetation restoration. In severely damaged zones, dolomite of 3.0 ton/ha and organic fertilizer of 1.5 ton/ha were recommended for soil amelioration. Plant species forming all layers of vegetation stratification were recommended for vegetation restoration, such as in the case of the very severely damaged zone. In moderately damaged zones, dolomite of 1.5 ton/ha and organic fertilizer of 0.75 ton/ha were recommended for soil amelioration. For restoring vegetation, plant species forming shrub and herb layers of vegetation stratification were recommended. Plant species forming shrub and herb layers were recommended in the moderately damaged zone. In the lightly damaged zone, dolomite of 1.0 ton/ha and organic fertilizer of 0.5 ton/ha were recommended for soil amelioration. Passive restoration was recommended for restoring vegetation.

In this restoration plan, we did not recommend for plant species forming the understory tree layer that vegetation to be introduced on the upper slope and ridge, as vegetation established there usually lacks the layer.

4. Discussion

4.1. The Effects of Air Pollution on Forest Ecosystems

Air pollution and atmospheric deposition emitted from industrial facilities have adverse effects on tree and forest health. Growing awareness of the air pollution effects on forests has, from the early 1980s on, led to intensive forest damage research and monitoring. This has fostered air pollution control, especially in developed countries, and also, to a smaller extent, in developing countries. At several forest sites, particularly in developed countries, there are the first indications of the recovery of forest soil and tree conditions that may be attributed to improved air quality [4,71]. This caused a decrease in the attention paid by the public to the air pollution effects on forests. However, air pollution continues to affect the structure and functioning of forest ecosystems just as when this study was conducted.

Air pollutants may impact trees as both wet and dry deposition. Wet deposition comprises rain, hail, and snow and is largely determined by atmospheric processes. Dry deposition consists of gases, aerosols, and dust and is largely influenced by the physical and chemical properties of the receptor surface. Forests receive higher deposition loads than open fields, depending on the tree species and canopy structure. A higher roughness of the canopy causes higher air turbulences and more intensive interactions between the air and the foliage. The interception of pollutants by the foliage in turn is determined by such factors as leaf area index, leaf shape, leaf surface roughness, and stomata size. Dry deposition accumulated on the foliage is washed off by precipitation and enhances the deposition under the canopy (throughfall) in comparison to deposition in an open field (bulk deposition). Moreover, throughfall is influenced by two components of canopy exchange: canopy leaching and canopy uptake of elements. The main air pollutants involved in forest damage are sulfur compounds, nitrogen compounds, ozone, and heavy metals [72–76].

Sulfur dioxide (SO_2) was the first air pollutant found to cause damage to trees [77]. Its air concentrations increased rapidly when it was released into the atmosphere by the combustion of fossil fuels during the course of industrialization. While damaging trees directly via their foliage, SO_2 also reacts with water in the atmosphere to form sulfurous acid (H_2SO_3) and sulfuric acid (H_2SO_4), thus contributing to the formation of acid precipitation and hence to the indirect damage of trees [72,73]. In Korea, forest decline has usually occurred around industrial areas, and SO_2 has played a leading role [7,8,13]. However, Korea has shown declining trends in SO_2 concentrations in recent years [4].

Nitrogen oxides (NOx) are released into the atmosphere in the course of various combustion processes in which nitrogen (N) in the air is oxidized mainly to nitrogen monoxide (NO), with a small admixture of nitrogen dioxide (NO_2). In daylight, NO is easily converted to NO_2 by photochemical reactions involving hydrocarbons present in the air. Both gases, especially NO, are also produced biologically by soil bacteria during nitrification, denitrification, and decomposition of nitrite (NO_2^-) [78]. These substances are gaseous and act on trees as dry deposition directly via the foliage. Some of them are acidifying and lead—by means of chemical reactions with water in the atmosphere—to acid precipitation. Acidifying compounds such as SO_2, NOx, and NH_3, however, enhance the concentrations of protons and form sulfuric acid, nitric acid (HNO_3), ammonium (NH_4), and nitrate (NO_3) [73,74,79].

Heavy metals result from most combustion processes and from many industrial production processes. They are released into the atmosphere by means of dust and, at high temperatures, also as gases. The main heavy metals considered to be detrimental to forest health are cadmium (Cd), lead (Pb), mercury (Hg), cobalt (Co), chromium (Cr), copper (Cu), nickel (Ni), and zinc (Zn). However, largely because of their impacts on human health, heavy metal emissions have been reduced greatly within the last 3 decades in many industrialized countries [80–82].

4.2. Damage Status to Forest Ecosystem around Seokpo Smelter

Vegetation in places close to smelters has been so severely damaged that only a few plants, such as *Athyrium yokoscense* (FR. Et SAV.) H.CHRIST, *Miscanthus sinensis* var. *purpurascens* RENDLE, and *Pteridium aquilinum* var. *latiusculum* (DESV.) UNDERW, sporadically exist; otherwise all of them have disappeared. As the distance from the smelter become farther, the damage decreases, resulting in shrubland and forest (Figures 4 and 5). The spatial distribution of vegetation in industrial areas usually reflects the degree of pollution damage [4,7,8,12]. If a forested ecosystem is being affected by air pollution, then the canopy stratum is generally impacted first and is stripped away. As canopy trees decline, shrubs and then the ground vegetation are affected. This syndrome of the sequential death of the horizontal strata of the terrestrial vegetation, described as a "peeling" of "layered vegetation effect" by [62], was observed clearly in this area (Figure 5).

Forests appeared as two types depending on the damage degree. Forests with moderate damage show poor development of vegetation stratum, while forests with light damage show visible damage in appearance and poor development of undergrowth (Figure 5). The damage was also reflected in species composition (Figure 8) and species diversity (Figure 9), showing a clear difference in species composition and species diversity compared to the reference area.

The soil is acidified and has a lower calcium and magnesium content compared to the reference area, while the aluminum content was higher in very severely damaged sites (Figure 6).

Synthesized results obtained from the diagnostic assessments on vegetation damage and soil acidification in the forest ecosystem around the Seokpo zinc smelter show that vegetation damage was so severe that denuded ground appeared throughout wide areas, and soil acidification was also relatively severe. In this respect, active restoration is required urgently to prevent follow-up damage, such as landslides [83–85]. However, the damage decreased with increasing distance from the pollution source and was restricted within the first ridge from the source. The results of this diagnostic assessment could help to determine the spatial range and level of restoration.

In the case of the Yeocheon industrial complex, passive restoration occurred in forests around industrial complexes [4], but the speed was slower than in the case of the Ulsan industrial complex where active restoration was applied [8]. In active restoration, dolomite and sludge treatment neutralized acidic soil and supplemented nutrients, thereby facilitating plant growth and contributing to the restorative effects [8]. In addition, the tolerant species, which was selected through field surveys and laboratory experiments, was well established and contributed to achieve successful restoration [8,36].

4.3. Necessity and Recommendation of Ecological Restoration

In Korea, most industrial complexes are located on the coastal areas [4,7,8]. However, the area where this study was conducted is uniquely located on a small mountain village. Most of this area is composed of a mountainous area with a steep slope, and it is therefore not easy to develop. Therefore, the nature is well conserved. The mountainous land of this area is very steep and shallow in soil depth. Thus, Korean red pine forest, which is adapted well to the dry and infertile environment, dominates the vegetation of this area. However, deciduous broad-leaved forests suitable for the climate condition of this region and the environmental characteristics of each site are well developed in lowlands below midslope, including mountainous valleys. The river that runs through this area corresponds to the upstream section of the Nakdong River, one of the longest rivers in Korea. As Korea is a mountainous nation where more than 65% of the total national territory is composed of mountainous areas, most riparian zones, including floodplains of rivers, are transformed into agricultural lands and urbanized areas, leaving few integrate rivers equipped with aquatic and riparian zones in the plains and lands with a gentle slope. However, as this area is not easy to develop due to the environmental characteristics composed of steep mountainous areas; the river also remains an almost completely intact status.

However, air pollutants emitted from the Seokpo smelter have destroyed forest vegetation established in the basin of this river, even leading to landslides. Therefore, its impact poses a great threat to the river ecosystem conserved so well. Considering these facts comprehensively, the restoration of the damaged forest ecosystem is absolutely necessary and urgently required. In particular, countermeasures such as restoring forests are important for ensuring future ecosystem services [7,8,13].

The continual growth of the human population and human land uses leads to declines in the quality of the environment. Further, the natural landscapes that provide many ecosystem services are being rapidly converted to agricultural, industrial, and urban areas, and even wastelands. The biodiversity and habitability of the planet are now more threatened than ever before. Therefore, it is imperative that degraded land be rehabilitated and that adjoining natural landscapes be protected. However, it is clear that degradation thresholds have been crossed in many habitats, and succession alone cannot restore viable and desirable ecosystems without intervention [86]. As was shown in our results (Figures 4 and 5), the forest has degraded through shrubland or grassland to denuded land, eventually producing continual landslides. In addition, the soil is acidified and becoming non-nutritive (Figure 6). Thus restoration actions are urgently required to prevent more land degradation.

To restore degraded ecosystems, in particular those degraded by pollution, we need to apply soil ameliorators, including dolomite [6,8,13,87,88]. Although these soil amendments contributed to improving the polluted environment and thereby achieved successful revegetation in the case of Sudbury [6], they may cause other problems, such as ground water pollution and eutrophication [73,89,90]. In this respect, we recommend planting tolerant plants or applying fertilizer plants rather than applying soil amendments as a restorative treatment in all cases [8].

4.4. Soil Amelioration for Restoration

The atmospheric environment in the industrial areas in Korea is improving due to a decrease in the emission of air pollutants [4]. However, the polluted soil has not improved so easily [7,8]. In fact, polluted soil usually provides the major challenge to most restoration programs [8,13,29,35]. Therefore, soil amendment was planned as a preparation for restoration. The soil amelioration focused on the neutralization of acidic soil and the enhancement of fertility. Previous research showed that dolomite was a superior ameliorator compared to lime [15], and it is also generally used [8,16]. The amount of dolomite applied was calculated with an equation being applied to determine the amount of dolomite required to improve varying soil pH to 5.5 [15]. In the present situation, the amounts of dolomite required were 4.5, 3.0, 1.5, and 1.0 ton/ha in very severely, severely, moderately, and lightly damaged zones, respectively (Table 1). These amounts were smaller than those used in the Sudbury region of Canada [16] and the Ulsan industrial area of Korea [13].

Soil neutralized by the addition of dolomite stimulates activities of soil microorganisms and enhances nutrient availability through the promoted decomposition of organic matter [8,91]. Organic fertilizer also may ameliorate acidified soil by raising pH and adding macronutrients, including phosphorus, which is often a limiting nutrient in acidic soil [13,88,92].

Aluminum toxicity results in rapid inhibition of root growth due to the impedance of both cell division and elongation [93–95], reduction of soil volume explored by the root system, and direct interference with the uptake of ions such as calcium and phosphate across the cell membrane of damaged roots [96,97]. The deficiency in soil nutrients, such as P, Ca^{2+}, and Mg^{2+}, exacerbate the problem of inefficient nutrient uptake due to restricted root growth and root damage [98,99].

Additions of undecomposed plant materials, such as pruning, to acid soils often increase soil pH, decrease Al^{3+} saturation, and improve conditions for plant growth generally [100–104]. Similarly, the addition of plant residue composts, urban waste compost, animal manure, and coal-derived organic products to acid soils increases soil pH, decreases Al^{3+} saturation, and improves conditions for plant growth [8,13,105–107]. The recycling of these waste products for soil amelioration has a double benefit for both the environment and the economy, provided that the waste materials are not contaminated with harmful impurities. These organic substances confer metal binding and pH buffering capacities, which are important determinants of the pH of the treated soil [104,108,109].

Treatment of sludge as a soil ameliorator contributed to the reduction of Al^{3+} content and resulted in increased plant growth [8]. This result suggests that the sludge is a chelating agent for Al^{3+} [104,109,110]. Although dolomite and sludge contribute to ameliorating the acidified soil, there are some serious concerns for the land application of dolomite and sewage sludge due to the potential for the contamination of ground water and eutrophication [73,88,89,110]. By stimulating the mineralization of soil organic matter, dolomitic liming causes ground water pollution by increasing nitrate release from the soil [8,89,90]. We, therefore, recommend restricting the use of those soil ameliorators.

Meanwhile, N-fixing plants have been used for enhancing soil fertility in revegetation projects elsewhere (e.g., [6,13,16]). Therefore, we recommend planting N-fixing plants as an initial step for revegetating in this area.

4.5. Selection of Plant Species for Restoration

This study corresponds to a diagnostic assessment which analyzed the damage status of the forest ecosystem as a preparatory step for realizing the ecological restoration of this area [19,20]. Furthermore, we carried out a vegetation survey to obtain the reference information in a conservation reserve with similar environmental conditions, which was designated as the forest genetic resource reserve from the Korea Forest Administration.

Studies for restoration have chosen the species for restoration on the basis of the following criteria: (1) species importance for restoring the ecosystem function, (2) species that are to be the main components of the final ecosystem, and (3) many plants that make up the final biodiversity of the ecosystem and should be able to recolonize by their own efforts [7,8,13,35].

In this study, we first selected the species lost in this study area by comparing and analyzing the vegetation data obtained from the damaged and natural reference areas.

Next, we selected tolerant species which can withstand the polluted environment. Tolerant plants were selected as species flourishing specifically in the damaged area and species showing higher frequency in the damaged area than in the reference area. Tolerant plants were selected by classifying the four vegetation layers of canopy tree, understory tree, shrub, and herb composing the vegetation profile.

Finally, we selected the pioneer species frequently invading the bare ground to restore the heavily damaged zone. Synthesized, the plant species to be introduced for restoration were selected by combining the tolerant species with the polluted environment, the pioneer species frequently invading with the bare ground, and the species disappeared from the damaged area compared with the species composition of the reference area (Table 6) [7,13,36].

5. Conclusions

The result of a diagnostic assessment for the damaged forests around the Seokpo zinc smelter showed that ecological restoration is required urgently, as vegetation damage and soil acidification are very severe within the first ridge and at a distance of about 5 km from the pollution source. Vegetation damage appeared variously, such as the level to which bare ground appears due to extreme damage and the landslides that follow, the degree to which some of the vegetation strata is lost, and the visible damage level. The degree of soil acidification tended to be proportional to vegetation damage. In relation to soil acidification, deficiencies in nutrients, such as Ca^{2+} and Mg^{2+}, and an increase in toxic ion concentration, such as Al^{3+}, were also identified. In particular, landslides continued around places where vegetation was severely destroyed, making ecological restoration urgent. The restoration plan was prepared by compiling the results of these diagnostic assessments and reference information collected from intact natural forests. The restoration plan was prepared in two directions: amelioration of acidified soil and vegetation restoration. In order to successfully complete this ecological restoration, the continuous monitoring and management of soil and air pollution need to be prepared, as well as the continuous monitoring of the establishment process of the vegetation that is to be introduced.

Author Contributions: Conceptualization, A.R.K., C.H.L. and C.S.L; methodology, A.R.K., C.H.L. and C.S.L; software, A.R.K. and C.H.L.; validation, B.S.L., J.S. and C.S.L.; formal analysis, A.R.K. and C.H.L.; investigation, A.R.K., B.S.L., J.S., C.H.L., W.S.L., Y.H.Y.. C.S.L.; resources, C.H.L. and C.S.L.; data curation, B.S.L and J.S.; writing—original draft preparation, A.R.K.; writing—review and editing, Y.H.Y. and C.S.L.; visualization, A.R.K., B.S.L.; supervision, C.S.L.; project administration, C.S.L.; funding acquisition, C.S.L. All authors have read and agreed to the published version of the manuscript.

Funding: This research received no external funding.

Conflicts of Interest: The authors declare no conflict of interest.

Appendix A

Table A1. Chemical Properties of Organic Fertilizer Planned as a Soil Ameliorator.

Environmental Factors	Content
Water content (%)	48.34
Organic matter (%)	33.76
Total Nitrogen (%)	1.24
Available Phosphorus (%)	1.04
Exchangeable Potassium (%)	0.26
C.E.C (cmol+/kg)	35.0
Sodium (%)	0.57

References

1. Bull, K.; Fench, G. International activities to reduce pollution impacts at the regional scale. In *IUFRO Research Series 1, Forest Dynamics in Heavily Polluted Regions*; Innes, J.L., Oleksyn, J., Eds.; CABI Publishing: Wallingford, UK, 2000.
2. Grennfelt, P.; Engleryd, A.; Forsius, M.; Hov, Ø.; Rodhe, H.; Cowling, E. Acid rain and air pollution: 50 years of progress in environmental science and policy. *Ambio* **2020**, *49*, 849–864. [CrossRef] [PubMed]
3. NIER (National Institute of Environmental Research). *National Air Pollutants Emission*; National Institute of Environmental Research: Gimpo, Korea, 2018.
4. Lee, H.; Lim, B.S.; Kim, D.U.; Kim, A.R.; Seol, J.W.; Lim, C.H.; Kil, J.H.; Moon, J.S.; Lee, C.S. Decline and passive restoration of forest vegetation around the Yeocheon industrial complex of Southern Korea. *Forests* **2020**, *11*, 674. [CrossRef]
5. Luttermann, A.; Freedman, B. *Risks to Forests in Heavily Polluted Regions. Forest Dynamics in Heavily Polluted Regions*; Report No. 1; IUFRO Task Force on Environmental Change; IUFRO: Vienna, Austria, 2000; pp. 9–26.
6. Winterhalder, K. Landscape degradation by smelter emissions near Sudbury, Canada, and subsequent amelioration and restoration. In *IUFRO Research Series 1. Forest Dynamics in Heavily Polluted Regions*; Innes, J.L., Oleksyn, J., Eds.; CABI Publishing: Wallingford, UK, 2000; pp. 87–119.
7. Lee, C.S.; Lee, K.S.; Hwangbo, J.K.; You, Y.H.; Kim, J.H. Selection of tolerant plants and their arrangement to restore a forest ecosystem damaged by air pollution. *Water Air Soil Pollut.* **2004**, *156*, 251–273. [CrossRef]
8. Lee, C.S.; Moon, J.S.; Cho, Y.C. Effects of soil amelioration and tree planting on restoration of an air-pollution damaged forest in South Korea. *Water Air Soil Pollut.* **2006**, *179*, 239–254. [CrossRef]
9. United Nations Environment Programme (UNEP). *Annual Report 2009: Seizing the Hreen Opportunity*; United Nations Environment Programme: Nairobi, Kenya, 2009.
10. Sicard, P.; Augustaitis, A.; Belyazid, S.; Calfapietra, C.; De Marco, A.; Fenn, M.E.; Bytnerowicz, A.; Grulke, N.E.; He, S.; Matyssek, R.; et al. Global topics and novel approaches in the study of air pollution, climate change and forest ecosystems. *Environ. Pollut.* **2016**, *213*, 977–987. [CrossRef]
11. Manisalidis, I.; Stavropoulou, E.; Stavropoulos, A.; Bezirtzoglou, E. Environmental and health impacts of air pollution: A review. *Front. Public Health* **2020**, *8*, 14. [CrossRef]
12. Lee, C.S.; Cho, Y.C.; Shin, H.C.; Lee, S.M.; Lee, C.H.; Eom, A.H. An evaluation of the effects of rehabilitation practiced in the coal mining spoils in Korea 2. An evaluation based on physic-chemical properties of soil. *J. Ecol. Field Biol.* **2008**, *31*, 25–30. [CrossRef]
13. Lee, C.S.; Cho, Y.C. Selection of pollution-tolerant trees for restoration of degraded forests and evaluation of the experimental restoration practices at the Ulsan Industrial Complex, Korea. In *Ecology, Planning, and Management of Urban Forests: International Perspectives*; Springer: New York, NY, USA, 2008; pp. 369–392.
14. Lee, C.S. Regeneration of Pinus densiflora Community around the Yeocheon Industrial Complex Disturbed by Air Pollution. *Korean J. Ecol.* **1993**, *16*, 305–316.
15. Ministry of Environment. *Selection and Breeding of Tolerant Species and Bio-Indicator to Air Pollution and Acid Rain*; Ministry of Environment: Seoul, Korea, 1996; p. 353.
16. Gunn, J.M. *Restoration and Recovery of an Industrial Region: Progress in Restoring the Smelter-Damaged Landscape Near Sudbury, Canada*; Springer: Berlin/Heidelberg, Germany, 1995.
17. Aronson, J.; Floret, C.; Le floc'h, E.; Ovalle, C.; Pontainer, P. Restoration and rehabilitation of degraded ecosystems in arid and semi-arid lands. A review from the South. *Restor. Ecol.* **1993**, *1*, 8–17. [CrossRef]
18. Berger, J.J. Ecological restoration and non indigenous plant species: A review. *Restor. Ecol.* **1993**, *1*, 74–82. [CrossRef]
19. SERI (Society Ecological Restoration International Science & Policy Working Group). *The SER International Primer on Ecological Restoration*; Society for Ecological Restoration International: Tucson, AZ, USA, 2004.
20. McDonald, T.; Gann, G.; Jonson, J.; Dixon, K. *International Standards for the Practice of Ecological Restoration—Including Principles and Key Concepts*; Society for Ecological Restoration: Washington, DC, USA, 2016.

21. Bradshaw, A. Ecological principles and land reclamation practice. *Landsc. Plan.* **1984**, *11*, 35–48. [CrossRef]
22. Bradshaw, A.D. The reclamation of derelict land and the ecology of ecosystems. In *Restoration Ecology: A Synthetic Approach to Ecological Research*; Jordan, W.R., Gilpin, M.E., Aber, A.D., Eds.; Cambridge University Press: Cambridge, UK, 1987; pp. 53–74.
23. Cairns, J. Is restoration ecology practical? *Restor. Ecol.* **1993**, *1*, 3–7. [CrossRef]
24. Cairns, J.; Heckman, J.R. Restoration ecology: The state of an emerging field. *Annu. Rev. Energy Environ.* **1996**, *21*, 167–189. [CrossRef]
25. Hobbs, R.; Norton, D.A. Towards a conceptual framework for restoration ecology. *Restor. Ecol.* **1996**, *4*, 93–110. [CrossRef]
26. Jordan, W.R.; Ii, R.L.P.; Allen, E.B. Ecological restoration as a strategy for conserving biological diversity. *Environ. Manag.* **1988**, *12*, 55–72. [CrossRef]
27. Naveh, Z. From biodiversity to ecodiversity: A landscape-ecology approach to conservation and restoration. *Restor. Ecol.* **1994**, *2*, 180–189. [CrossRef]
28. Turner, F. The invented landscape. In *Beyond Preservation: Restoring and Inventing Landscapes*; Baldwin, A.D.J., De Luce, J., Pletsch, C., Eds.; University of Minnesota Press: Minneapolis, MN, USA, 1994; pp. 35–66.
29. Bradshaw, A.D. The biology of land restoration. In *Applied Population Biology*; Jain, S.K., Botsford, J.W., Eds.; Kluwer: Dordrecht, The Netherlands, 1992; pp. 25–44.
30. Temperton, V.M.; Zirr, K. Order of arrival and availability of safe sites: An example of their importance for plant community assembly in stressed ecosystems. In *Assembly Rules and Restoration Ecology-Bridging the Gap between Theory and Practice*; Temperton, V.M., Hobbs, R., Nuttle, T., Halle, S., Eds.; Island Press: Washington, DC, USA, 2004; pp. 285–303.
31. Hough, M. *City Form and Natural Process*; Croom Helm: London, UK, 1984; p. 279.
32. Aber, J.D. Restored forests and the identification of critical factors in species-site interactions. In *Restoration Ecology: A Synthetic Approach to Ecological Research*; Jordan, W.R., Gilpin, M.E., Aber, J.D., Eds.; Cambridge University Press: Cambridge, UK, 1987; pp. 241–250.
33. MacMahon, J.A. Disturbed lands and ecological theory: An essay about a mutualistic association. In *Restoration Ecology*; Jordan, W.R., Gilpin, M.E., Aber, J.D., Eds.; Cambridge University Press: Cambridge, UK, 1987; pp. 221–240.
34. Gunn, J.M. Restoring the smelter-damaged landscape near Sudbury, Canada. *Ecol. Restor.* **1996**, *14*, 129–136. [CrossRef]
35. Dobson, A.P.; Bradshaw, A.D.; Baker, A.J.M. Hopes for the future: Restoration ecology and conservation biology. *Science* **1997**, *277*, 515–522. [CrossRef]
36. Kim, G.S.; Pee, J.H.; An, J.H.; Lim, C.H.; Lee, C.S. Selection of air pollution tolerant plants through the 20-years-long transplanting experiment in the Yeocheon industrial area, southern Korea. *Anim. Cells Syst.* **2015**, *19*, 208–215. [CrossRef]
37. Kercher, J.; Axelrod, M.; Bingham, G. Forecasting effects of S02 pollution on growth and succession in a Western conifer forest. In Proceedings of the Symposium on Effects of Air Pollutants on Mediterranean and Temperate Forest Ecosystems, Riverside, CA, USA, 22–27 June 1980; p. 200.
38. Kozlowski, T.T. Impacts of air pollution on forest ecosystems. *BioScience* **1980**, *30*, 88–93. [CrossRef]
39. Healey, M.; Raine, A.; Parsons, L.; Cook, N. *River Condition Index in New South Wales: Method Development and Application*; NSW Office of Water: Sydney, Australia, 2012.
40. Munné, A.; Solà, C.; Rieradevall, M.; Prat, N. Índex QBR. Mètode per a l'avaluació de la qualitat dels ecosistemes de ribera. *Estud. Qual. Ecol.* **1998**, *4*, 28. (In Catalan)
41. Webb, B.W.; Petts, G.E.; Möller, H.; Roux, A.L. Historical change of large alluvial rivers: Western Europe. *Geogr. J.* **1990**, *156*, 91. [CrossRef]
42. Mant, J.; Janes, M. Restoration of rivers and floodplains. In *Restoration Ecology*; Blackwell Publishing: Malden, MA, USA, 2005; pp. 141–157.
43. Kuemmerlen, M.; Reichert, P.; Siber, R.; Schuwirth, N. Ecological assessment of river networks: From reach to catchment scale. *Sci. Total Environ.* **2019**, *650*, 1613–1627. [CrossRef] [PubMed]
44. Kim, A.R.; Kim, D.U.; Lim, B.S.; Seol, J.W.; Lee, C.S. An evaluation on restoration effect in the restored Yangjae stream and the improvement plan based on the result. *Korean J. Ecol. Environ.* **2020**, *53*, 390–407. [CrossRef]
45. Lee, C.S.; Cho, Y.C.; Shin, H.C.; Moon, J.S.; Lee, B.C.; Bae, Y.S.; Byun, H.G.; Yi, H. Ecological response of streams in Korea under different management regimes. *Water Eng. Res.* **2005**, *6*, 131–147.
46. An, J.H.; Lim, C.H.; Lim, Y.K.; Nam, K.B.; Lee, C.S. A review of restoration project evaluation and post management for ecological restoration of the river. *J. Restor. Ecol.* **2014**, *4*, 15–34.
47. Jung, S.H.; Kim, A.R.; Seol, J.W.; Lim, B.S.; Lee, C.S. Characteristics and reference information of riparian vegetation for realizing ecological restoration classified by reach of the river in Korea. *J. Korean Soc. Water Environ.* **2018**, *34*, 447–461.
48. Lee, C.S.; Jeong, Y.M.; Kang, H.S. Concept, direction, and task of ecological restoration. *J. Restor. Ecol.* **2011**, *2*, 59–71.
49. Reif, D.M.; Martin, M.T.; Tan, S.W.; Houck, K.A.; Judson, R.S.; Richard, A.M.; Knudsen, T.B.; Dix, D.J.; Kavlock, R.J. Endocrine profiling and prioritization of environmental chemicals using ToxCast data. *Env. Health Perspect.* **2010**, *118*, 1714–1720.
50. Lee, C.S. Role and task of restoration ecology in changing environment. *Proc. Natl. Acad. Sci. USA* **2015**, *2015*, 481–527.
51. Korea Forest Service. *Study on the Cause of Forest Damage and Restoration Plan in Seokpo*; Korea Forest Service: Daejeon, Korea, 2019; p. 357.
52. Jury, M.R. Meteorology of air pollution in Los Angeles. *Atmos. Pollut. Res.* **2020**, *11*, 1226–1237. [CrossRef]
53. Bonan, G. *Ecological Climatology: Concepts and Applications*, 1st ed.; Cambridge University Press: Cambridge, UK, 2002; p. 678.

54. Korea Metrological Administration. Climatological Normals 1981–2010. Available online: https://www.weather.go.kr/weather/climate/average_30years.jsp?yy_st=2011&stn=271&norm=Y&x=9&y=18&obs=TA (accessed on 10 April 2021).
55. Korea Forest Service. Forest Soil Map. Available online: https://www.forest.go.kr/newkfsweb/html/HtmlPage.do?pg=/fgis/UI_KFS_5002_020200.html&mn=KFS_02_04_03_04_02&orgId=fgis (accessed on 10 April 2021).
56. National Institute of Ecology. National Ecosystem Survey. Available online: https://www.nie-ecobank.kr/spceinfo/main.do (accessed on 14 February 2019).
57. Küchler, A.W.; Zonneveld, I.S. *Vegetation Mapping*; Kluwer Academic Publishers: Boston, MA, USA, 1988.
58. ESRI. *Arcview GIS*; Environmental System Research Institute: Redlands, CA, USA, 2005.
59. ESRI. *Image Classification Using the ArcGIS Spatial Analyst Extension, 10.3*; ESRI: Redlands, CA, USA, 2014.
60. Richards, J.A. *Remote Sensing Digital Image Analysis*; Springer: Berlin, Germany, 1999; Volume 3, pp. 10–38.
61. Haque, M.I.; Basak, R. Land cover change detection using GIS and remote sensing techniques: A spatio-temporal study on Tanguar Haor, Sunamganj, Bangladesh. *Egypt. J. Remote Sens. Space Sci.* **2017**, *20*, 251–263. [CrossRef]
62. Gordon, A.G.; Gorham, E. Ecological aspects of air pollution from an iron-sintering plant at Wawa, Ontario. *Can. J. Bot.* **1963**, *41*, 1063–1078. [CrossRef]
63. Ellenberg, D.; Mueller-Dombois, D. *Aims and Methods of Vegetation Ecology*; Wiley: New York, NY, USA, 1974.
64. Jackson, M.L. *Soil Chemical Analysis*; Prentice-Hall: New Delhi, India, 1967; p. 498.
65. Allen, S.E.; Grimshaw, H.M.; Rowland, A.P. Chemical analysis. In *Methods in Plant Ecology*; Moore, P.D., Chapman, S.B., Eds.; Blackwell: Oxford, UK, 1986.
66. SAS Institute. *PROC User's Manual*, 6th ed.; SAS Institute: Cary, NA, USA, 2001; p. 956.
67. Jeong, J.H.; Koo, K.S.; Lee, C.H.; Kim, C.S. Physic-chemical properties of Korean forest soils by regions. *J. Korean Soc.* **2002**, *91*, 694–700. (In Korean)
68. Rietkerk, M.; van den Bosch, F.; van de Koppel, J. Site-specific properties and irreversible vegetation changes in semi-arid grazing systems. *Oikos* **1997**, *80*, 241–252. [CrossRef]
69. National Institute of Ecology. Ecology and Nature Map. Available online: https://www.nie-ecobank.kr/ (accessed on 10 April 2021).
70. Kim, H.Y.; Cho, H.J. Vegetation composition and structure of Sogwang-ri forest genetic resources reserve in Uljin-gun, Korea. *Korean J. Environ. Ecol.* **2017**, *31*, 188–201. [CrossRef]
71. Lorenz, M.; Clarke, N.; Paoletti, E.; Bytnerowicz, A.; Grulke, N.; Lukina, N.; Sase, H.; Staelens, J. *Air Pollution Impacts on Forests in a Changing Climate*; Forest and Society—Responding to Global Drivers of Change, IUFRO World Series; International Union of Forest Research Organizations: Vienna, Austria, 2010; Volume 25, pp. 55–74.
72. Smith, W.H. *Air Pollution and Forests: Interactions between Air Contaminants and Forest Ecosystems*; Springer Science & Business Media: Berlin, Germany, 1990.
73. Freedman, B. *Environmental Ecology: The Ecological Effects of Pollution, Disturbance, and Other Stresses*, 2nd ed.; Academic Press: San Diego, CA, USA, 1995.
74. Longhurst, J.W.S.; Owen, P.S.; Conlan, D.E.; Watson, A.F.R.; Raper, D.W. Atmospheric pollution: Components, mechanisms, control and remediation. In *Clean Technology and the Environment*; Kirkwood, R.C., Longley, A.J., Eds.; Springer: Dordrecht, The Netherlands, 1995.
75. Zawar-Reza, P.; Spronken-Smith, R. Air pollution climatology. In *Encyclopedia of World Climatology. Encyclopedia of Earth Sciences Series*; Oliver, J.E., Ed.; Springer: Dordrecht, The Netherlands, 2005.
76. Akimoto, H.; Luangjame, J.; Hara, H.; Gromov, S.; Khummongkol, P.; Carandang, W. The second periodic report on the state of acid deposition in East Asia. In *Part I. Regional Assessment*; Acid Deposition Monitoring Network in East Asia: Bangkok, Thailand, 2011; p. 270.
77. Rau, H. *Das Papsttum: Seine Entstehung, Seine Blüthe und Sein Verfall*; G. Stöckhardt: Stuttgart, Germany, 1871. (In German)
78. Saliba, N.A.; Mochida, M.; Finlayson-Pitts, B.J. Laboratory studies of sources of HONO in polluted urban atmospheres. *Geophys. Res. Lett.* **2000**, *27*, 3229–3232. [CrossRef]
79. Ramadan, A.E.K. Acid deposition phenomena. *TESCE* **2004**, *30*, 1369–1389.
80. Wuana, R.A.; Okieimen, F.E. Heavy metals in contaminated soils: A review of sources, chemistry, risks and best available strategies for remediation. *Int. Sch. Res. Netw. Isrn Ecol.* **2011**, *2011*, 402647. [CrossRef]
81. Masindi, V.; Muedi, K.L. Environmental contamination by heavy metals. *Heavy Met.* **2018**, 115–133. [CrossRef]
82. Hervás, J. *Lessons Learnt from Landslide Disasters in Europe*; European Communities: Luxembourg, 2003; p. 91.
83. Gupta, A.K.; Nair, S.S. *Ecosystem Approach to Disaster Risk Reduction*; National Institute of Disaster Management: New Delhi, India, 2012; p. 202.
84. Ryu, S.R.; Choi, H.T.; Lim, J.H.; Lee, I.K.; Ahn, Y.S. Post-fire restoration plan for sustainable forest management in South Korea. *Forests* **2017**, *8*, 188. [CrossRef]
85. Briffa, J.; Sinagra, E.; Blundell, R. Heavy metal pollution in the environment and their toxicological effects on humans. *Heliyon* **2020**, *6*, e04691. [CrossRef] [PubMed]
86. Lee, C.S.; Kim, J.Y.; You, Y.H. Amelioration of soil acidified by air pollutant around the industrial complexes. *Korean J. Ecol.* **1998**, *21*, 313–320.
87. Edmeades, D.C.; Ridley, A.M. Using lime to ameliorate topsoil and subsoil acidity. In *Handbook of Soil Acidity*; Rengel, Z., Ed.; Marcel Dekker: New York, NY, USA, 2003; pp. 297–336.

88. Kaupenjohann, M.; Hantschel, R.; Zech, W.; Horn, R. Ergebnisse von Düngungsversuchen mit Magnesium and vermutlich immissionsgeschädigten Fichten [*Picea abies* (L.) Karst.] im Fichtelgebirge. *Forstwiss Cent.* **1987**, *106*, 78–84. (In German) [CrossRef]
89. Kreutzer, K. Effects of forest liming on soil processes. In *Nutrient Uptake and Cycling in Forest Ecosystems*; Springer: Dordrecht, The Netherlands, 1995; pp. 447–470.
90. Matzner, E.; Meiwes, K.J. Effects of liming and fertilization on soil solution chemistry in North German forest ecosystems. *Water Air Soil Pollut.* **1991**, *54*, 377–389.
91. Borgegard, S.S.; Rydin, H. Utilization of waste products and inorganic fertilizer in the restoration of iron mine tailings. *J. Appl. Ecol.* **1989**, *26*, 1083–1088. [CrossRef]
92. Blamey, E.P.C.; Edwards, D.G. Limitations to food crop production in tropical acid soils. In *Nutrient Management for Food Crop Production in Tropical Farming Systems*; van der Heide, J., Ed.; Institute for Soil Fertility: Haren, The Netherlands, 1989; pp. 73–94.
93. Yang, Z.B.; Rao, I.M.; Horst, W.J. Interaction of aluminum and drought stress on root growth and crop yield on acid soils. *Plant Soil* **2013**, *372*, 3–25. [CrossRef]
94. Awad, K.M.; Salih, A.M.; Khalaf, Y.; Suhim, A.A.; Abass, M.H. Phytotoxic and genotoxic effect of Aluminum to date palm (*Phoenix dactylifera* L.) in vitro cultures. *J. Genet. Eng. Biotechnol.* **2019**, *17*, 7. [CrossRef]
95. Kochain, L.V. Cellular mechanisms of aluminum toxicity and resistance in plants. *Annu. Rev. Plant Physiol. Plant Mol. Biol.* **1995**, *46*, 237–260. [CrossRef]
96. Rahman, M.A.; Lee, S.H.; Ji, H.C.; Kabir, A.H.; Jones, C.S.; Lee, K.W. Importance of mineral nutrition for mitigating aluminum toxicity in plants on acidic soils: Current status and opportunities. *Int. J. Mol. Sci.* **2018**, *19*, 3073. [CrossRef]
97. Sumner, M.E.; Fey, M.V.; Noble, A.D. Nutrient status and toxicity problems in acid soils. In *Soil Acidity*; Ulrich, B., Sumner, M.E., Eds.; Springer: Berlin, Germany, 1991; pp. 149–182.
98. Song, Z.; Wan, F.; Chang, X.; Zhang, J.; Sun, M.; Liu, Y. Effects of nutrient deficiency on root morphology and nutrient allocation in *Pistacia chinensis* bunge seedlings. *Forests* **2019**, *10*, 1035. [CrossRef]
99. Hoyt, P.B.; Turner, R.C. Effects of organic materials added to very acid soils on pH, aluminum, exchangeable NH4, and crop yields. *Soil Sci.* **1975**, *119*, 227–237. [CrossRef]
100. Asghar, M.; Kanehiro, Y. Effects of sugar-cane trash and pineapple residue on soil pH, redox potential, extractable Al, Fe and Mn. *Trop. Agric.* **1980**, *57*, 245–258.
101. Ahmad, F.; Tan, K.H. Effect of lime and organic matter on soybean seedlings grown in aluminum-toxic soil. *Soil Sci. Soc. Am. J.* **1986**, *50*, 656–661. [CrossRef]
102. Bessho, T.; Bell, L.C. Soil solid and solution phase changes and mung bean response during amelioration of aluminium toxicity with organic matter. *Plant Soil* **1992**, *140*, 183–196. [CrossRef]
103. Wong, M.T.F.; Swift, R.S. Role of organic matter in alleviating soil acidity. In *Handbook of Soil Acidity*; Rengel, Z., Ed.; Marcel Dekker: New York, NY, USA, 2003; pp. 337–358.
104. Hue, N.V.; Amien, I. Aluminum detoxification with green manures. *Commun. Soil Sci. Plant Anal.* **1989**, *20*, 1499–1511. [CrossRef]
105. Alter, D.; Mitchell, A. Use of vermicompost extract as an aluminium inhibitor in aqueous solutions. *Commun. Soil Sci. Plant Anal.* **1992**, *23*, 231–240. [CrossRef]
106. Hue, N.V. Correcting soil acidity of a highly weathered Ultisol with chicken manure and sewage sludge. *Commun. Soil Sci. Plant Anal.* **1992**, *23*, 241–264. [CrossRef]
107. Rowley, M.C.; Grand, S.; Verrecchia, É.P. Calcium-mediated stabilisation of soil organic carbon. *Biogeochemistry* **2018**, *137*, 27–49. [CrossRef]
108. Wei, H.; Liu, Y.; Xiang, H.; Zhang, J.; Li, S.; Yang, J. Soil pH responses to simulated acid rain leaching in three agricultural soils. *Sustainability* **2020**, *12*, 280. [CrossRef]
109. de la Fuente, J.M.; Verenice, R.R.; Jose Luis, C.P.; Luis, H.E. Aluminum tolerance in transgenic plants by alteration of citrate synthesis. *Science* **1997**, *276*, 1566–1568. [CrossRef] [PubMed]
110. Singh, R.P.; Singh, P.; Ibrahim, M.H.; Hashim, R. Land application of sewage sludge: Physic-chemical and microbial response. *Rev. Environ. Contam. Toxicol.* **2011**, *214*, 41–61. [CrossRef]

 forests

MDPI

Article

SWAT Model Adaptability to a Small Mountainous Forested Watershed in Central Romania

Nicu Constantin Tudose [1,*], Mirabela Marin [1], Sorin Cheval [2], Cezar Ungurean [1], Serban Octavian Davidescu [1], Oana Nicoleta Tudose [1], Alin Lucian Mihalache [1,3] and Adriana Agafia Davidescu [1]

1 National Institute of Research and Development in Forestry 'Marin Drăcea', EroilorBulevard, No. 128, 077190 Voluntari, Romania; mirabelamarin@yahoo.com (M.M.); ucezar@yahoo.com (C.U.); serydavidro@yahoo.com (S.O.D.); oanatodoni@yahoo.com (O.N.T.); mihalache.alin.93@gmail.com (A.L.M.); agafiadavidescu@yahoo.com (A.A.D.)

2 National Meteorological Administration, 013686 Bucharest, Romania; sorincheval@yahoo.com

3 Faculty of Silviculture and Forest Engineering, Transilvania University of Brașov, 29 EroilorBulevard, 500036 Brasov, Romania

* Correspondence: cntudose@yahoo.com

Abstract: This study aims to build and test the adaptability and reliability of the Soil and Water Assessment Tool hydrological model in a small mountain forested watershed. This ungauged watershed covers 184 km^2 and supplies 90% of blue water for the Brașov metropolitan area, the second largest metropolitan area of Romania. After building a custom database at the forest management compartment level, the SWAT model was run. Further, using the SWAT-CUP software under the SUFI2 algorithm, we identified the most sensitive parameters required in the calibration and validation stage. Moreover, the sensitivity analysis revealed that the surface runoff is mainly influenced by soil, groundwater and vegetation condition parameters. The calibration was carried out for 2001–2010, while the 1996–1999 period was used for model validation. Both procedures have indicated satisfactory performance and a lower uncertainty of model results in replicating river discharge compared with observed discharge. This research demonstrates that the SWAT model can be applied in small ungauged watersheds after an appropriate parameterisation of its databases. Furthermore, this tool is appropriate to support decision-makers in conceiving sustainable watershed management. It also guides prioritising the most suitable measures to increase the river basin resilience and ensure the water demand under climate change.

Keywords: SWAT; hydrological model; sensitivity analysis; calibration; validation; small forested watershed

Citation: Tudose, N.C.; Marin, M.; Cheval, S.; Ungurean, C.; Davidescu, S.O.; Tudose, O.N.; Mihalache, A.L.; Davidescu, A.A. SWAT Model Adaptability to a Small Mountainous Forested Watershed in Central Romania. *Forests* **2021**, *12*, 860. https://doi.org/10.3390/f12070860

Academic Editors: Ovidiu Badea, Alessandra De Marco, Pierre Sicard, Mihai A. Tanase and Timothy A. Martin

Received: 22 April 2021
Accepted: 24 June 2021
Published: 29 June 2021

1. Introduction

Watershed behaviour is influenced by multiple factors such as its geomorphologic characteristics (e.g., slope, soil, land use) and climate conditions [1]. Evaluating the watershed response to these stressors is pivotal for achieving environmental sustainability [2], considering that worldwide, meaningful changes are projected by the Intergovernmental Panel on Climate Change (IPCC) in rainfall, temperatures and extreme events [3]. Additionally, for European regions, an increased risk of droughts and floods is projected [4]. Besides, flood events generated by faster snowmelt or compounded rain-snow events due to increased temperatures will be more frequent, particularly in the mountainous regions [5–7]. Those changes will jeopardise the future sustainability of natural resources and, accordingly, all activity sectors [8], particularly water resources, through changes in flow regime [9,10]. Alongside land use modifications due to urban development, water resources are more vulnerable to additional pressures [11,12], especially its quality and quantity [13]. It is noteworthy that there is an intensification of hydrological processes in urban watersheds simultaneous with increments in the degree of urbanisation [14]. Furthermore, as a climate change consequence, increments in water demand are forecasted [15].

Considering the resource interlinkages, the entire ecosystem and humanity's well-being are jeopardised by individual shifts with a spillover effect [16]. Hydrological modelling is a useful and valuable approach for understanding these interconnections at the watershed level and to assess the impact of multiple drivers (e.g., climate, land use, socio-economic) on ecosystems. Hydrological processes within different sizes and scales of watersheds can be understood, described and explored using hydrologic models [17]. Lately, many researchers have employed various models (like the Soil and Water Assessment Tool (SWAT), Distributed Hydrology Soil Vegetation Model (DHSVM), Hydrologic Engineering Center's Hydrologic Modelling System (HEC-HMS), Variable Infiltration Capacity (VIC), European Hydrological System Model (MIKE SHE) and so on) to investigate these cumulative impacts on hydrological processes within the watersheds aiming to anticipate and mitigate multiple challenges [18–21]. This action is crucial for the appropriate planning and management of natural resources in various environments [22]. Almost 40% of the worldwide population is located in mountainous watersheds [23]. The mountainous regions and urban areas are characterised by a high vulnerability to climate change [23]. Therefore, natural resources can be endangered by those changes [23]. Those environments assure up to 80% of freshwater resources [24] and are susceptible to numerous shortcomings related to water resources, particularly water supply reservoirs located in urban areas [25]. To capture the local or regional specificity of watersheds with high accuracy, the simulation must be performed under the regional climate model [26,27]. Moreover, small watersheds are more numerous than large ones and receive less attention globally [6,28]. Starting from a small scale is essential for accurate hydrological assessments [29,30], even if, unfortunately, limited hydrological data are available for local levels [17,31,32]. Therefore, considering the orographic influence and assessing the long-term impacts at an appropriate resolution is fundamental for mountainous watersheds [7,33]. For watersheds located in mountainous regions, as is the studied watershed, the topography and snowmelt significantly influence streamflow [33,34]. Consequently, not only heavy rainfall but also the snow melt process, amplified by increased temperatures [35], will generate higher downstream river flows [5,36], events that were already confirmed at the national level, particularly for the winter months and early spring [37,38].

Unfortunately, the National Strategy for Flood Risk Management is currently designed for large watersheds only; for small watersheds, no nationwide action plan exists [39]. Small watersheds, mostly without conventional gauges, have short response times and are therefore more vulnerable to flash flood events [40,41]. Hence, a new approach focused on small watersheds is mandatory for developing appropriate response strategies for these watersheds. In this respect, investigating small watersheds' behaviour under multiple challenges by assessing the negative impact on the local environment, and thus on the local society, is fundamental. Further, short-, medium- and long-term stream flow prediction is necessary to inform decision-makers and support them in achieving sustainable water management [42,43]. Amongst the wide range of hydrologic models developed to date, for this study, we chose the SWAT hydrological model due to its high adaptability and flexibility to investigate a wide range of water-related issues and supportive user groups that can be easily accessed. Constantly improved since the 1990s, SWAT is a physical open-sources model that, even if it was initially developed for large river basins, has also been proved to be suitable for watersheds up to 1000 km^2 [13]. Additionally, the model is recognised as suitable for investigating long-term impacts, particularly in watersheds without conventional gauges [1,44]. SWAT is considered a valuable tool that assists decision-makers and enables them to project a series of impacts and, hence, identify and prioritise measures needed to alleviate future risks.

To our knowledge, the application and validation of the SWAT model for a small watershed represent a novelty for both the region being studied and the entire country. In this respect, the specific objectives of this research are: (1) to personalise the SWAT model databases and (2) to test its adaptability to the local specificity of a small mountain forested watershed. Given that a large local population depends on its reservoir, the calibrated and validated SWAT model represents a valuable tool for local and national decision-makers, supporting them in designing new sustainable water resources management strategies, particularly because small watersheds are usually seen with reservoirs that ensure downstream water demand [15]. In this context, considering the multiple challenges that society faces nowadays, a new integrated approach for investigating the possible changes realistically and advocating for achieving sustainable management of those changes is required [15,25].

2. Materials and Methods

2.1. Study Area

The watershed is located in the central part of Romania (Figure 1) at 45°30′56″ N and 25°48′13″ E. Our research was performed in the Tărlung watershed upstream of the Săcele reservoir.

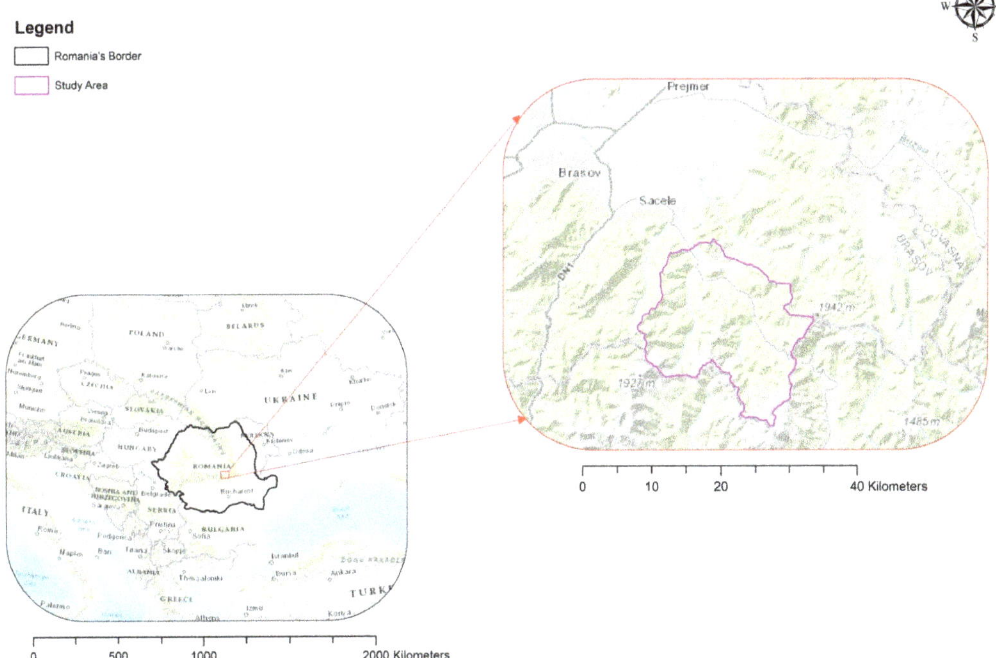

Figure 1. Study area location.

The watershed upstream of the Săcele reservoir covers 184 km^2 and represents the main source of water (90%) for the Brașov metropolitan area. The watershed elevation ranges between 724 and 1899 m. The study area is characterised by a continental climate that receives an annual precipitation of 700-800 mm and records an average temperature of 4–5 °C. The main land use within the watershed is forests (73%), followed by mountain meadows (12% of the area), pastures with scattered trees (8%), pastures (2%), meadows (4%) and water bodies (1%). Regarding the soil types, 84% of the watershed soils are included in the cambisoil class, followed by spodisols (11%), cernisols (2%) and protisols (1%).

2.2. SWAT Hydrological Model

SWAT is a basin-scale model that operates at a daily time step and is extensively used in gauged and ungauged watersheds to simulate long-term hydrological processes under different drivers [45]. The model divides watersheds into sub-basins, which subsequently are delineated into multiple hydrological response units (HRUs) in agreement with homogeneous characteristics of soils, slopes and land use [46]. Thus, a more accurate physiographic description of the watershed will be ensured [1]. The model has a default database, but it also enables users to create a personalised database for the request inputs: soil, land use and weather database [47]. The flowchart to run the SWAT model is highlighted in Figure 2.

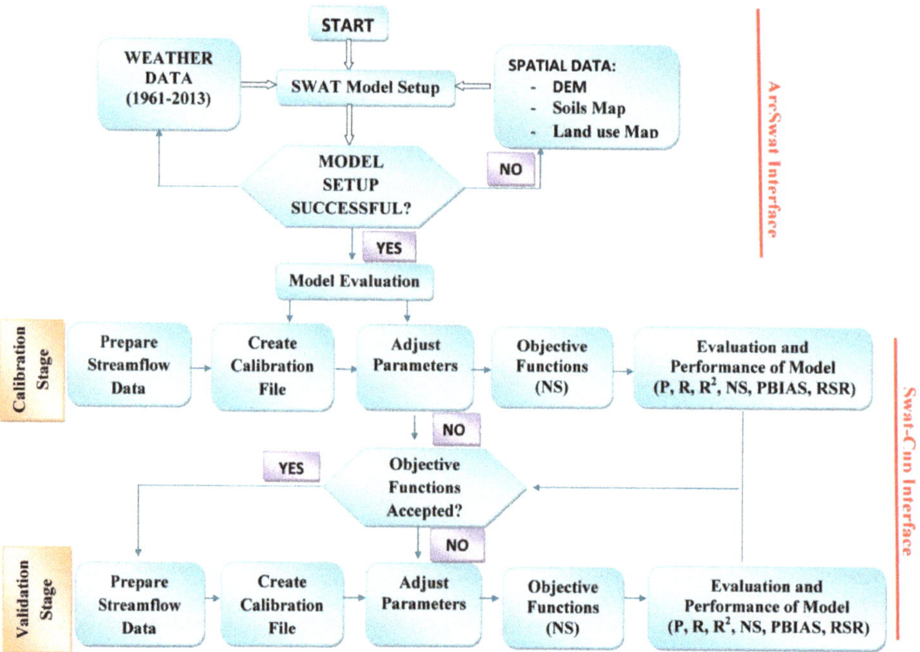

Figure 2. Diagram of SWAT model [48].

The model is an open-source software with low parameter requirements that enables users to customise their database and define elevation bands to adjust the orographic effect on precipitation and temperature, particularly for watersheds located in mountainous regions [46]. Moreover, for each elevation band, SWAT estimates accumulation, sublimation and snow melt [35] parameters with a large influence on hydrological processes within those river basins [36,49].

2.3. Model Parameterisation

To setup SWAT, four components are needed: The digital elevation model (DEM), weather, soil and a land use database. All model input data in vector and raster format (namely DEM, land use and soil) are in the EPSG 3844 projection (the projected coordinate system for Romania), datum Pulkovo 1942 (58)/Stereo70. DEM is the first and most important input considering that defining all the watershed characteristics relies on this component. We used a DEM with a 10-meter spatial resolution for our study, characterised by a 10-meter horizontal resolution and 5-meter vertical resolution. DEM has been supplied by the National Institute of Hydrology and Water Management (INHGA database). Using the ArcSWAT interface (an ArcGIS extension tool), the Tărlung watershed was delineated. Afterwards, we continued with HRU delineation by overlapping three spatial characteristics: land use, soil maps and slope. This procedure is based on similar characteristics of land use, soil and slopes that are lumped together after a threshold set by the user. For the Tărlung watershed, we established a threshold level of 10% each for soil, slope and land use to minimise errors due to multiple HRUs covering minimal surfaces. The action allows the reallocation at the sub-basin level of those three basic characteristics, which cover areas lower than the threshold value [50]. In doing so, the studied watershed was delineated into 169 sub-basins and 2419 HRUs. Moreover, to encapsulate the orographic influence and obtain accurate results, we defined ten elevation bands. After stream delineation, the morphological parameters and flow direction were obtained at the sub-watershed level.

Weather data are the second input requested by the SWAT model. For our research, we utilised data retrieved from the ROCADA dataset V 1.0 [51,52] and covered the 1961–2013 period. ROCADA represents a state-of-the-art homogenised gridded climatic dataset encompassing Romania at a spatial resolution of 0.1°. This database has been used and its accuracy has been confirmed in many studies [53,54]. We also two used other patchy datasets regarding precipitation (1988–2010) and river discharge (1974–2015) that were recorded inside of the watershed (Babarunca and Săcele Reservoir hydrometric stations). The river discharge measurements were used to calibrate and validate the model and minimise the model's uncertainty. These two hydrometric stations belong to the INHGA that provided us with the river discharge datasets. The INHGA is empowered to provide hydrological data for different types of research and development projects. The weather database comprises the precipitation, minimum and maximum temperature, average wind speed, solar radiation and relative humidity and was conceived in a particular format accepted by SWAT, and afterwards embedded in the model and used for performing simulations.

The soil database was updated based on the information retrieved from the forest and pastoral management plans (Forest Management Plan 2009 and 2013, Silvopastoral Management Plan 1989) compiled for the Tărlung watershed by the National Institute for Research and Development in Forestry (INCDS) database. The maps enclosed in the aforementioned studies were used to identify the spatial distribution of the soil types within the studied watershed (Figure 3). The database was developed at the forest management compartment level in vector format and subsequently converted into raster format.

Due to time and money constraints, we did not have information regarding some soil characteristics like bulk density (SOL_BD), hydraulic conductivity (SOL_K) and water content (SOL_AWC) required when building the SWAT model. Instead, we used the soil-plant-atmosphere-water model (SPAW), an open-source software [55]. The SPAW program automatically determines those parameters according to certain soil properties like organic matter, sand and clay percentage. The value of each soil characteristic was inserted into the SPAW application, which automatically delivered water content, bulk density and hydraulic conductivity for each soil layer. Other parameters like soil albedo (SOL_ALB) and soil erodibility factor (K_USLE) were computed considering the research performed by [56,57], respectively. After determining all the required parameters regarding soil characteristics, the user soil table was completed (Table S1). To connect the default database and the raster of soil types at the watershed level, we created a table (user soil .txt format) with codes for each soil type. The codes can be found both at the raster level and in the SWAT default database. Finally, the user soil table was fed into the model and soils were reclassified in agreement with the SWAT codes. Subsequently, soil types were classified by hydrological groups. This classification was made considering the research performed by [58] in accordance with sand and clay percentage and soil layer depth. The soils within the watershed were framed in two hydrological groups, namely Group B (90.57%) and Group C (9.43%), which are characterised by medium and low infiltration capacity, respectively [59].

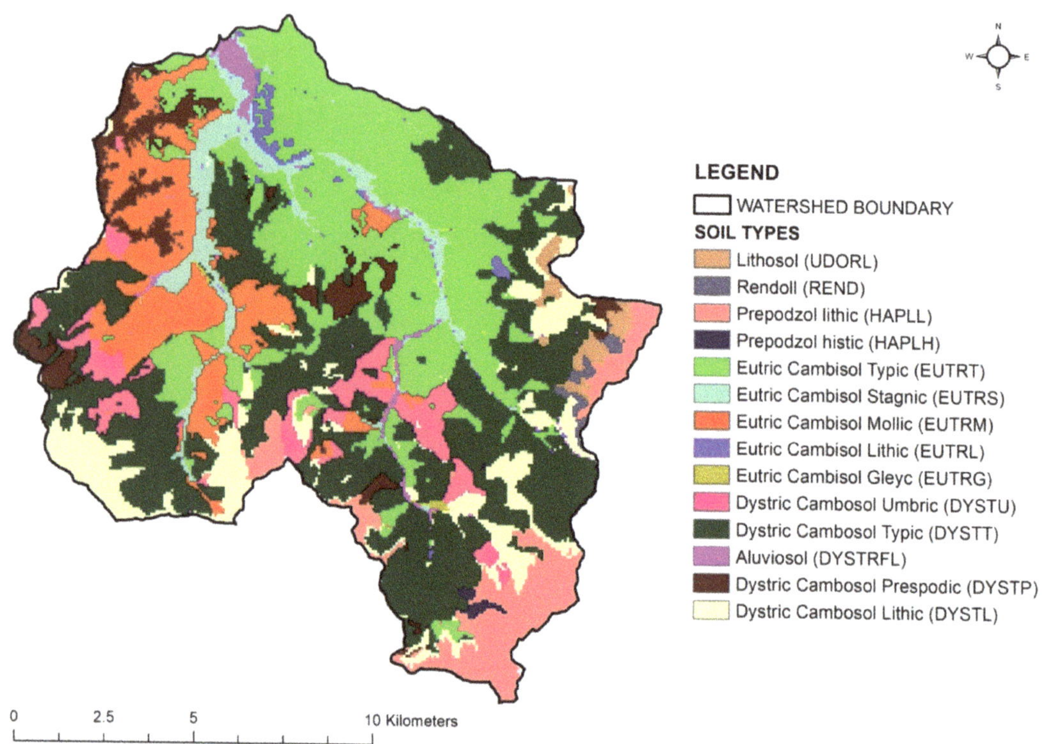

Figure 3. Soil types within the Tărlung watershed.

The land use database was updated using the information collected from the management plans alluded to above and observations on satellite images regarding roads, buildings and water bodies. The dominant land use categories (Figure 4) identified in the watershed were forests (73%), followed by mountain meadows (12%) and pastures with scattered trees (8%). Small percentages within the watershed area were occupied by meadows (4%), pastures (2%) and water bodies (1%).

The land use look-up table was designed in the requested format *(.txt file)* and was uploaded in the model, and afterwards, the land use was reclassified accordingly with the codes defined in ArcSWAT. Soil and land databases were developed at the forest management unit level. After building the requested databases and feeding them into the model, we set SWAT to run at a monthly time step for the 1961–2013 period (i.e., 53 years). This procedure also implied setting a warm-up period, namely 1961–1965, a length of time following the recommendations regarding the warm-up period setting for hydrologic models [60]. Hence, we obtained the hydrological parameters at the sub-basin level for 48 years and were also able to identify potential errors.

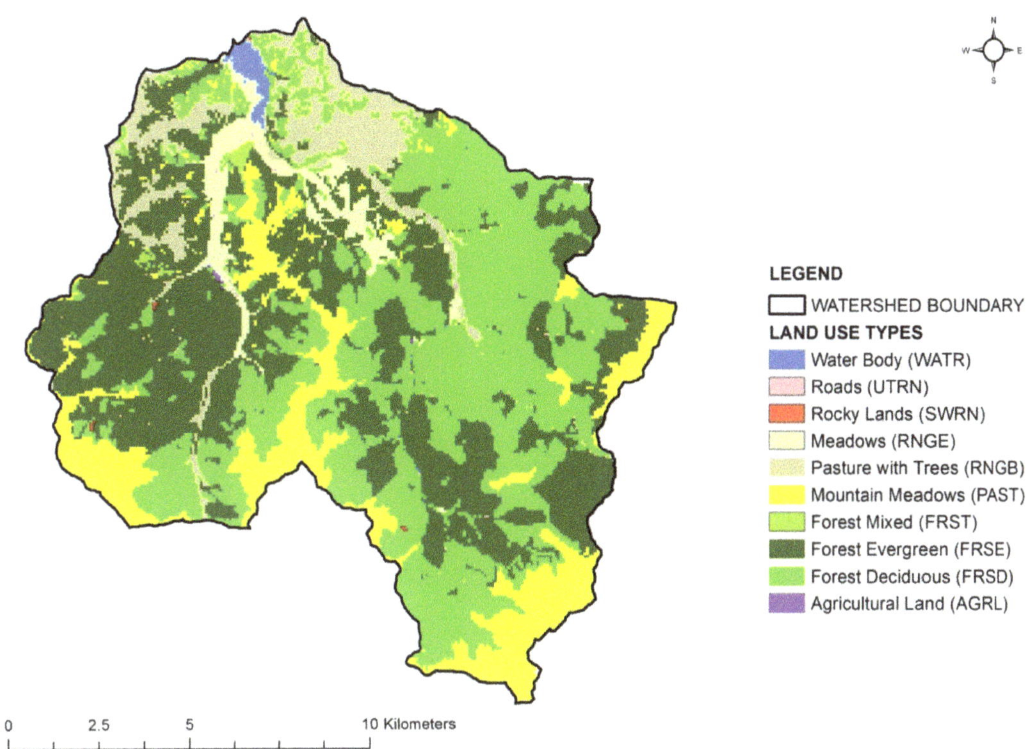

Figure 4. Land use at the Tărlung watershed level.

2.4. Model Performance Evaluation Criteria

The performance of the SWAT model was automatically carried out using the SWAT-CUP software [61]. We selected the SUFI-2 (Sequential Uncertainty Fitting version 2) algorithm from the four distinct procedures provided by SWAT-CUP due to its ability to optimise the parameters with minimal repetitions [44]. Another advantage is that this procedure considers both the model uncertainty and the uncertainty between the SWAT parameters and those that are measured [61]. The following widely applied parameters in hydrological studies were used for evaluating the model performance [62]: The coefficient of determination (R^2), percent bias (PBIAS), standard deviation rate (RSR) and Nash Sutcliffe Model Efficiency (NSE). Choosing a multiple statistics indicator has to "increase the likelihood of mixed interpretation of model performance" [63]. R^2 reflects the degree of co-linearity amongst simulated and observed values and is computed using Equation (1) [64]. This index ranges between 0 and 1, where 0 describes no correlation, while 1 shows a good agreement:

$$R^2 = \frac{\left[\sum_{i=1}^{n}\left(Q_{obs} - Q_{obs,m}\right)\left(Q_{sim} - Q_{sim,m}\right)\right]^2}{\left[\sum_{i=1}^{n}\left(Q_{obs} - Q_{obs,m}\right)^2 \sum_{i=1}^{n}\left(Q_{sim} - Q_{sim,m}\right)\right]^2} \quad (1)$$

where Q_{obs} is the discharge measured, Q_{sim} is the discharge simulated, $Q_{obs,\,m}$ is the mean of measured discharge, and $Q_{sim,\,m}$ is the mean of simulated discharge.

PBIAS calculates the model errors [65]. Expressed in percentage after using Equation (2), the good fit of the model is indicated through values close to 0 [66]. The underestimation of the model results is highlighted by positive simulated PBIAS values, while negative simulated values suggest overestimation [63]:

$$\text{PBIAS} = \frac{\sum_{i=1}^{n}\left(Y_i^{obs} - Y_i^{sim}\right) * (100)}{\sum_{i=1}^{n}\left(Y_i^{obs}\right)} \tag{2}$$

where Y^{obs} is the measured value of considered variable, Y^{sim} is the simulated value of considered variable.

RSR is computed as the ratio between root mean square error (RMSE) and standard deviation of observed values ($STEDEV_{obs}$) using Equation (3) [63]. A value close to 0 of this parameter indicates a perfect model simulation [63]:

$$\text{RSR} = \frac{\text{RMSE}}{\text{STEDEV}_{obs}} = \frac{\left[\sqrt{\sum_{i=1}^{n}\left(Y_i^{obs} - Y_i^{sim}\right)^2}\right]}{\left[\sqrt{\sum_{i=1}^{n}\left(Y_i^{obs} - Y_i^{mean}\right)^2}\right]} \tag{3}$$

where Y^{obs} is the measured value of considered variable, Y^{sim} is the simulated value of considered variable, Y^{mean} is the mean of the measured and simulated value.

NSE highlights the 1:1 fit between observed and simulated values using Equation (4) [67]:

$$\text{NSE} = \frac{\left[\sum_{i=1}^{n}(Q_{sim} - Q_{obs})\right]^2}{\left[\sum_{i=1}^{n}(Q_{obs} - Q_{obs,m})\right]^2} \tag{4}$$

where Q_{sim} is the discharge simulated, Q_{obs} is the discharge measured and $Q_{obs,\,m}$ is the mean of measured discharge.

Additionally, the model performance was evaluated using the *p-factor* and *r-factor*. The *p-factor* indicates the fraction of data bracketed by the 95PPu band, while the *r-factor* represents the ratio of the average width of the 95PPu band and the standard deviation of the measured variable [68–70]. For *p-factor*, better values are higher than 0.7, while for *r-factor* values between 0.7–1.5 are recommended [68–70].

3. Results

3.1. Sensitivity Analysis

The sensitivity analysis aims to identify the parameters with the largest influence on model outputs, thus influencing its successful application. Undertaken before calibration, this procedure has the role of identifying key parameters that subsequently will be used in model calibration [71]. The sensitivity analysis is a mathematical technique applied to enable users to examine how variations in the outputs of a numerical model can be attributed to variations of its inputs [45]. Alongside calibration and validation, this procedure is decisive for minimising the output uncertainty and efficiently perform the simulations [71]. The sensitivity analysis uses a *t*-test to assess the relative parameter significance, while the *p*-value indicates the sensitivity rank. After performing the global sensitivity analysis, the parameters with large *t*-test values and smallest *p*-values are the most sensitive [72]. We considered 12 parameters (defined in Table 1) with the largest influence on model outputs: CN2, REVAPMN, GW_DELAY, SOL_K, ESCO, GWQMN, CH_N2, CH_K2, GW_REVAP, ALPHA_BF, LAT_TIME and SOL_BD.

Table 1. The default range and adjusted values of parameters included in the calibration procedure.

Parameter	Description	Variation Method	Minimum and Maximum Value	Adjusted Value
	First calibration performed for parameters that insert water into the system			
SFTMP.bsn	Snowfall temperature	Replace	−20 ... 20	−4.791781
SMFMX.bsn	Maximum melt rate for snow during year	Replace	0 ... 20	13.605089
SMFMN.bsn	Minimum melt rate for snow during the year	Replace	0 ... 20	6.092970
SMTMP.bsn	Snow melt base temperature	Replace	−20 ... 20	2.299827
CANMX.hru_FRSE	Maximum canopy storage for forest evergreen	Replace	0 ... 100	2.149979
CANMX.hru_FRSD	Maximum canopy storage for forest deciduous	Replace	0 ... 100	4.746581
CANMX.hru_PAST	Maximum canopy storage for pastures	Replace	0 ... 100	4.563951
	Second calibration performed for chosen parameters			
CN2.mgt	SCS runoff curve number (-)	Multiply	−0.20 ... 0.20	0.120750
ESCO.hru	Soil evaporation compensation factor	Replace	0 ... 1	0.506750
EPCO.hru	Plant uptake compensation factor (-)	Replace	0 ... 1	0.337250
HRU_SLP.hru	Average slope steepness (m/m)	Multiply	0 ... 1	0.597250
OV_N.hru	Manning's "n" value for overland flow (-)	Multiply	−0.20 ... 0.00	−0.078850
GW_REVAP.gw	Coefficient for groundwater revap (days)	Replace	0.02 ... 0.2	0.165935
GW_DELAY.gw	Groundwater delay time (days)	Replace	0 ... 500	496.875000
ALPHA_BF.gw	Base flow alpha factor (1/days)	Replace	0 ... 1	0.640750
RCHRG_DP.gw	Deep aquifer percolation fraction (-)	Multiply	0 ... 1	0.899750
REVAPMN.gw	Threshold depth of water in the shallow aquifer for revap or percolation (mm)	Replace	0 ... 500	132.875000
GWQMN.gw	Threshold depth of water in the shallow aquifer for return flow (mm)	Replace	0 ... 5000	288.750000
SURLAG.bsn	Surface runoff lag time	Replace	0.05 ... 24	10.847938
SOL_BD(1).sol	Moist bulk density	Multiply	0.9 ... 2.5	0.047175
SOL_K(1).sol	Saturated hydraulic conductivity (mm/hr)	Multiply	−0.80 ... 0.80	−0.410800
SOL_AWC(1).sol	Available water capacity of the soil layer (mmH_2O/mm soil)	Multiply	−0.20 ... 0.10	−0.175625
CH_N2.rte	Manning's "n" value for the main channel	Replace	−0.01 ... 0.3	0.119475
CH_K2.rte	Effective hydraulic conductivity in main channel alluvium	Replace	−0.01 ... 500	172.625000

3.2. Model Calibration

After sensitivity analysis, we performed the calibration procedure to minimise the discrepancies amongst simulated data and recorded values [73]. The automatic calibration was also conducted using the SWAT-CUP program under the SUFI-2 algorithm. The model performance was assessed in agreement with the model performance evaluation criteria alluded to above.

The calibration was done for 2001–2010. This period was chosen due to continuous measurements and the dry, average and wet years necessary to ensure a high model performance with a lower uncertainty in the predictions [74]. Previously, we set up a five-year warm-up period (1996–2000) requested for model initialisation [61]. In doing so, we obtained the monthly river discharge for 10 years (Figure 5).

To obtain the best estimates between simulated and observed flow (Figure 6), we used the parallel processing module and performed seven iterations of 2000 simulations each. The process stopped when the model achieved a good performance rating indicated by the values of the statistical parameters recommended by [63], which can be accepted and used for assessing future impacts.

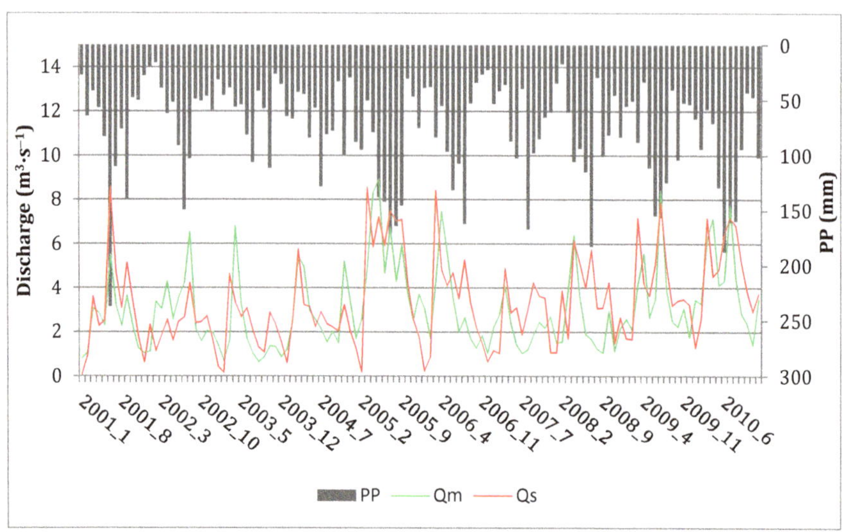

Figure 5. Simulated river discharge (Qs), measured river discharge (Qm) and precipitations (PP) for the Tărlung watershed for the 2001-2010 period.

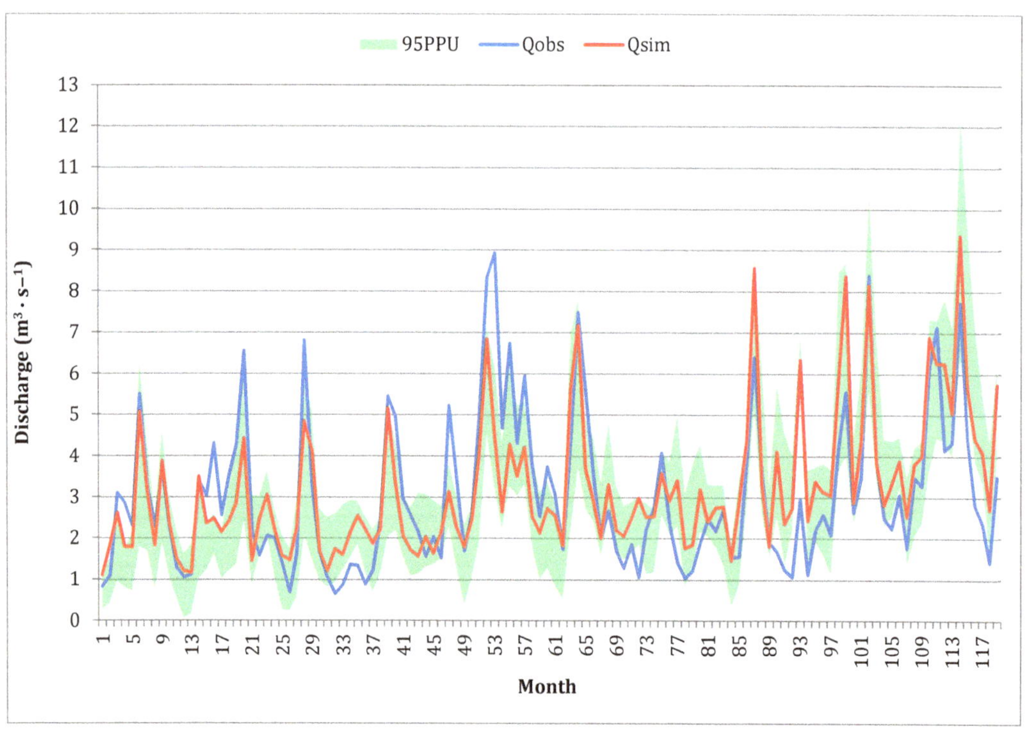

Figure 6. The 95PPU plot between observed and best simulated discharges after the calibration procedure.

The parameters that insert water into the system (e.g., snowmelt or canopy storage parameters) should be calibrated independently from the other parameters [73]. Therefore, we performed the first calibration, including only SFTMP, SMTMP, SMFMX, SMFMN,

TIMP and CANMX parameters, and ran the model until the statistical indices reached the performance rating recommended [63]. After those parameters were adjusted and fixed, they were subsequently excluded for the following calibration simulations. The second calibration was done independently for the first one and for 17 parameters that concerned only the parameters related to soil, groundwater, watershed and management characteristics. The selected parameters, the default range and their adjusted values are given in Table 1. Similar to the first calibration, the procedure was repeated until the statistical indices met the performance level that proves model acceptance.

Overall, the calibration procedure revealed a satisfactory SWAT performance, indicated by the statistical parameter values, appraised after [63], namely: R^2 = 0.61 (satisfactory), NSE = 0.59 (satisfactory), RSR = 0.64 (satisfactory), PBIAS = −5.7, *p-factor* = 0.72, and *r-factor* = 1.22. Hence, the SWAT performance was satisfactory to very good, and the obtained values revealed the model acceptance for simulating hydrological processes within the Tărlung watershed.

3.3. Model Validation

The validation confirms the results obtained after calibration [73]. This stage is important for ensuring the accuracy of the outputs considering that these will be further used in the decision process [75]. In our study, the validation was carried out for the same parameters used in calibration and considering the 1996–1999 period after previously setting up a five-year period for model warm-up. The period adopted for validation followed the same characteristics as in the calibration, namely continuous river discharges measurements and the presence of wet, dry and average years. For obtaining the best estimates between simulated and observed river discharge during validation, we performed a single iteration of 2000 simulations (Figure 7).

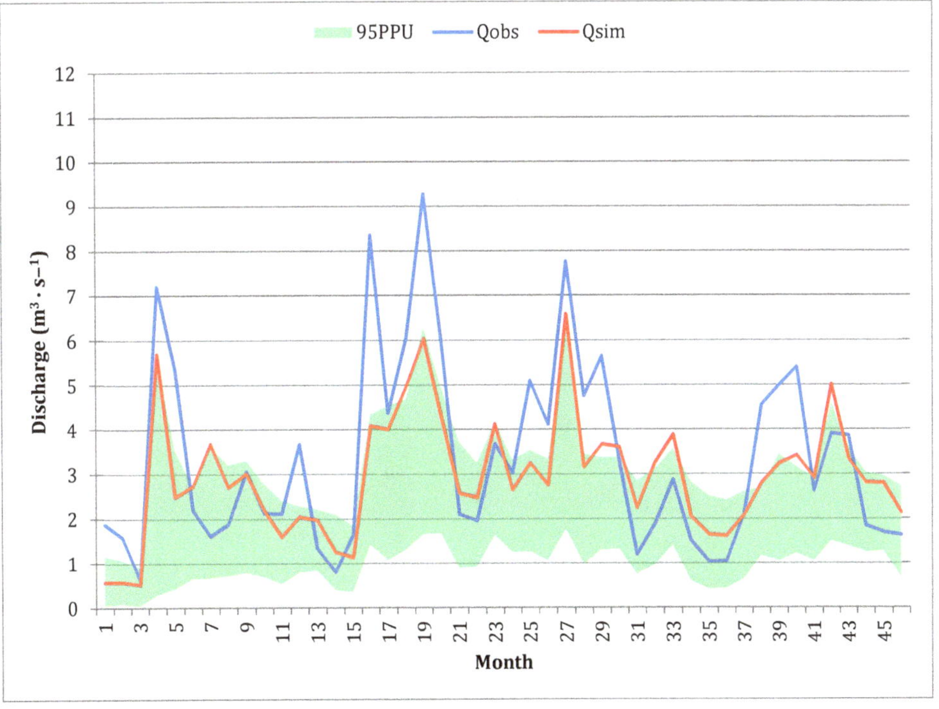

Figure 7. The 95PPU plot between observed and best simulated discharges after the validation procedure.

Nonetheless, if the user performs more than one iteration, it will increase the uncertainty of the model output due to the iterative character of the SUFI-2 algorithm [69]. The model efficiency was assessed using the same statistical indices as in the calibration. For those indices we obtained the following values appraised in accordance with the recommended performance rating [63]: $R^2 = 0.78$ (very good), NSE = 0.62 (satisfactory), RSR = 0.62 (satisfactory), *p-factor* = 0.67 and *r-factor* = 1.22. Overall, the model performance was satisfactory to very good, and the validation procedure results indicate that SWAT is suitable for assessing future impacts within the Tărlung watershed.

4. Discussion

The SWAT model developed particularly for large watersheds was applied, for the first time, both to the case study area and nationwide for a small-sized watershed. In this respect, we first customised the SWAT database to the local specificity of the studied region. In the next step, we performed the sensitivity analysis procedure that reduces the time required for calibration. During this stage, the parameters with the largest influences on hydrological processes are identified. In doing so, it was revealed that snowmelt and canopy retention are parameters with large influences on water balance within the Tărlung watershed. Those parameters have triggered lately, in the mountainous area, perilous floods during the spring months [48], particularly when snowmelt is overlapped with rainfall [38]. Due to their meaningful influence on hydrological process parameters that directly insert water into the system, they should not be calibrated together with other parameters (e.g., groundwater delay time, the coefficient for groundwater revap, base flow alpha factor and so on) because, as [73] states, they can generate identifiability issues. Therefore, snowmelt and canopy retention parameters were calibrated separately from the rest of the parameters that describe the watershed characteristics. In this respect, the first calibration includes only the snowmelt and canopy retention parameters, and the second calibration was made only for parameters that illustrate the watershed characteristics. Comparing the maximum canopy storage (CANMX) for evergreen forests and deciduous forests, the lower value was obtained for evergreen forests (see Table 1). A similar situation was also reported by [76]. However, the maximum canopy storage of deciduous forests is quite similar to the value obtained for pasture (see Table 1). This result agrees with the findings reported by [77], who obtained for pastures a maximum canopy storage even higher than those obtained for forests. In the case study area, an extension of pasture will affect the water quality due to the turbidity increments. These increments are also favoured by the main soil types from the watershed (Eutric Cambisol and Dystric Cambisol), which have high percentages of clay and silt (see Table S1), particles that are retained longer in suspension and affect the quality of water [1]. Thus, the water treatment capacity of the water plant will be exceeded and the water demand will not be covered (as has previously happened in the case study area). To prevent turbidity increments, the decision-makers should consider promoting "close to nature" forest management. This management practice will help preserve biodiversity and achieve the objectives highlighted and promoted in the EU's Biodiversity Strategy for 2030 [78].

Afterwards, the model performance was appraised through calibration and validation procedures that provided a satisfactory rating. This result indicates that the hydrological processes within the Tărlung watershed are well captured. After performing both procedures we noticed that the values of R^2 and NSE parameters increased in the validation compared with calibration. This is an unusual situation because the optimisation of parameters occurs during the first procedure, but this circumstance has been reported in other research [22,66,79–81]. This situation may be due to the symmetry regression of the SWAT model [80], the number of wet or dry years included in both procedures or most likely due to the iterative character of the model [79–81]. The model uncertainties were assessed through *p-factor* and *r-factor*. The values obtained for the *p-factor* showed that the 95PPU band envelops 72% of the measured river discharges in the calibration and 67% during validation. Those results indicate a minimum uncertainty for calibration compared with validation. Although the *p-factor* value during validation was 0.67 (slightly less than the lower limit of the interval recommended in the literature), we preserved this result. We did not perform another iteration because this repetition would have increased the uncertainty of the model results [73]. The *r-factor* represents the thicknesses of the 95PPU envelope and was 1.22 both for calibration and validation. According to [61], the values obtained for these two indices during both procedures revealed lower uncertainties in model results. Overall, the SWAT performance evaluated using the R^2, NSE, RSR, and PBIAS showed a satisfactory model performance.

After running the SWAT model, both overestimations and underestimations at the monthly level were revealed (see Figure 6). The most meaningful overestimations were observed during the spring season (e.g., March 2003, 2005, 2006, 2009) and can be attributed to the fast snowmelt process [17,38,80,82–85]. Overestimations were also noticed during the summer season after heavy rainfall events, with similar results being reported by other authors [11,86,87]. The most significant underestimations were noticed during May in 2003, 2005, 2006 and 2010. These deficiencies can be generated by rainfall spatial variability within the watershed [86,88] and underline the necessity of research infrastructure installation that is properly spatially distributed to capture the spatial variability of rainfalls inside the watershed with high accuracy. Another consequence can be an inaccurate simulation of some parameters included in the water balance equation like groundwater and evapotranspiration [89], highlighting the importance of using field measurements.

Nevertheless, the SWAT model proved its performance and reliability and is suitable scientific support for decision-makers in planning activities, particularly in watersheds located in mountainous regions. These environments are important sources of freshwater, food, energy and biodiversity, and therefore enhancing their resilience is imperative under climate and land use change [24]. This task is a priority mentioned in the SDG 15: "Protect, restore and promote sustainable use of terrestrial ecosystems, sustainably manage forests, combat desertification, and halt and reverse land degradation and halt biodiversity loss". SDG's target is in 15.4 "By 2030, ensure the conservation of mountain ecosystems, including their biodiversity, in order to enhance their capacity to provide benefits that are essential for sustainable development" [90]. Considering that mountain areas host 23% of the total forests [24], their protection will alleviate climate change effects [78]. Furthermore, the "Framework convention on the protection and sustainable development of the Carpathians," signed by our country, also highlights the importance of mountainous regions for all ecosystems and the role of the local community to achieve an integrated and balanced sustainability of these environments [91]. Therefore, the calibrated and validated SWAT model can be considered a valuable planning tool for designing action plans for small watersheds, which are currently neglected.

5. Conclusions

This research is an effort that can be considered a novel step for future studies investigating the hydrological behaviour of small watersheds. We presented the methodology used for customising the SWAT model to the local specificity for testing its ability to simulate the hydrological processes within a small forested ungauged watershed located in a mountainous region. The studied watershed has meaningful importance for Brașov city and its surrounding areas because it represents the main drinking and industrial water source. Future climate change projections published for the 21st century underline the importance of conducting such hydrological assessments to investigate watershed behaviour under climate-related risks. Therefore, we focused on testing, for the first time (nationwide and for a small forested watershed), the applicability of the SWAT hydrological model in a small watershed located in a mountainous area. Given that we built a detailed and customised database, the calibration and validation procedures revealed that SWAT meets the requirements and is adequate to simulate the hydrological processes within the Tărlung watershed. The model was developed for large river basins and had certain deficiencies reported in the literature. Nevertheless, this study stresses the importance of several factors (e.g., the accuracy of input data, choosing the proper interval for performing the model's calibration and validation procedures, and carefully selecting the parameters to perform those procedures) that contribute and ensure the SWAT's suitability for application in small ungauged watersheds. After running the SWAT model for 53 years, we noticed a good agreement in mirroring the hydrological process, which is accurately captured within the watershed. The contribution of this paper enables the local upgraded SWAT model to be further used as a guidance tool for management decisions that pursue sustainable and integrated watershed management under multiple challenges (climate, environmental and societal).

Supplementary Materials: The following are available online at https://www.mdpi.com/article/10.3390/f12070860/s1, Table S1 presents the physicochemical characteristics across soils type under study case (Tărlung watershed).

Author Contributions: Conceptualisation, methodology, investigation, formal analysis, writing—original draft, writing—review & editing, validation (N.C.T.); methodology, investigation, formal analysis, writing—review & editing (M.M.); review & editing, formal analysis (S.C., C.U., S.O.D., A.L.M., O.N.T. and A.A.D.). All authors have read and agreed to the published version of the manuscript.

Funding: The present research benefits from funding from the project Climate Services for Water-Energy-Land-Food Nexus (CLISWELN) funded by ERA4CS. ERA4CS is an ERA-NET initiated by JPI Climate, and CLISWELN is funded by BundesministeriumfürBildung und Forschung (BMBF-Germany), Executive Agency for Higher Education, Research, Development and Innovation Funding (UEFISCDI -Romania), BundesministeriumfürfürBildung, Wissenschaft und Forschung and ÖsterreichischeForschungsförderungsgesellschaft (BMBWF and FFG -Austria), and Ministerio de Economía y Competitividad (MINECO-Spain), with co-funding from the European Union's Horizon 2020 under Grant Agreement No 690462.This paper and the content included in it do not represent the opinion of the European Union, and the European Union is not responsible for any use that might be made of its content.

Institutional Review Board Statement: Not applicable.

Informed Consent Statement: Not applicable.

Data Availability Statement: Datasets generated and/or analysed during the current study are available from the corresponding authors on request.

Acknowledgments: The authors would like to express their gratitude to Roger Cremades, Hermine Mitter, Anabel Sanches, Annelies Broekman and Bernadette Kropf for comments and discussions that helped to improve the manuscript. We would also like to thank the editor and anonymous reviewers for their useful advice that helped to improve the manuscript.

Conflicts of Interest: The authors declare no conflict of interest.

References

1. Jabbar, F.K.; Grote, K. Evaluation of the predictive reliability of a new watershed health assessment method using the SWAT model. *Environ. Monit. Assess.* **2020**, *192*. [CrossRef] [PubMed]
2. Deshmukh, A.; Singh, R. Physio-climatic controls on vulnerability of watersheds to climate and land use change across the US. *Water Resour. Res.* **2016**, *52*, 8775–8793. [CrossRef]
3. IPCC. Summary for Policymakers: Climate Change 2014: Impacts, Adaptation and Vulnerability. Part A: Global and Sectoral Aspects. Contributions of the Working Group II to the Fifth Assessment Report. 2014.
4. Guerreiro, S.B.; Dawson, R.J.; Kilsby, C.; Lewis, E.; Ford, A. Future heat-waves, droughts and floods in 571 European cities. *Environ. Res. Lett.* **2018**, *13*. [CrossRef]
5. Miller, J.D.; Immerzeel, W.W.; Rees, G. Climate change impacts on glacier hydrology and river discharge in the Hindu Kush-Himalayas. *Mt. Res. Dev.* **2012**, *32*, 461–467. [CrossRef]
6. Song, Y.; Park, Y.; Lee, J.; Park, M.; Song, Y. Flood forecasting and warning system structures: Procedure and application to a small urban stream in South Korea. *Water* **2019**, *11*, 1571. [CrossRef]
7. Duan, Y.; Meng, F.; Liu, T.; Huang, Y.; Luo, M.; Xing, W.; De Maeyer, P. Sub-daily simulation of mountain flood processes based on the modified soil water assessment tool (Swat) model. *Int. J. Environ. Res. Public Health* **2019**, *16*, 3118. [CrossRef] [PubMed]

8. Weiskopf, S.R.; Rubenstein, M.A.; Crozier, L.G.; Gaichas, S.; Griffis, R.; Halofsky, J.E.; Hyde, K.J.W.; Morelli, T.L.; Morisette, J.T.; Muñoz, R.C.; et al. Climate change effects on biodiversity, ecosystems, ecosystem services, and natural resource management in the United States. *Sci. Total Environ.* **2020**, *733*, 137782. [CrossRef]
9. Cui, T.; Yang, T.; Xu, C.Y.; Shao, Q.; Wang, X.; Li, Z. Assessment of the impact of climate change on flow regime at multiple temporal scales and potential ecological implications in an alpine river. *Stoch. Environ. Res. Risk Assess.* **2018**, *32*, 1849–1866. [CrossRef]
10. Senent-Aparicio, J.; Liu, S.; Pérez-Sánchez, J.; López-Ballesteros, A.; Jimeno-Sáez, P. Assessing Impacts of Climate Variability and Reforestation Activities on Water Resources in the Headwaters of the Segura River Basin (SE Spain). *Sustainability* **2018**, *10*, 3277. [CrossRef]
11. Grey, O.P.; St G Webber, D.F.; Setegn, S.G.; Melesse, A.M. Application of the Soil and Water Assessment Tool (SWAT Model) on a small tropical island (Great River Watershed, Jamaica) as a tool in Integrated Watershed and Coastal Zone Management. *Rev. Biol. Trop. Int. J. Trop. Biol. Conserv.* **2014**, *62*, 293–305.
12. Orozco, I.; Martínez, A.; Ortega, V. Assessment of the water, environmental, economic and social vulnerability of a watershed to the potential effects of climate change and land use change. *Water* **2020**, *12*, 1682. [CrossRef]
13. Marin, M.; Clinciu, I.; Tudose, N.; Ungurean, C.; Adorjani, A.; Mihalache, A.; Davidescu, A.; Davidescu, Ş.O.; Dinca, L.; Cacovean, H. Assessing the vulnerability of water resources in the context of climate changes in a small forested watershed using SWAT: A review. *Environ. Res.* **2020**, *184*, 109330. [CrossRef]
14. Berezowski, T.; Chybicki, A. High-resolution discharge forecasting for snowmelt and rainfall mixed events. *Water* **2018**, *10*, 56. [CrossRef]
15. Jurik, Ľ.; Húska, D.; Halászová, K.; Bandlerová, A. Small water reservoirs—Sources of water or problems? *J. Ecol. Eng.* **2015**, *16*, 22–28. [CrossRef]
16. Cai, X.; Wallington, K.; Shafiee-Jood, M.; Marston, L. Understanding and managing the food-energy-water nexus—Opportunities for water resources research. *Adv. Water Resour.* **2018**, *111*, 259–273. [CrossRef]
17. Mengistu, A.G.; van Rensburg, L.D.; Woyessa, Y.E. Techniques for calibration and validation of SWAT model in data scarce arid and semi-arid catchments in South Africa. *J. Hydrol. Reg. Stud.* **2019**, *25*, 100621. [CrossRef]
18. Osei, M.A.; Amekudzi, L.K.; Wemegah, D.D.; Preko, K.; Gyawu, E.S.; Obiri-Danso, K. The impact of climate and land-use changes on the hydrological processes of Owabi catchment from SWAT analysis. *J. Hydrol. Reg. Stud.* **2019**, *25*, 100620. [CrossRef]
19. Chen, Q.; Chen, H.; Wang, J.; Zhao, Y.; Chen, J.; Xu, C. Impacts of Climate Change and Land-Use Change on Hydrological Extremes in the Jinsha River Basin. *Water* **2019**, *11*, 1398. [CrossRef]
20. Yan, R.; Cai, Y.; Li, C.; Wang, X.; Liu, Q. Hydrological Responses to Climate and Land Use Changes in a Watershed of the Loess Plateau, China. *Sustainability* **2019**, *11*, 1443. [CrossRef]
21. Phung, Q.A.; Thompson, A.L.; Baffaut, C.; Costello, C.; Sadler, E.J.; Svoma, B.M.; Lupo, A.; Gautam, S. Climate and Land Use Effects on Hydrologic Processes in a Primarily Rain-Fed, Agricultural Watershed. *JAWRA J. Am. Water Resour. Assoc.* **2019**, *55*, 1196–1215. [CrossRef]
22. Esmali, A.; Golshan, M.; Kavian, A. Investigating the performance of SWAT and IHACRES in simulation streamflow under different climatic regions in Iran. *Atmósfera* **2021**, *34*, 79–96. [CrossRef]
23. Beniston, M. Climatic change in mountain regions: A review of possible impacts. *Clim. Change* **2003**, *59*, 5–31. [CrossRef]
24. Mountain Partnership. Mountains and the Sustainable Development Goals. 2014. Available online: http://www.fao.org/fileadmin/templates/mountain_partnership/doc/POLICY_BRIEFS/Mountains_and_the_Sustainable_Development_Goals_-_NY_-_8Jan.2014.pdf (accessed on 15 March 2021).
25. BIO Intelligence Service. Literature Review on the Potential Climate Change Effects on Drinking Water Resources across the EU and the Identification of Priorities among Different Types of Drinking Water Supplies, Final Report—ADWICE Project Prepared for European Commission DG Environment, 20–22 Villa Deshayes, 75014 Paris. 2012. Available online: https://ec.europa.eu/environment/archives/water/adaptation/pdf/ADWICE_FinalReport.pdf (accessed on 15 March 2021).
26. Jacob, D.; Petersen, J.; Eggert, B.; Alias, A.; Christensen, O.B.; Bouwer, L.M.; Braun, A.; Colette, A.; Déqué, M.; Georgievski, G.; et al. EURO-CORDEX: New high-resolution climate change projections for European impact research. *Reg. Environ. Chang.* **2014**, *14*, 563–578. [CrossRef]
27. Kotlarski, S.; Keuler, K.; Christensen, O.B.; Colette, A.; Déqué, M.; Gobiet, A.; Goergen, K.; Jacob, D.; Lüthi, D.; Van Meijgaard, E.; et al. Regional climate modeling on European scales: A joint standard evaluation of the EURO-CORDEX RCM ensemble. *Geosci. Model Dev.* **2014**, *7*, 1297–1333. [CrossRef]
28. Downing, J.A. Emerging global role of small lakes and ponds: Little things mean a lot. *Limnetica* **2010**, *29*, 9–24. [CrossRef]
29. Hackenbruch, J.; Kunz-Plapp, T.; Müller, S.; Schipper, J.W. Tailoring climate parameters to information needs for local adaptation to climate change. *Climate* **2017**, *5*, 25. [CrossRef]
30. Bayabil, H.K.; Dile, Y.T. Improving hydrologic simulations of a small watershed through soil data integration. *Water* **2020**, *12*, 2763. [CrossRef]
31. Irving, K.; Kuemmerlen, M.; Kiesel, J.; Kakouei, K.; Domisch, S.; Jähnig, S.C. Data descriptor: A high-resolution streamflow and hydrological metrics dataset for ecological modeling using a regression model. *Sci. Data* **2018**, *5*, 1–14. [CrossRef]
32. Zhao, Q.; Zhu, Y.; Wan, D.; Yu, Y.; Lu, Y. Similarity Analysis of Small- and Medium-Sized Watersheds Based on Clustering Ensemble Model. *Water* **2019**, *12*, 69. [CrossRef]

33. Rahman, K.; Maringanti, C.; Beniston, M.; Widmer, F.; Abbaspour, K.; Lehmann, A. Streamflow Modeling in a Highly Managed Mountainous Glacier Watershed Using SWAT: The Upper Rhone River Watershed Case in Switzerland. *Water Resour. Manag.* **2013**, *27*, 323–339. [CrossRef]
34. Arora, M.; Kumar, R.; Kumar, N.; Malhotra, J. Hydrological modeling and streamflow characterization of Gangotri Glacier. In *Geostatistical and Geospatial Approaches for the Characterization of Natural Resources in the Environment: Challenges, Processes and Strategies, IAMG 2014*; Capital Publishing Company: New Delhi, India, 2016.
35. Jain, S.; Jain, S.; Jain, N.; Xu, C.-Y. Hydrologic modeling of a Himalayan mountain basin by using the SWAT mode. *Hydrol. Earth Syst. Sci. Discuss.* **2017**. [CrossRef]
36. Mateo-Lázaro, J.; Castillo-Mateo, J.; Sánchez-Navarro, J.Á.; Fuertes-Rodríguez, V.; García-Gil, A.; Edo-Romero, V. Assessment of the role of snowmelt in a flood event in a gauged catchment. *Water* **2019**, *11*, 506. [CrossRef]
37. Romanescu, G.; Jora, I.; Stoleriu, C. The most important high floods in Vaslui river basin -causes and consequences. *Carpathian J. Earth Environ. Sci.* **2011**, *6*, 119–132.
38. Birsan, M.V.; Zaharia, L.; Chendes, V.; Branescu, E. Recent trends in streamflow in Romania (1976–2005). *Rom. Reports Phys.* **2012**, *64*, 275–280.
39. The World Bank. Romania, Climate Change and Low Carbon Green Growth Program, Component B Sector Report, Forest Sector Rapid Assessment. 2014. Available online: https://openknowledge.worldbank.org/bitstream/handle/10986/17570/842620WP0P14660Box0382136B00PUBLIC0.pdf?sequence=1&isAllowed=y (accessed on 1 March 2021).
40. Yang, L.; Smith, J.A.; Baeck, M.L.; Zhang, Y. Flash flooding in small urban watersheds: Storm event hydrologic response. *Water Resour. Res.* **2016**, *52*, 4571–4589. [CrossRef]
41. Kvočka, D.; Ahmadian, R.; Falconer, R.A. Predicting Flood Hazard Indices in Torrential or Flashy River Basins and Catchments. *Water Resour. Manag.* **2018**, *32*, 2335–2352. [CrossRef]
42. Petroselli, A.; Grimaldi, S. Design hydrograph estimation in small and fully ungauged basins: A preliminary assessment of the EBA4SUB framework. *J. Flood Risk Manag.* **2018**, *11*, S197–S210. [CrossRef]
43. Apollonio, C.; Bruno, M.F.; Iemmolo, G.; Molfetta, M.G.; Pellicani, R. Flood Risk Evaluation in Ungauged Coastal Areas. *Water* **2020**, *5*, 1466. [CrossRef]
44. Ha, L.; Bastiaanssen, W.; van Griensven, A.; van Dijk, A.; Senay, G. SWAT-CUP for Calibration of Spatially Distributed Hydrological Processes and Ecosystem Services in a Vietnamese River Basin Using Remote Sensing. *Hydrol. Earth Syst. Sci. Discuss.* **2017**. [CrossRef]
45. Arnold, J.G.; Moriasi, D.N.; Gassman, P.W.; Abbaspour, K.C.; White, M.J.; Srinivasan, R.; Santhi, C.; Harmel, R.D.; Van Griensven, A.; Van Liew, M.W.; et al. SWAT: Model use, calibration, and validation. *Trans. ASABE* **2012**, *55*, 1491–1508. [CrossRef]
46. Neitsch, S.; Arnold, J.; Kiniry, J.; Williams, J. Neitsch Grassland, Soil, Water Research Laboratory; Agricultural Research Service Blackland Research Center; Texas AgriLife Research. Soil & Water Assessment Tool, Theoretical Documentation Version 2009. Texas Water Resources Institute. exas Water Resources Institute Technical Report No. 365 Texas A&M University System College Station, Texas 77843-2118. 2011. [CrossRef]
47. Neitsch, S.L.; Arnold, J.G.; Kiniry, J.R.; Williams, J.R. *Soil and Water Assessment Tool Theoretical Documentation Version 2005*; Grassland, Soil and Water Research Laboratory, Agricultural Research Service: Temple, TX, USA, 2005.
48. Tudose, N.C.; Davidescu, S.O.; Cheval, S.; Chendes, V.; Ungurean, C.; Babata, M. Integrated Model of River Basin, Land Use and Urban Water Supply. Deliverable 3.4. CLISWELN Project. 2018. Available online: https://ms.hereon.de/imperia/md/content/csc/projekte/projekte/clisweln_d3.4_romania_study_case_final-2.pdf (accessed on 9 November 2020).
49. Noor, H.; Vafakhah, M.; Taheriyoun, M.; Moghadasi, M. Hydrology modelling in Taleghan mountainous watershed using SWAT. *J. Water L. Dev.* **2014**, *20*, 11–18. [CrossRef]
50. Winchell, M.; Srinivasan, R.; Di Luzio, M.; Arnold, J. ArcSWAT 2.3.4 Interface for SWAT2005: User's Guide, Version September 2009. Texas Agricultural Experiment Station and Agricultural Research Service- US Department of Agriculture, Temple. 2009.
51. Bîrsan, M.-V.; Dumitrescu, A. ROCADA: Romanian daily gridded climatic dataset (1961-2013) V1.0. *Nat. Hazards* **2014**, *78*, 1045–1063. [CrossRef]
52. Dumitrescu, A.; Birsan, M.V. ROCADA: A gridded daily climatic dataset over Romania (1961–2013) for nine meteorological variables. *Nat. Hazards* **2015**, *78*, 1045–1063. [CrossRef]
53. Popa, I.; Badea, O.; Silaghi, D. Influence of climate on tree health evaluated by defoliation in the ICP level I network (Romania). *IForest* **2017**, *10*, 554–560. [CrossRef]
54. Sfîcă, L.; Croitoru, A.E.; Iordache, I.; Ciupertea, A.F. Synoptic conditions generating heatwaves and warm spells in Romania. *Atmosphere* **2017**, *8*, 50. [CrossRef]
55. Karim, T.H.; Fattah, M.A. Efficiency of the SPAW model in estimation of saturated hydraulic conductivity in calcareous soils. *J. Univ. Duhok* **2020**, *23*, 189–201. [CrossRef]
56. Post, D.F.; Fimbres, A.; Matthias, A.D.; Sano, E.E.; Accioly, L.; Batchily, A.K.; Ferreira, L.G. Predicting Soil Albedo from Soil Color and Spectral Reflectance Data. *Soil Sci. Soc. Am. J.* **2000**, *64*, 1027–1034. [CrossRef]
57. Williams, J.R. The EPIC Model. In *Computer Models of Watershed Hydrology*; Singh, V.P., Ed.; Water Resources Publications: Highlands Ranch, CO, USA, 1995; ISBN 0918334918.
58. Gijsman, A.J.; Thornton, P.K.; Hoogenboom, G. Using the WISE database to parameterize soil inputs for crop simulation models. *Comput. Electron. Agric.* **2007**, *56*, 85–100. [CrossRef]

59. Arnold, J.G.; Kiniry, J.R.; Srinivasan, R.; Williams, J.R.; Haney, E.B.; Neitsch, S.L. Input/Output Documentation. 2012. Available online: https://swat.tamu.edu/media/69296/swat-io-documentation-2012.pdf (accessed on 20 November 2020).
60. Daggupati, P.; Pai, N.; Ale, S.; Douglas-Mankin, K.R.; Zeckoski, R.W.; Jeong, J.; Parajuli, P.B.; Saraswat, D.; Youssef, M.A. A recommended calibration and validation strategy for hydrologic and water quality models. *Trans. ASABE* **2015**, *58*, 1705–1719. [CrossRef]
61. Abbaspour, K.C. SWAT-CUP. SWAT Calibration and Uncertainty Programs. 2015, p. 100. Available online: https://swat.tamu.edu/media/114860/usermanual_swatcup.pdf (accessed on 6 December 2020).
62. Harmel, R.D.; Smith, P.K.; Migliaccio, K.W.; Chaubey, I.; Douglas-Mankin, K.R.; Benham, B.; Shukla, S.; Muñoz-Carpena, R.; Robson, B.J. Evaluating, interpreting, and communicating performance of hydrologic/water quality models considering intended use: A review and recommendatrions. *Environ. Model. Softw.* **2014**, *21*, 40–51.
63. Moriasi, D.N.; Arnold, J.G.; Van Liew, M.W.; Bingner, R.L.; Harmel, R.D.; Veith, T.L. Model evaluation guidelines for systematic quantification of accuracy in watershed simulations. *Trans. ASABE* **2007**, *50*, 885–900. [CrossRef]
64. Moriasi, D.N.; Gitau, M.W.; Pai, N.; Daggupati, P. Hydrologic and water quality models: Performance measures and evaluation criteria. *Trans. ASABE* **2015**, *58*, 1763–1785. [CrossRef]
65. Gupta, H.V.; Sorooshian, S.; Yapo, P.O. Status of Automatic Calibration for Hydrologic Models: Comparison with Multilevel Expert Calibration. *J. Hydrol. Eng.* **1999**, *4*, 135–143. [CrossRef]
66. da Silva, M.G.; de Aguiar Netto, A.O.; de Jesus Neves, R.J.; do Vasco, A.N.; Almeida, C.; Faccioli, G.G. Sensitivity Analysis and Calibration of Hydrological Modeling of the Watershed Northeast Brazil. *J. Environ. Prot.* **2015**, *6*, 837–850. [CrossRef]
67. Nash, J.E.; Sutcliffe, J.V. River flow forecasting through conceptual models part I - A discussion of principles. *J. Hydrol.* **1970**, *10*, 282–290. [CrossRef]
68. Abbaspour, K.C.; Johnson, C.A.; Th van Genuchten, M. Estimating Uncertain Flow and Transport Parameters Using a Sequential Uncertainty Fitting Procedure. *Vadose Zone J.* **2004**, *3*, 1340–1352. [CrossRef]
69. Abbaspour, K.C.; Rouholahnejad, E.; Vaghefi, S.; Srinivasan, R.; Yang, H.; Kløve, B. A continental-scale hydrology and water quality model for Europe: Calibration and uncertainty of a high-resolution large-scale SWAT model. *J. Hydrol.* **2015**, *524*, 733–752. [CrossRef]
70. Thavhana, M.P.; Savage, M.J.; Moeletsi, M.E. SWAT model uncertainty analysis, calibration and validation for runoff simulation in the Luvuvhu River catchment, South Africa. *Phys. Chem. Earth* **2018**, *105*, 115–124. [CrossRef]
71. Qiu, L.J.; Zheng, F.L.; Yin, R.S. SWAT-based runoff and sediment simulation in a small watershed, the loessial hilly-gullied region of China: Capabilities and challenges. *Int. J. Sediment Res.* **2012**, *27*, 226–234. [CrossRef]
72. Emam, R.A.; Kappas, M.; Hoang Khanh Nguyen, L.; Renchin, T. Hydrological Modeling in an Ungauged Basin of Central Vietnam Using SWAT Model. *Hydrol. Earth Syst. Sci. Discuss.* **2016**, 1–33. [CrossRef]
73. Abbaspour, K.C.; Vaghefi, S.A.; Srinivasan, R. A guideline for successful calibration and uncertainty analysis for soil and water assessment: A review of papers from the 2016 international SWAT conference. *Water* **2017**, *10*, 6. [CrossRef]
74. Abbaspour, K.C.; Yang, J.; Maximov, I.; Siber, R.; Bogner, K.; Mieleitner, J.; Zobrist, J.; Srinivasan, R. Modelling hydrology and water quality in the pre-alpine/alpine Thur watershed using SWAT. *J. Hydrol.* **2007**, *333*, 413–430. [CrossRef]
75. Wallace, C.W.; Flanagan, D.C.; Engel, B.A. Evaluating the effects ofwatershed size on SWAT calibration. *Water* **2018**, *10*, 898. [CrossRef]
76. Miralles, D.G.; Gash, J.H.; Holmes, T.R.H.; De Jeu, R.A.M.; Dolman, A.J. Global canopy interception from satellite observations. *J. Geophys. Res. Atmos.* **2010**, *115*, 1–8. [CrossRef]
77. Cui, Y.; Jia, L.; Hu, G.; Zhou, J. Mapping of interception loss of vegetation in the heihe river basin of china using remote sensing observations. *IEEE Geosci. Remote Sens. Lett.* **2015**, *12*, 23–27. [CrossRef]
78. European Commission. Communication from the Commission to the European Parliament, the Council, the European Economic and Social Committee and the Committee of the Regions. 2020. Available online: https://eur-lex.europa.eu/resource.html?uri=cellar:a3c806a6-9ab3-11ea-9d2d-01aa75ed71a1.0001.02/DOC_1&format=PDF (accessed on 13 January 2021).
79. Brouziyne, Y.; Abouabdillah, A.; Bouabid, R.; Benaabidate, L. SWAT streamflow modeling for hydrological components' understanding within an agro-sylvo-pastoral watershed in Morocco. *J. Mater. Environ. Sci.* **2018**, *9*, 128–138. [CrossRef]
80. Aawar, T.; Khare, D. Assessment of climate change impacts on streamflow through hydrological model using SWAT model: A case study of Afghanistan. *Model. Earth Syst. Environ.* **2020**, *6*, 1427–1437. [CrossRef]
81. Leng, M.; Yu, Y.; Wang, S.; Zhang, Z. Simulating the hydrological processes of a meso-scalewatershed on the Loess Plateau, China. *Water* **2020**, *12*, 878. [CrossRef]
82. Amatya, D.M.; Jha, M.K. Evaluating the SWAT model for a low-gradient forested watershed in coastal South Carolina. *Trans. ASABE* **2011**, *54*, 2151–2163. [CrossRef]
83. Mapes, K.L.; Pricope, N.G. Evaluating SWAT model performance for runoff, percolation, and sediment loss estimation in low-gradientwatersheds of the Atlantic Coastal Plain. *Hydrology* **2020**, *7*, 21. [CrossRef]
84. Tolera, M.B.; Chung, I.M.; Chang, S.W. Evaluation of the climate forecast system reanalysis weather data for watershed modeling in Upper Awash Basin, ethiopia. *Water* **2018**, *10*, 725. [CrossRef]
85. Busico, G.; Colombani, N.; Fronzi, D.; Pellegrini, M.; Tazioli, A.; Mastrocicco, M. Evaluating SWAT model performance, considering different soils data input, to quantify actual and future runoff susceptibility in a highly urbanized basin. *J. Environ. Manag.* **2020**, *266*. [CrossRef]

86. Abbas, T.; Nabi, G.; Boota, M.W.; Hussain, F.; Azam, M.I.; Jin, H.; Faisal, M. Uncertainty analysis of runoff and sedimentation in a forested watershed using sequential uncertainty fitting method. *Sci. Cold Arid Reg.* **2016**, *8*, 297–310. [CrossRef]
87. Briones, R.U.; Ella, V.B.; Bantayan, N.C. Hydrologic impact evaluation of land use and land cover change in Palico Watershed, Batangas, Philippines Using the SWAT model. *J. Environ. Sci. Manag.* **2016**, *19*, 96–107.
88. Narsimlu, B.; Gosain, A.K.; Chahar, B.R.; Singh, S.K.; Srivastava, P.K. SWAT Model Calibration and Uncertainty Analysis for Streamflow Prediction in the Kunwari River Basin, India, Using Sequential Uncertainty Fitting. *Environ. Process.* **2015**, *2*, 79–95. [CrossRef]
89. Adhikary, P.P.; Sena, D.R.; Dash, C.J.; Mandal, U.; Nanda, S.; Madhu, M.; Sahoo, D.C.; Mishra, P.K. Effect of Calibration and Validation Decisions on Streamflow Modeling for a Heterogeneous and Low Runoff–Producing River Basin in India. *J. Hydrol. Eng.* **2019**, *24*, 05019015. [CrossRef]
90. United Nations Transforming Our World: The 2030 Agenda for Sustainable Development: Sustainable Development Knowledge Platform. Available online: https://sustainabledevelopment.un.org/post2015/transformingourworld (accessed on 12 June 2020).
91. Carpathian Framework Convention. Framework Convention on the Protection and Sustainable Development of the Carpathians. Available online: http://www.carpathianconvention.org/tl_files/carpathiancon/Downloads/01TheConvention/1.1.1.1_CarpathianConvention.pdf (accessed on 19 February 2021).

 forests

MDPI

Article

Intraspecific Growth Response to Drought of *Abies alba* in the Southeastern Carpathians

Georgeta Mihai [1,*], Alin Madalin Alexandru [1,2,*], Emanuel Stoica [1] and Marius Victor Birsan [3]

1 Department of Forest Genetics and Tree Breeding, "Marin Dracea" National Institute for Research and Development in Forestry, 077190 Voluntari, Romania; emanuel_s_96@yahoo.com
2 Faculty of Silviculture and Forest Engineering, Transilvania University of Brasov, 500123 Brasov, Romania
3 Department of Research and Infrastructure Projects, National Meteorological Administration, 013686 Bucharest, Romania; marius.birsan@gmail.com
* Correspondence: gmihai_2008@yahoo.com (G.M.); alexandru.alin06@yahoo.com (A.M.A.)

Abstract: The intensity and frequency of drought have increased considerably during the last decades in southeastern Europe, and projected scenarios suggest that southern and central Europe will be affected by more drought events by the end of the 21st century. In this context, assessing the intraspecific genetic variation of forest tree species and identifying populations expected to be best adapted to future climate conditions is essential for increasing forest productivity and adaptability. Using a tree-ring database from 60 populations of 38-year-old silver fir (*Abies alba*) in five trial sites established across Romania, we studied the variation of growth and wood characteristics, provenance-specific response to drought, and climate-growth relationships during the period 1997–2018. The drought response of provenances was determined by four drought parameters: resistance, recovery, resilience, and relative resilience. Based on the standardized precipitation index, ten years with extreme and severe drought were identified for all trial sites. Considerable differences in radial growth, wood characteristics, and drought response parameters among silver fir provenances have been found. The provenances' ranking by resistance, recovery, and resilience revealed that a number of provenances from Bulgaria, Italy, Romania, and Czech Republic placed in the top ranks in almost all sites. Additionally, there are provenances that combine high productivity and drought tolerance. The correlations between drought parameters and wood characters are positive, the most significant correlations being obtained between radial growth and resilience. Correlations between drought parameters and wood density were non-significant, indicating that wood density cannot be used as indicator of drought sensitivity. The negative correlations between radial growth and temperature during the growing season and the positive correlations with precipitation suggest that warming and water deficit could have a negative impact on silver fir growth in climatic marginal sites. Silvicultural practices and adaptive management should rely on selection and planting of forest reproductive material with high drought resilience in current and future reforestation programs.

Keywords: silver fir; radial growth; wood characteristics; drought response; climate change

Citation: Mihai, G.; Alexandru, A.M.; Stoica, E.; Birsan, M.V. Intraspecific Growth Response to Drought of *Abies alba* in the Southeastern Carpathians. *Forests* **2021**, *12*, 387. https://doi.org/10.3390/f12040387

Academic Editors: Alessandra De Marco, Ovidiu Badea, Pierre Sicard and Mihai A. Tanase

Received: 21 February 2021
Accepted: 22 March 2021
Published: 24 March 2021

1. Introduction

Climate change is a major threat to forests in the 21st century. According to the reports of the Intergovernmental Panel on Climate Change [1,2], temperatures have increased globally, and the highest rates of warming have taken place in the last decades. Furthermore, recent evidence has shown a significant increase in the frequency of extreme weather events (prolonged droughts, heat waves, cold snaps, and floods) related to global climate change [3,4].

Among the extreme meteorological events, drought is considered to have the largest detrimental impact on forest ecosystems. Drought and heat stress associated with climate change could fundamentally alter the productivity, genetic diversity, and distribution of forest ecosystems [5–7]. In recent years, it was observed that drought frequency, severity, and duration increased in many regions in Europe [8]. The most affected regions have been southern Europe with the Mediterranean region as a hotspot [9,10] and South-eastern Europe, particularly the Carpathian region [11–13].

The moderate scenario projections (RCP4.5) show that southern Europe, western Europe, and northern Scandinavia will be affected by a substantial increase in drought frequency by the end of the 21st century. However, the extreme emission scenario (RCP8.5) suggests that the entirety of Europe will be affected by more frequent and severe droughts compared to the last century. Under both scenarios, drought frequency is projected to increase in spring and summer everywhere in Europe, but especially in southern Europe [14,15].

The climate changes will also enhance the action of the new biotic (pest and disease) and abiotic disturbance factors (fire, windstorm) with major consequences for forest ecosystems. Increasing the extreme events, such as drought and disturbance factors, in the near future will pose serious threats to the growth and persistence of forest species than gradual climate changes [16,17]. There is a consensus that the ability of forest ecosystems to provide multiple goods and services will be impacted [18]. The mountain ecosystems and those located at the edges of forest species' distribution will be the most vulnerable.

Silver fir (*Abies alba* Mill.) is one of the main species of mountain ecosystems in Europe with multiple functions, including ecological, economic, and soil protection roles. European silver fir is a shade-tolerant species and can grow in an array of soil conditions, with various amounts of nutrients, but prefers humid and deep soils [19]. Results so far regarding the potential of silver fir to thrive under expected warmer and drier conditions are optimistic. The species distribution models (SDMs) suggest that the suitable distribution area of silver fir will decrease by the end of the century, particularly in the southern and eastern parts of its distribution, but the lowest decrease is projected for silver fir compared to other coniferous species [20,21]. On the other hand, paleoecological studies, as well as dynamic models accounting for biotic and abiotic disturbances, suggest that this species has a high potential to cope with the expected climate change [22] and can even expand in regions with summer water deficit from central and eastern Europe [23,24]. Additionally, other recent studies showed that European silver fir has high phenotypic plasticity [25] and is less vulnerable to drought stress than other conifers of temperate forests [26–31]. However, a possible decline may occur in the driest and warmest areas at the distribution edge [32–34]. Therefore, European silver fir could be one of the future species for consideration under changing climate conditions, particularly at lower altitudes in the mixed vegetation layer.

Considering that drought events will become more frequent and intense in the near future, the strategies to cope with climate change have to prepare forests by increasing the adaptive capacity of tree populations [35]. Recent research shows that selecting and transferring forest reproductive material adapted to the new environmental conditions of the planting site could increase genetic diversity in those areas and could facilitate the adaptation of forest species [36–38]. Therefore, assessment of intraspecific genetic variation and identifying populations expected to be best adapted to the future climate conditions is essential for increasing forest productivity and adaptability in the context of climate changes.

Many classical studies in the field of dendrochronology have investigated the potential impact of climate changes on tree growth [28,29,32]. Unfortunately, these studies do not take into account the existence of intraspecific genetic variation, considering genetically homogeneous species. Provenance trials, where tree populations throughout the entire distribution of a species are tested in different site conditions, can provide important data concerning intraspecific adaptive capacity and selection of suitable populations for reforestation programs. These genetic tests facilitate the identification of climatic variables that exert strong selective pressure on studied populations and the developing of the models that can be used to predict species' response to future climates [39,40].

Provenance trials have been used to analyze intraspecific variation in climate growth response in several tree species, such as *Pinus contorta* [39,41], *Pinus sylvestris*, *Fagus sylvatica* and *Quercus petraea* [42], *Quercus robur* [43], *Picea glauca* [44,45], *Pseudotsuga menziesii* var. *menziesii* [46,47], *Picea abies* [42,48–51], and *Abies alba* [34,52]. However, there are still gaps in our knowledge concerning the intraspecific genetic response of forest species to extreme climate events, such as severe or extreme droughts, because they require long-term experiments and growth and climate assessments over several decades. Recent studies have shown that there are significant genetic variations in drought response both within and among trees populations [53–56]. The provenance-specific drought response for silver fir has been investigated in even fewer studies [30,57,58].

The Romanian Carpathians represent the southeastern distribution limit of silver fir in Europe. Meteorological records show a warming trend and increasingly severe and extreme summer droughts in recent decades in this region. Considering that negative effects of predicted climate change will be more pronounced, especially at the xeric edge of species distribution range [14,15,34], knowing the adaptive capacity of silver fir becomes of major importance. In this context, the aim of this study was to investigate the genetic adaptive capacity and response of silver fir provenances originated from nine European countries to extreme drought events that have occurred in this region in the last 22 years. Understanding of the population's performance in relation to climate stress, selection of the best adapted seed sources to future climate conditions, and using them in reforestation programs (i.e., assisted migration) is essential for increasing forest productivity and adaptability.

Based on the assumption that drought will significantly impact silver fir ecosystems in southeastern Europe, in the near future, the objectives of this study were to (1) assess the genetic variation of radial growth and wood characteristics among silver fir provenances, (2) evaluate the provenances-specific drought response, (3) establish the climate–growth relationships, (4) determine correlations between radial growth and wood characteristics, and drought parameters, and (5) provide practical information for sustainable forest management in a changing climate context.

2. Materials and Methods

2.1. Trial Site and Plant Material

The study was conducted in a series of five provenances trials established in 1980 in Romania. The provenance trials were established in five geographic regions with different climatic conditions (Figure 1 and Table 1). Two trials are located outside of the natural range of silver fir in Romanian Carpathians, in the European beech zone, while three are within the natural range.

Table 1. Geographic and climatic variables for silver fir trial sites.

Trials	Prov. Region	Altitude m	TMA °C	TM_{VEG} °C	SAP mm	PM_{VEG} °C	SP_{VEG} mm	De Martonne Index	De Martonne $Index_{VEG}$
Bucova	D2	650	7.37	13.46	878	92	550	50	23
Domnesti	C2	880	6.68	12.83	916	104	622	55	27
Moinesti	A2	815	7.72	13.71	868	91	544	49	23
Sacele	B1	1225	4.92	10.91	1001	113	678	67	32
Strambu-Baiut	A1	890	6.90	13.24	879	87	521	52	22

In these trials are tested 60 populations originating from the entire species distribution range in Europe (Figure 1). They were grouped as core, western, eastern, northern, southeastern, and southern according to their location within the natural distribution range (Table A1 in Appendix A). Forty-three provenances are common in all trials, and 17 provenances are tested additionally at the Sacele trial only. The silver fir tested provenances range from 38°33′ to 51°07′ N and from 4°00′ to 26°40′ E, and include both lower as well as higher mountain regions (altitudes between 130–1600 m above sea level). In four sites, the field layout was the randomized square lattice, type 7 × 7, with three repetitions and 25 trees per plot planted at 1.0 × 2.0 m, while at the Sacele trial, the field layout was the randomized square lattice, type 8 × 8, also with tree repetitions. The five field trials were established with six-year-old bare root seedlings, which have been produced in the Sinaia nursery situated in the mountain beech zone, at 45°29′ N latitude, 25°59′ E longitude, and at 695 m a.s.l.

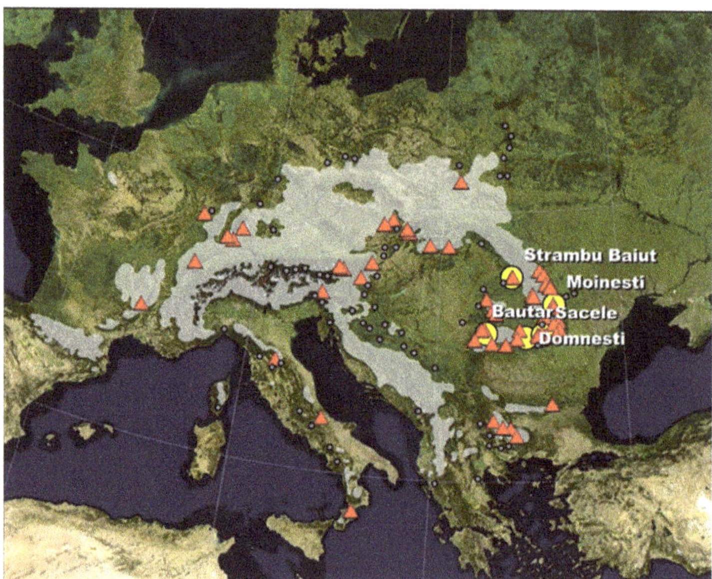

Figure 1. Location of silver fir provenances (triangles) and trials (circles). Gray area/dots—the natural distribution of silver fir (by EUFORGEN).

2.2. Phenotypic and Climatic Data

In each provenance trial, four dominant or (co)dominant trees per provenance and repetition (12 trees in total for each provenance) have been cored at 1.3 m breast height using 5 mm increment borers (Haglof, Sweden), from slope-parallel stem radii, to avoid tension and compression wood. In order to avoid tree damage, only one core per tree was taken. Cores were dried and progressively sanded [59]. Then, the core samples were scanned at 1200 dpi, using an Epson Expression 10,000 XL, and the ring width (RW), earlywood width (EW), and latewood width (LW) were measured using the Ligno Vision software package to the nearest 0.001 mm. Additionally, latewood proportion (LWP) was calculated as an indicator of wood quality.

For each trial, the tree-ring series has been cross-dated using COFECHA [60] to avoid dating errors due to missing or false rings, which could be present in an increment radial core. Only the tree-ring series that presented intercorrelation values > 0.328 ($p < 0.01$) were included in final tree-ring data. All tree-ring time series were standardized to a mean value of one to obtain a width index (RWI) [61,62]. The negative exponential regression in the R package dplR [63,64] was applied for each raw measurement series, because it is deterministic, meaning that it follows a model of tree growth. The final tree ring data set comprised 2699 tree-ring series, 669 from the Sacele trial and 500 from each of the other four trials. The analyzed period was 1997–2018, being the common interval for all tree-ring series.

Additionally, wood density (WD) in g/cm^3 was determined for each core sample and whole analyzed period using the [65] formula:

$$\rho_c = 1/[(M_{max}/M_o) - 1 + 1/\rho_{ml}] \quad (1)$$

where: ρ_c = conventional density (g/cm^3), M_{max} = weight of saturated sample (g), M_o = weight of dried sample (g), ρ_{ml} = wood density (1.53 g/cm^3).

The climatic data have been calculated using a daily gridded climatic dataset covering the Romanian territory (ROCADA). The dataset used herein consists of a higher spatial resolution (1 × 1 km) for improved reproduction of the climatic spatial variability and has been made using state-of-the-art interpolation techniques [66]. The following climatic variables have been calculated for each trial site over the period 1997–2018: mean annual temperature (MAT); mean temperature during the growing season (April to September) (MT_{VEG}); mean temperature for January (MT_{JAN}) and July (MT_{JUL}) (the coldest and the warmest months, respectively); mean temperature from October to December of the previous year ($MT_{OCT\text{-}DEC}$); mean temperature from October of the previous year to March of the current year ($MT_{OCT\text{-}MAR}$); mean temperature from January to March of the current year ($MT_{JAN\text{-}MAR}$); mean annual precipitation amount (MAP); mean precipitation during the growing season (MP_{VEG}); mean precipitation of the coldest (MP_{JAN}) and the warmest (MP_{JUL}) months; mean precipitation from October to December of the previous year ($MP_{OCT\text{-}DEC}$); mean precipitation from January to March of the current year ($MP_{JAN\text{-}MAR}$); mean precipitation from October of the previous year to March of the current year ($MP_{OCT\text{-}MAR}$); annual precipitation amount (SAP); precipitation amount in the growing season (SP_{VEG}); precipitation amount in the autumn-winter of the previous year ($SP_{OCT\text{-}DEC}$); precipitation amount from January to March of the current year ($SP_{JAN\text{-}MAR}$); precipitation amount from October of the previous year to March of the current year ($SP_{OCT\text{-}MAR}$) (Table 2).

Table 2. Description of the wood and climatic characteristics.

Abbreviation	Wood and Climate Characteristics
RW	Ring width
EW	Earlywood width
LW	Latewood width
LWP	Latewood proportion
DW	Wood density
MAT	Mean annual temperature
MT_{VEG}	Mean temperature during growing season (April to September)
MT_{JAN}	Mean temperature for January (the coldest month)
MT_{JUL}	Mean temperature for July (the warmest month)
$MT_{OCT\text{-}DEC}$	Mean temperature from October to December of previous year
$MT_{OCT\text{-}MAR}$	Mean temperature from October of previous year to March of current year
$MT_{JAN\text{-}MAR}$	Mean temperature from January to March of current year
MAP	Mean annual precipitation amount
MP_{VEG}	Mean precipitation during growing season
MP_{JAN}	Mean precipitation of the coldest month
MP_{JUL}	Mean precipitation of the warmest month
$MP_{OCT\text{-}DEC}$	Mean precipitation from October to December of previous year
$MP_{OCT\text{-}MAR}$	Mean precipitation from October of previous year to March of current year
$MP_{JAN\text{-}MAR}$	Mean precipitation from January to March of the current year
SAP	Annual precipitation amount
$SP_{OCT\text{-}DEC}$	Precipitation amount in autumn-winter of previous year
$SP_{OCT\text{-}MAR}$	Precipitation amount from October of previous year to March of current year
$SP_{JAN\text{-}MAR}$	Precipitation amount from January to March of current year

2.3. Determination of Drought Events and Drought Response Parameters

As an indicator for the meteorological droughts, we calculated the standardized precipitation indices (SPI) [67], which account for anomalous low rainfall, over the period 1989–2018. Given that an extreme drought event obviously lasts two to three months, to identify drought years within the analyzed period, we calculated SPI for three consecutive drought months in each trial site. That allowed us to detect both seasonal and annual variation of drought events during the analyzed period. The drought years have been classified as follows: SPI ≤ -2—extreme drought year, SPI between -1.99 to -1.50—severe drought year, SPI between −1.49 to −1.0—moderate drought year, SPI between −1.0 to +1.0—normal precipitation year [57].

The response of provenances to drought events was evaluated by four drought parameters [68]: resistance (Res), recovery (Rec), resilience (Rsl), and relative resilience (rRsl). Resistance was calculated as the ratio between ring width during (Dr) and before the drought event (preDr): Res = Dr/preDr and indicates how much the radial growth decreased during drought (Res ≥ 1 means high tolerance, Res < 1 means low tolerance). Recovery was calculated as the ratio between the ring width after drought event (postDr) and during drought (Rc = postDr/Dr) and indicates the revitalization capacity after a drought period. Resilience (Rsl) represents the ratio of the ring width after drought (postDr) and pre-drought (preDr): Rsl = postDr/preDr and describes the capacity of a provenance to reach pre-drought increment after a drought event (Rsl ≥ 1 means full restoration, Rs < 1 means long-term growth reductions). Relative resilience (rRsl) was calculated by rRsl = (postDr − Dr)/preDr. Pre-drought and post-drought ring widths were calculated as average values for three-year period before or after the drought year.

2.4. Data Analysis

Analyses of variance were performed at two levels, each trial site and among sites, using the GLM procedure (SPSS v20). The total amount of variation was divided into the following sources of variation: provenance, site, year, and the interaction between them. All effects were considered random, except for the trial location, which was considered fixed.

The following mixed model was applied:

$$Z_{ijkln} = \mu + P_i + S_l + B_j + Y_k + PS_{il} + PY_{ik} + SY_{lk} + e_{ijkln} \quad (2)$$

where: Z_{ijkln} = the trait (wood characters, drought parameters), μ = the overall mean, P_i, S_l, B_j, Y_k, PS_{il}, PY_{ik}, SY_{lk}, and e_{ijkln} are the effect due to the ith provenance, lth site, jth repetition (block), kth year, interaction due to ith provenance and lth site, interaction due to ith provenance and kth year, interaction due to lth site and kth year, and random error associated with the ijklnth trees.

In order to investigate to what extent the local adaptation to climate conditions of origin location influence traits variation, Pearson correlations based on provenance means were computed between the wood characters, the drought parameters, and the geographical coordinates of the provenances' origin for each trial site.

Relationships between the wood characters and the climatic variable of trial sites were investigated by regression analysis. The growth response functions were developed to assess the impact of climate at trial sites on provenances radial growth. The quadratic models based on both temperature and precipitation were used to develop growth response functions, considering them more suitable [34,40,41,44]. We used seven temperature variables and 12 precipitation variables, and the best models were chosen based on the R^2 coefficient (SPSS program, stepwise selection method).

3. Results

3.1. Identification of Drought Years

Large variation in mean annual temperature and annual precipitation amount were recorded in each trial site (Figure 2). Based on SPI values we have identified the moderate, severe, and extreme drought years in each trial site in the period 1989–2018 (Figure 3). Ten years with extreme and severe drought during the analyzed period have been identified, in all trial sites. Most of the extreme and severe drought events occurred after the year 2000. Additionally, during this period, the most consecutive drought years were recorded, such as 2002–2003, 2011–2012, and 2013–2015.

The number of extreme drought years during the analyzed period (1997–2018) have varied among sites and ranged between three at Moinesti and Domnesti trials to five at the Strambu Baiut trial. However, 2000, 2002, and 2011 were the common extreme drought years in all testing sites.

The most significant drought event occurred in 2000 when the highest number of months with severe and extreme drought (nine at Moinesti, seven at Bucova, five at Sacele, four at Stambu Baiut, and three at Domnesti) has been recorded (Table 3). Among all extreme droughts, the 2000 drought had the longest duration in almost all sites. Furthermore, the drought overlapped with the growing season, in four of the five trial sites. The 2011 drought was characterized by the highest intensity and generally by two peaks, while the 2002 drought by lower duration and intensity (five to two months with extreme and severe drought).

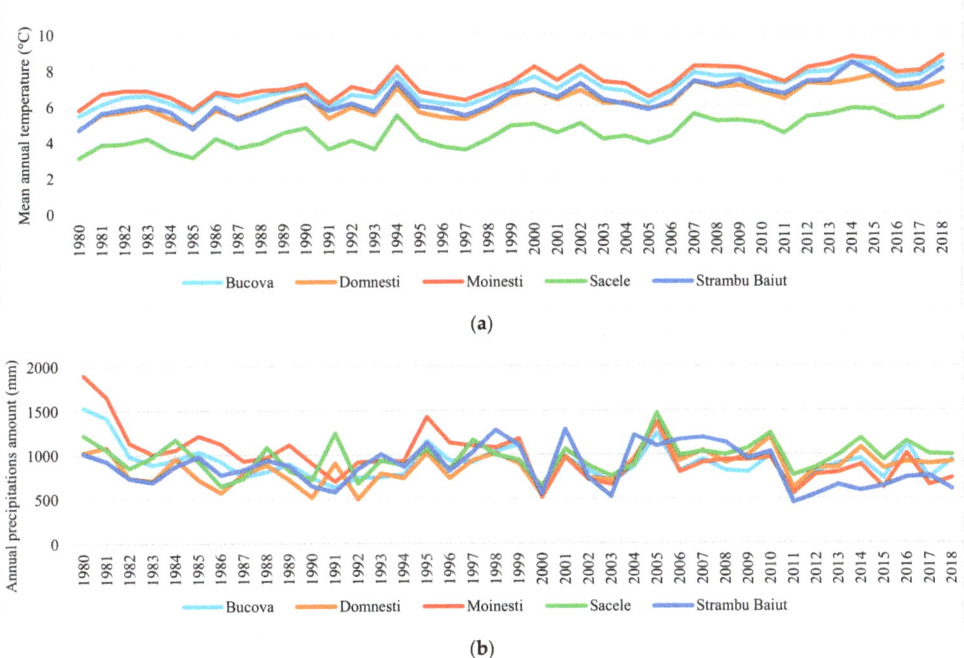

Figure 2. Variation of the mean annual temperature (**a**) and annual precipitation amount (**b**) in trial sites.

3.2. The Effect of Provenance, Site, and Year on Radial Growth and Wood Characteristics

The analysis of variance for each trial site and analyzed period was presented in Table 4. Results have highlighted that both provenance and year effects were significant for studied traits in all trial sites. Provenance x year interaction was also significant in three testing sites. Multifactorial analysis of variance across sites highlighted significant provenance, year, and also site effects. Provenance x site and site x year interactions were very significant also (Table 5).

Results indicate that, during the analyzed period, the studied characters have varied significantly among sites (Figure 4). The highest values of average on experiment for RW were obtained at Strambu Baiut trial (3.8 mm) followed by Sacele (3.7 mm) and Domnesti (3.7 mm) trials. The lowest values were recorded at Moinesti trial (3.4 mm). Regarding the LWP and WD, the highest value of average on experiment was obtained at Domnesti trial (47% LWP and 0.36 g/cm^3 WD). The lowest values for LWP (41%) and WD (0.34 g/cm^3) have been recorded at Sacele and Bucova, respectively.

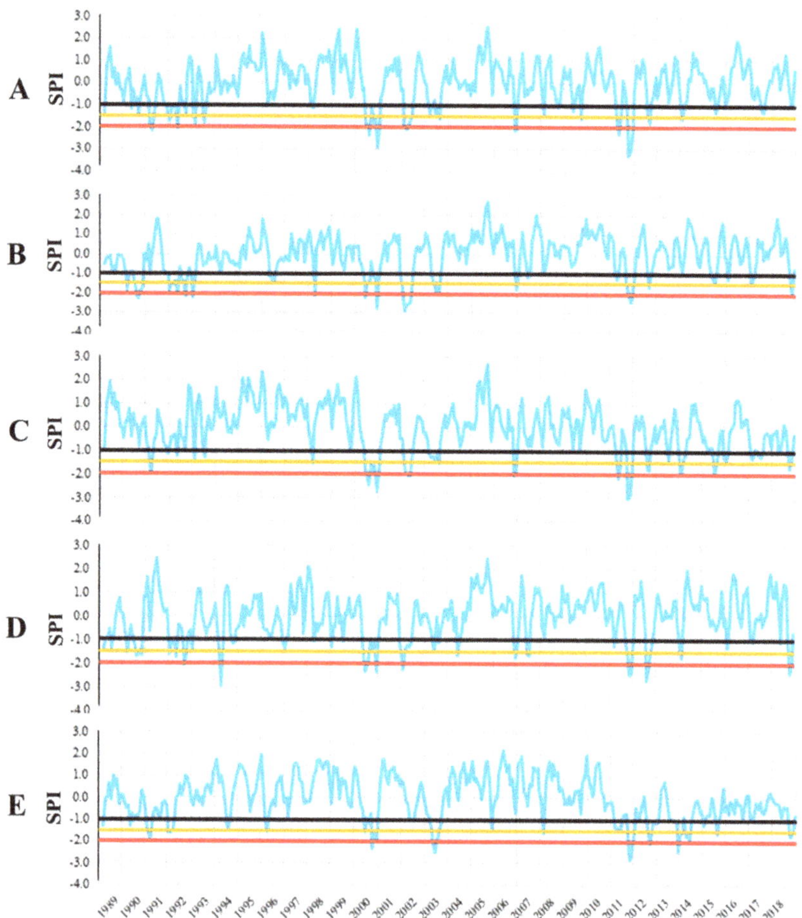

Figure 3. Standardized precipitation index (SPI) for the trial sites. Trials are Bucova (**A**), Domnesti (**B**), Moinesti (**C**), Sacele (**D**), and Strambu Baiut (**E**).

Table 3. The months when occurred extreme and severe droughts during 2000, 2002, 2003, and 2011.

Trial	2000			2002/2003 (1)			2011		
	Extreme Drought	Severe Drought	Total Months	Extreme Drought	Severe Drought	Total Months	Extreme Drought	Severe Drought	Total Months
Bucova	I, II, VIII, XII	VI, VII, X	7	III, IV	II, V	4	V, X, XI, XII	IV	5
Domnesti	VI, XII	VII	3	II, III, IV, V	I	5	X, XI, XII	-	3
Moinesti	I, VII, VIII, XII	II, VI, IX, X, XI	9	IV, V	II, III	4	V, XI	VI, XII	4
Sacele	VI, VII, XII	VIII, X	5	II	I	2	XI, XII	X	3
Strambu Baiut	X, XII	VI, XI	4	V, VI, VII	VIII	4	X, XI, XII	-	3

(1) Explanatory note: the year 2003 was extremely dry at Strambu Baiut only.

Table 4. Analysis of variance of wood traits for the period 1997–2018.

Trial	Source of Variation	DF	Variance (s^2)				
			RW	LW	EW	LWP	WD
SACELE	Provenance (P)	59	8750.95 ***	2054.92 ***	5973.12 ***	0.071 ***	0.002 ***
	Repetition (B)	2	0.010	0.079	0.115	0.098	0.001
	Year (Y)	21	1.880 ***	0.289 ***	1.422 ***	0.290 ***	-
	Interaction P × Y	1239	0.038 ***	0.011	0.018 *	0.004	-
	Error	2640	0.032	0.010	0.017	0.004	0.001
DOMNESTI	Provenance (P)	42	5568.98 ***	1543.09 ***	2011.95 ***	0.015 ***	0.003 **
	Repetition (B)	2	0.003	0.036	0.035	0.036	0.009
	Year (Y)	21	1.868 ***	0.497 ***	0.623 ***	0.088 ***	-
	Interaction P × Y	882	0.023 ***	0.007 ***	0.007 ***	0.001	-
	Error	2838	0.015	0.005	0.006	0.002	0.001
BUCOVA	Provenance (P)	42	4107.42 ***	1495.19 ***	3182.79 ***	0.072 ***	0.001
	Repetition (B)	2	0.002	0.234	0.265	0.244	0.001
	Year (Y)	21	1.080 ***	0.387 ***	0.568 ***	0.201 ***	-
	Interaction P × Y	882	0.024	0.008	0.012	0.003	-
	Error	2838	0.023	0.009	0.011	0.004	0.001
STRAMBU BAIUT	Provenance (P)	42	3543.39 ***	1139.49 ***	1930.62 ***	0.036 ***	0.005
	Repetition (B)	2	0.003	0.136	0.100	0.125	0.006
	Year (Y)	21	1.426 ***	0.467 ***	0.708 ***	0.213 ***	-
	Interaction P × Y	882	0.020 ***	0.008	0.011 **	0.004	-
	Error	2838	0.014	0.007	0.010	0.005	0.004
MOINESTI	Provenance (P)	42	5347 ***	1137 ***	2497 ***	0.023 ***	0.002 *
	Repetition (B)	2	0.006	0.304	0.329	0.289	0.001
	Year (Y)	21	1.241 ***	0.305 ***	0.617 ***	0.133 ***	-
	Interaction P × Y	882	0.013	0.005	0.009	0.003	-
	Error	2838	0.015	0.006	0.010	0.004	0.001

The level of significance is represented as follows: * $p < 0.05$; ** $p < 0.01$; *** $p < 0.001$.

Table 5. Multifactorial analysis of variance of wood traits for the period 1997–2018.

Source of Variation	DF	Variance (s^2)				
		RW	LW	EW	LWP	WD
Provenance (P)	42	5894.39 ***	1476.69 ***	2743.90 ***	0.032 ***	0.003 **
Site (S)	4	62,418.47 ***	44,090.87 ***	30,738.85 ***	1.616 ***	0.009 ***
Year (Y)	21	5.405 ***	1.369 ***	2.860 ***	0.471 ***	-
Interaction P × S	168	50,470.00 ***	14,551.71 ***	27,404.64 ***	0.042 ***	0.001
Interaction P × Y	882	0.024	0.008	0.011	0.004	-
Interaction S × Y	84	0.396 ***	0.118 ***	0.153 ***	0.056 ***	-
Error	13,200	0.023	0.009	0.013	0.005	0.001

The level of significance is represented as follows: ** $p < 0.01$; *** $p < 0.001$.

In all trials, it can be seen a strong relationship between the RW variation and severe and extreme drought years. The tree-ring pattern of provenances showed a strong increment drop in those years (Figure 4). The descriptive statistics of silver dendrochronology in each trial were presented in Table A2.

The average radial growth has varied between 4.83 mm (provenance 24-Devin at Sacele trial) to 2.91 mm (provenance 25-Kitilovo at Moinesti trial), latewood percentage between 53% (provenance 45-Le Joux at Bucova trial) to 33% (provenance 37-Liezen at Sacele trial), whereas wood density has varied from 0.45 g/cm3 (provenance 45-Le Joux at Strambu Baiut trial) to 0.31 g/cm3 (provenance 59-Banska Bystrica at Strambu Baiut trial) (Figure A1). However, despite this high variability among sites, there were some provenances that have obtained RW values over the average of the experiment in all trial sites: 63-Zarovice, 54-Strambu Baiut, 16-Toplita, 55-Valea Iadului, 21-Azuga, 26-St.

Dimitrov, 51-Gura Putnei, 48-Pangarati, 47-Moinesti, and 14-Asau. The ranking by LWP has shown that the highest spatial stability had the provenances 43-Greseuss, 26-St. Dimitrov, 29-Vallombrosa, 45-Le Joux, 41-Enzklosterle, 42-Sulzburg, and 44-Lepilat.

3.3. Genetic Variation in Drought Response

The analysis of variance for all extreme drought years, taken together, revealed significant variation in drought response among silver fir provenances in all trial sites, except the Bucova trial (Table 6). Additionally, significant differences were obtained for the year's effect and provenance x year interaction. The highest variation among provenances was obtained for the resilience to drought, in four of the five trials. Significant differences for resistance capacity were obtained at Sacele and Stambu Baiut, while for recovery only at Domnesti.

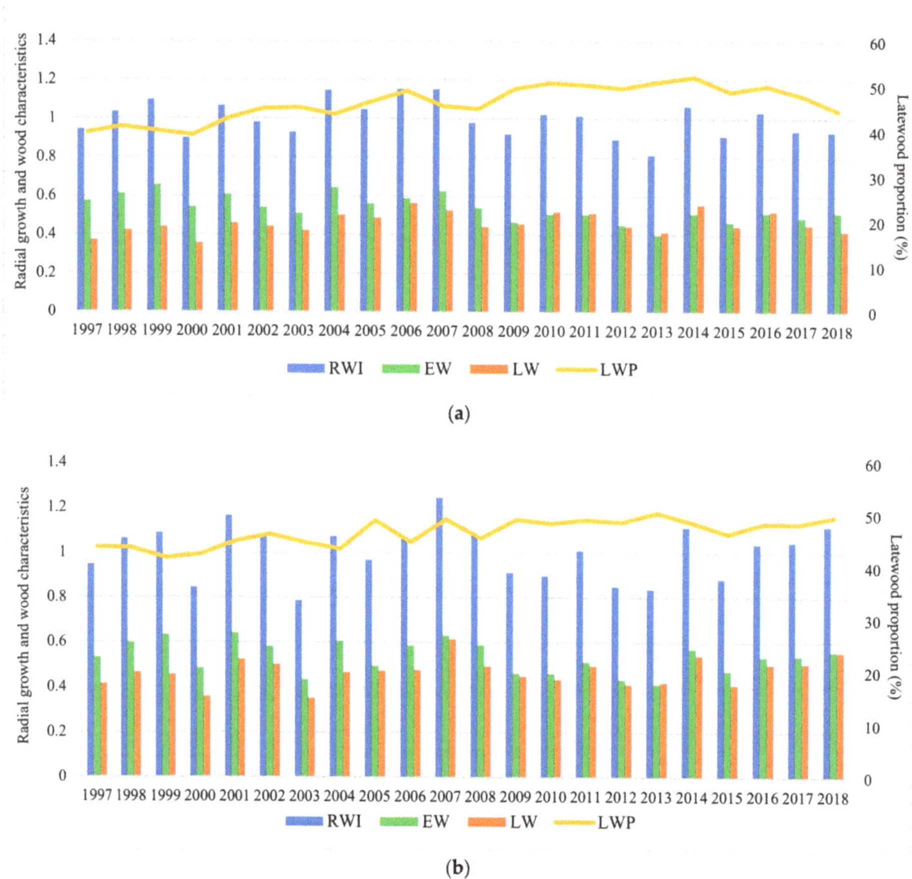

Figure 4. *Cont.*

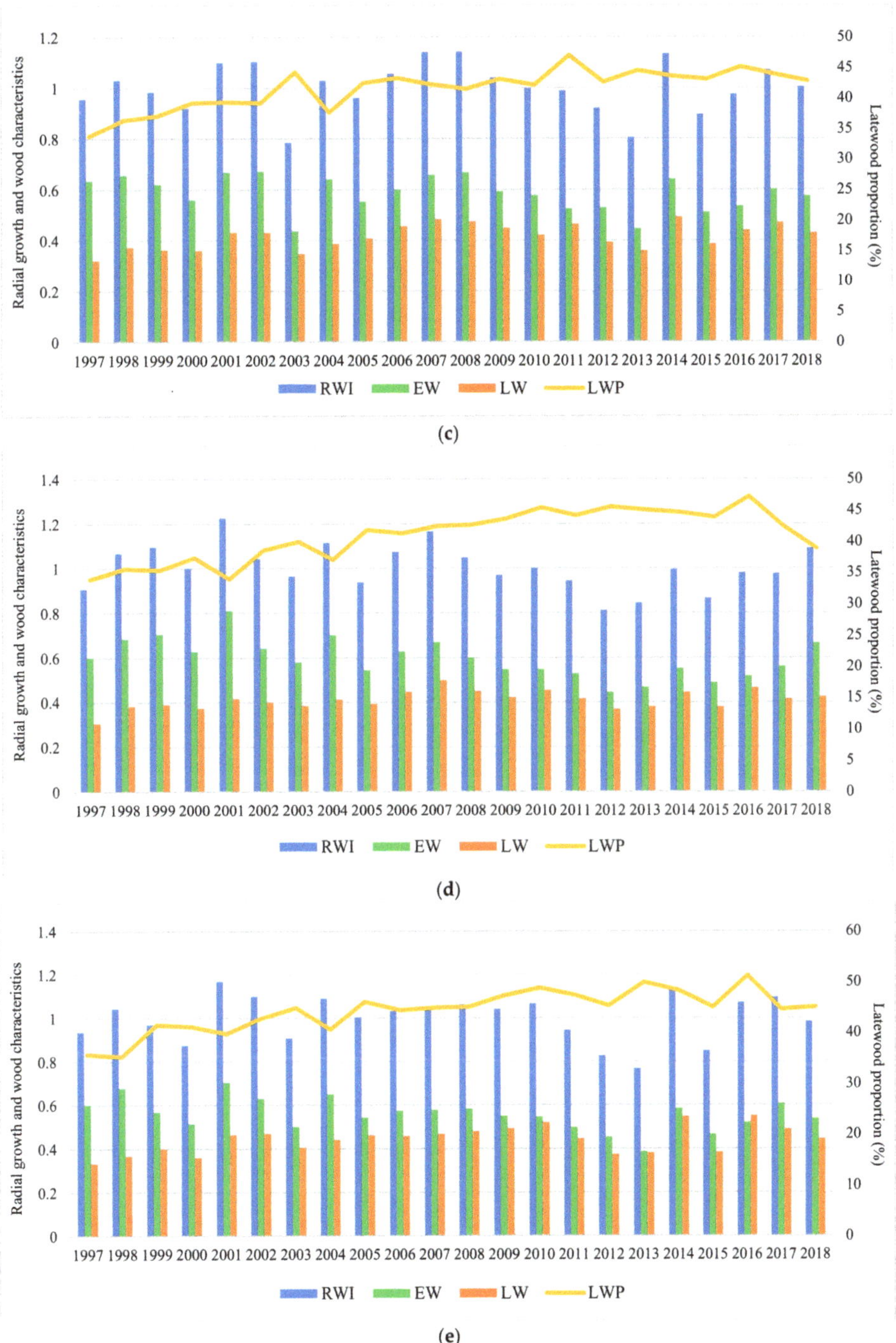

Figure 4. Variation of the radial growth and wood characteristics in trial sites during the analyzed period. Trials are Bucova (**a**), Domnesti (**b**), Moinesti (**c**), Sacele (**d**), and Strambu Baiut (**e**).

Table 6. Analysis of variance for drought parameters of silver fir provenances in all extreme drought years and each trial site.

Trial	Source of Variation	DF	Variance (s^2)			
			Res	Rec	Rsl	rRsl
SACELE	Provenance (P)	59	0.119 ***	0.100	0.138 ***	0.049
	Repetition (B)	2	0.288	0.622	0.602	0.209
	Extreme drought year (DrY)	3	2.418 ***	4.893 ***	4.272 ***	2.863 ***
	Interaction P x DrY	177	0.078	0.093	0.110 **	0.050
	Error	2436	0.068	0.103	0.079	0.052
DOMNESTI	Provenance (P)	42	0.038	0.108*	0.087 **	0.033
	Repetition (B)	2	0.260	0.154	0.110	0.082
	Extreme drought year (DrY)	2	6.318 ***	16.716 ***	0.520 ***	9.681 ***
	Interaction P x DrY	86	0.080 ***	0.119 **	0.066 *	0.061 ***
	Error	1368	0.049	0.081	0.052	0.034
BUCOVA	Provenance (P)	42	0.074	0.121	0.071	0.049
	Repetition (B)	2	0.304	0.858	0.034	0.148
	Extreme drought year (DrY)	3	3.873 ***	8.484 ***	0.307 ***	4.845 ***
	Interaction P x DrY	126	0.099 *	0.162	0.089 *	0.075
	Error	1856	0.081	0.185	0.073	0.063
STRAMBU BAIUT	Provenance (P)	42	0.081 *	0.125	0.081 ***	0.063
	Repetition (B)	2	0.330	0.168	0.342	0.108
	Extreme drought year (DrY)	4	0.911 ***	9.110 ***	3.391 ***	3.856 ***
	Interaction P x DrY	168	0.052	0.102	0.037	0.050
	Error	2055	0.061	0.098	0.042	0.054
MOINESTI	Provenance (P)	42	0.029	0.048	0.065 **	0.027
	Repetition (B)	2	0.069	0.111	0.291	0.076
	Extreme drought year (DrY)	2	4.277 ***	6.144 ***	0.950 ***	6.171 ***
	Interaction P x DrY	86	0.044	0.065 **	0.035	0.048 **
	Error	1368	0.035	0.043	0.039	0.030

The level of significance is represented as follows: * $p < 0.05$; ** $p < 0.01$; *** $p < 0.001$.

Considering only the common extreme drought years in all testing sites 2000, 2002 (2003 at Strambu Baiut), and 2011, significant differences were found among drought parameters of silver fir provenances (Table 7). The provenance-specific drought response depended on the trial site and drought year. Thus, significant differences for all parameters and all extreme drought years were obtained at Domnesti and Moinesti trials. The highest differences in drought response were found in the year 2000, in all trial sites. Additionally, the 2011 drought caused a significant genetic variation in the drought response of silver fir provenances.

The ranking of silver fir provenances by drought parameters in the year 2000, as the most significant drought year, and in all sites, revealed a certain variation pattern (Figure 5). Thus, the provenances ranking by resistance, recovery, and resilience, taken together, have highlighted a best performing group placed at the top ranks in almost all sites. This group include the following silver fir provenances: 23-Rakitovo, 30-Paularo, 7-Vadul Dobri, 53-Botiza, 55-Valea Iadului, and 63-Zarovice. In terms of resistance and resilience, the most valuable provenances were 12-Naruja I, 25-Kitilovo, 33-Abeti Soprani, 43-Greseuss, and 45- Le Joux. Regarding the resistance and recovery, the most valuable provenances were 51-Gura Putnei, 54-Strambu Baiut, and 56-Ilisoara Mures, while regarding the recovery and resilience the provenance 6-Bucium obtained good results. Additionally, there are provenances that obtained a good response and high spatial stability for only one drought parameter. For instance, 26- St. Dimitrov revealed high resistance capacity; 21-Azuga, 50-Malini, 52-Solca, and 59-Banska Bystrica revealed high recovery; while 22-Vallombrosa, 41-Enzklosterle, 44-Lepilat, 47-Moinesti, and 4-Avrig showed high resilience (Figure 5).

Making the ranking of drought parameters for all extreme drought years, the highest values of resistance were observed at Moinesti and Bucova trials (the drought-prone environments), while for recovery and resilience at the Strambu Baiut trial.

Table 7. ANOVA of drought parameters of silver fir provenances for the common extreme drought years during the analyzed period.

Trial	Drought Parameters	s^2			
		2000	2002/2003 [(1)]	2011	2012
SACELE	Resistance	0.143 *	0.082	0.066 *	0.064
	Recovery	0.123	0.102	0.082	0.074
	Resilience	0.207 ***	0.085	0.078 *	0.098 *
	Rel. resilience	0.071	0.044	0.039	0.044
DOMNESTI	Resistance	0.060	0.072 **	0.067 **	-
	Recovery	0.272 *	0.023	0.050 *	-
	Resilience	0.109 *	0.048 *	0.063 *	-
	Rel. resilience	0.090 **	0.022	0.042 *	-
BUCOVA	Resistance	0.203 **	0.049	0.046	-
	Recovery	0.251	0.217	0.059	-
	Resilience	0.159 *	0.072	0.062	-
	Rel. resilience	0.084	0.058	0.051	-
STRAMBU BAIUT	Resistance	0.085	0.035	0.053	0.033
	Recovery	0.174 *	0.078	0.060	0.081
	Resilience	0.054	0.043	0.032	0.044
	Rel. resilience	0.090	0.025	0.042	0.046
MOINESTI	Resistance	0.027	0.061 *	0.028	
	Recovery	0.080 **	0.051	0.049 *	
	Resilience	0.041	0.049	0.047 *	
	Rel. resilience	0.050 *	0.039	0.035	

Explanatory note: [(1)] the year 2003 was extremely dry at Strambu Baiut only. The level of significance is represented as follows: * $p < 0.05$; ** $p < 0.01$; *** $p < 0.001$.

3.4. Phenotypic Correlations

The correlations between wood characters and WD were negative in all trial sites, although statistically significant correlations were obtained in few trials only (Table A3).

Additionally, the correlations between wood characters and geographic coordinates of the provenances were few and indicate low pattern of local adaptation. The most significant correlations were found with LONG and, generally, they have been negative in almost all trials, except for RW and EW at Bucova trial, and for LW and LWP at Sacele trial, where they were positive. Statistically, correlations with LAT and ALT of seed origin were few and only at Moinesti trials and Strambu Baiut, respectively.

The correlations between drought parameters and wood characters were positive, and the most were obtained between RW and resilience (Table 8). Correlations between drought parameters and wood density were non-significant.

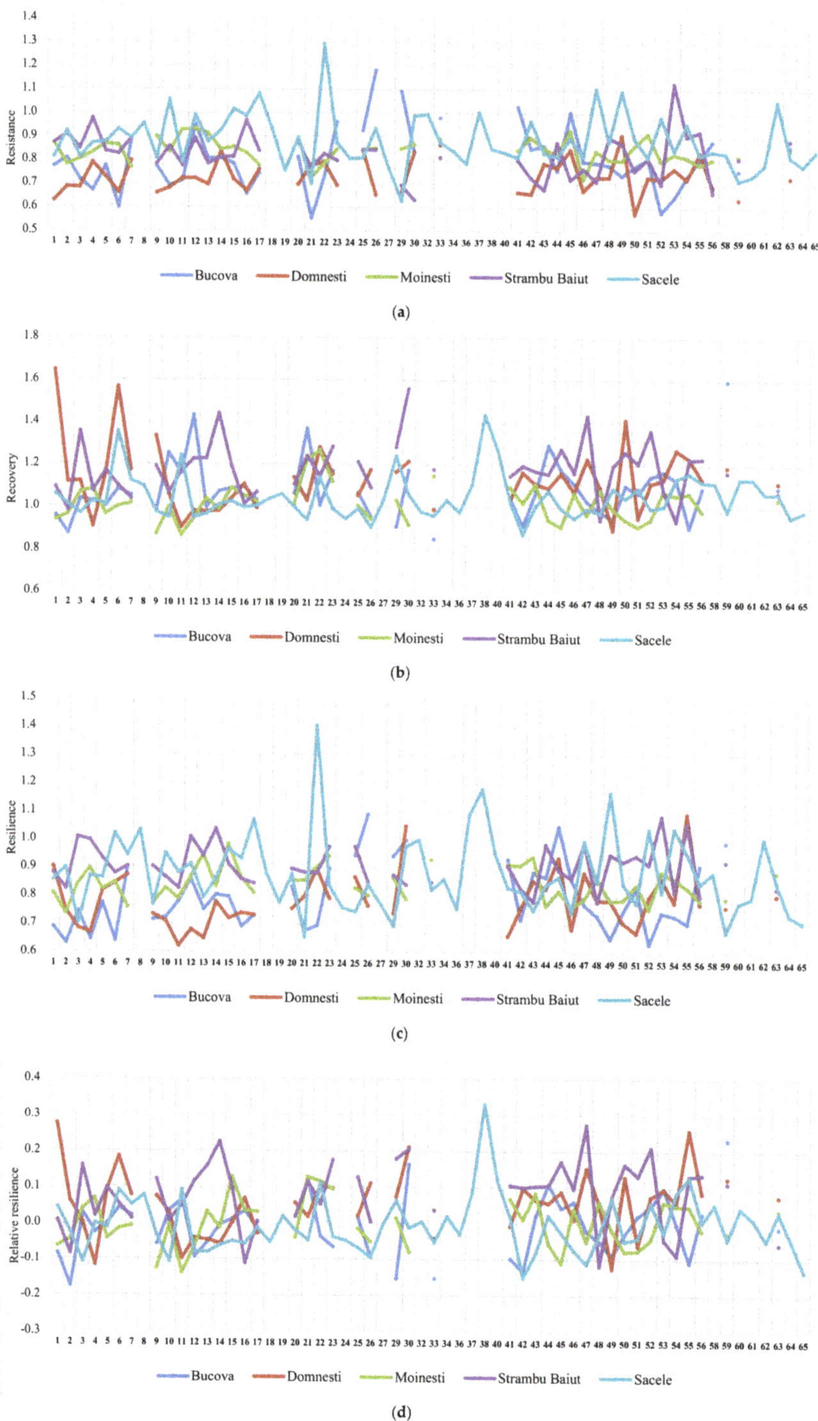

Figure 5. Variation of the drought parameters of silver fir provenances calculated for 2000 drought year in each trial site. Drought parameters are resistance (**a**), recovery (**b**), resilience (**c**), and relative resilience (**d**).

Table 8. Phenotypic correlations between wood characters and drought parameters of silver fir provenances.

Trial	Trait	RW	LW	EW	LWP	WD
SACELE	Resistance	0.211	0.301 *	0.097	0.188	0.262
	Recovery	0.077	0.068	0.063	−0.014	0.065
	Resilience	0.261 *	0.320 *	0.144	0.139	0.145
	Rel. resilience	0.129	0.091	0.100	−0.053	0.096
DOMNESTI	Resistance	0.301 *	0.317 *	0.255	0.121	0.137
	Recovery	0.244	0.206	0.207	0.014	−0.101
	Resilience	0.389 **	0.362 *	0.337 *	0.090	0.052
	Rel. resilience	0.295 *	0.214	0.281	−0.038	−0.086
BUCOVA	Resistance	0.094	0.513 ***	−0.242	0.465 **	0.094
	Recovery	0.131	0.170	0.030	0.033	−0.167
	Resilience	0.316 *	0.698 ***	−0.115	0.473 **	−0.123
	Rel. resilience	0.255	0.162	0.181	−0.058	−0.243
STRAMBU BAIUT	Resistance	−0.170	−0.150	−0.122	−0.084	0.265
	Recovery	0.225	0.252	0.106	0.126	0.027
	Resilience	0.130	0.223	−0.004	0.119	0.249
	Rel. resilience	0.308 *	0.402 **	0.106	0.231	0.043
MOINESTI	Resistance	0.336 *	0.240	0.329 *	−0.084	0.106
	Recovery	0.126	−0.095	0.249	−0.365 *	0.094
	Resilience	0.298 *	0.109	0.363 *	−0.263	0.114
	Rel. resilience	0.110	−0.055	0.198	−0.262	0.068

The level of significance is represented as follows: * $p < 0.05$; ** $p < 0.01$; *** $p < 0.001$.

3.5. Growth Response Functions

The influence of climate on the RW and LWP in each trial site has been investigated using quadratic regressions, and the best models obtained were presented in Tables 9 and 10. The main climatic drivers explaining the radial growth of silver fir were MT_{VEG}, $MP_{OCT\text{-}MAR}$, and $MP_{JAN\text{-}MAR}$. The growth–climate relationship was moderate, R^2 ranging between 0.37 and 0.50, indicating that a substantial amount of the radial growth variation can be explained by these climatic factors. Partial R^2 indicates that silver fir is less sensitive to precipitation than to temperature. The influence of temperature during the growing season accounted for 29% and 48% of the total variation of RW. For latewood percentage, the response models founded were modest (R^2 varied between 0.09 and 0.15). MAT, MT_{VEG}, MAP, and MT_{VEG} are the main climatic factors that influence LWP of silver fir (Table 10). The temperature variables accounted for the greater part of LWP variation.

Table 9. Climatic response models for radial growth of silver provenances. MT_{VEG}—the mean temperature of the growing season, $MP_{OCT\text{-}MAR}$—the mean precipitation from October of the previous year to March of the current year, $MP_{JAN\text{-}MAR}$—the mean precipitation from January to March of the current year.

Trial	Growth Response Model	Signif.	R^2	Partial R^2	
				Temp.	Precip.
Bucova	10,632.399—41.943 MT^2_{VEG} + 0.154 $MP^2_{JAN\text{-}MAR}$	***	0.458	0.412	0.043
Domnesti	15,399.754—55.908 MT^2_{VEG}—46.317 $MP_{OCT\text{-}MAR}$	***	0.496	0.475	0.154
Moinesti	8780.790—29.468 MT^2_{VEG} + 0.060 $MP^2_{JAN\text{-}MAR}$	***	0.372	0.292	0.017
Sacele	11,985.713—49.258 MT^2_{VEG}—41.649 $MP_{OCT\text{-}MAR}$	***	0.415	0.326	0.093
S. Baiut	9154.895—36.635 MT^2_{VEG} + 17.794 $MP_{OCT\text{-}MAR}$	***	0.503	0.326	0.062

The level of significance is represented as follows: *** $p < 0.001$.

Table 10. Climatic response models for late wood percentage of silver provenances. MAT—the mean annual temperature, MT_{VEG}—the mean temperature of the growing season, MAP—the mean annual precipitation amount, MT_{VEG}—the mean precipitation, during the growing season.

Trial	Growth Response Model	Signif.	R^2	Partial R^2	
				Temp.	Precip.
Bucova	$32.271 + 0.212\ MAT^2 + 0.027\ MP_{VEG}$	***	0.091	0.092	0.008
Domnesti	$26.927 + 0.087\ MT^2_{VEG} + 0.052\ MP_{VEG}$	***	0.222	0.206	0.111
Moinesti	$30.329 + 0.069\ MT^2_{VEG} - 0.002\ MAP$	***	0.136	0.066	0.006
Sacele	$16.864 + 0.147\ MT^2_{VEG} + 0.057\ MP_{VEG}$	***	0.154	0.152	0.052
S. Baiut	$32.411 + 0.081\ MT^2_{VEG} - 0.003\ MAP$	***	0.150	0.058	0.011

The level of significance is represented as follows: * $p < 0.05$; ** $p < 0.01$; *** $p < 0.001$.

4. Discussion

In this study, we have analyzed the radial growth, wood characteristics, and drought response of 60 silver fir provenances tested in five long-term trials established in different geographic regions and climatic conditions across Romania. Considerable differences in radial growth and wood characteristics among silver fir provenances were found. The influence of the local site conditions of each experiment and interaction of provenance with site and year were also significant in three testing sites, suggesting that the stability over time of silver fir radial growth and wood characteristics depends on site conditions.

The analysis of climate data during the period 1997–2018 showed large variations in terms of temperature and precipitation at site and time scale too. Results revealed a warming trend and a decreasing in the sum of annual precipitation during the analyzed period. The De Martonne aridity index, calculated for each trial site at the entire year level and entire period, had a value between 50 to 67 indicating that climatic conditions of the trial sites fall into the wet category. However, the values of De Martonne aridity index calculated for the growing season ranged between 22 to 27 that classifies the sites climate into silvostepic.

Extreme drought events have increased their frequency during the last two decades and among all extreme droughts, the most significant in duration and intensity have been the 2000, 2002, and 2011 droughts, in all trial sites. Abrupt growth changes were detected in tree ring chronologies related to these drought events. The losses in RW caused by drought have varied depending on the site, drought year, and provenance. The highest losses in RW have occurred in 2011, characterized by the highest drought intensity and two peaks in all sites, ranging between 18% at Moinesti to 27% at the Sacele trial. The year 2011 exerted the highest water stress on vegetation over the half-century in many regions of Europe [1].

Results revealed significant genetic variation in drought response among tested provenances. The drought reaction of silver fir provenances varied significantly depending on the extreme drought year and site conditions. The highest response by almost all drought parameters was found in the year 2000 when a consistent pattern in provenances drought response across the sites was observed. Thus, the provenances ranking by resistance, recovery, and resilience revealed several provenances placed in the top ranks in almost all sites. This group include provenances from Bulgaria, Italy, Romania, and Czech Republic. Additionally, some provenances had a good tolerance and high spatial stability for two or only one drought parameter. The remarkable performance combining superior growth with high tolerance to drought events had the provenances 63-Zarovice, 53-Botiza, 54-Strambu Baiut, 55-Valea Iadului from core distribution range, 47-Moinesti, 50-Malini, 51-Gura Putnei from eastern edge, and 7-Vadul Dobrii and 26-St. Dimitrov from southeastern edge. The highest values of resistance to drought were observed at Moinesti and Bucova trials, in the drought-prone environments.

It is notable that our results highlight higher genetic variation in drought response among silver fir provenances compared to previous studies. Thus, George et al. [57], studying drought sensitivity of ten provenances of silver fir and four Mediterranean fir species in eastern Austria, found both intra- and inter-specific variation to drought.

However, his results indicated that genetic variation in drought response among silver fir provenances is more reduced than among *Abies* species. Additionally, Sagnard et al. [69], analyzing growth traits and drought resistance of silver fir seedlings in France, found a low variation among provenances in drought response, while Sindelar and Beran [70] found little genetic differentiation among silver fir provenances for drought resilience. The high genetic differentiation of drought response revealed in our study can be explained by the broad geographic amplitude of the provenances tested in these trials. This geographic area comprises two putative glacial refugia in southern Europe where silver fir survived during the last glaciation: in the Appenines and in the Balkan Peninsula of southeastern Europe. The remarkable growth performances and drought resilience of some provenances from the eastern distribution range (48, 52, 53) and southeastern (7,23, 25, 26) and southern edge (33) indicate that these populations, most of them peripheral, possess high adaptive potential, most likely as a consequence of the selection pressure.

Forest species hold different adaptive capacity to withstand the impacts of drought according to their ecophysiological characteristics and evolutionary adaptation. For instance, Arend et al. [54] showed that *Quercus robur* needs a prolonged recovery phase after the drought, indicating a lower fitness for drought tolerance. Forner et al. [71] found that *Pinus nigra* was able to recover after the extreme event while *Quercus faginea* was not. Additionally, Gazol et al. [72] revealed that *Pinus ponderosa* and *Pseudotsuga menziesii* displayed greater plasticity in resistance to a drought that the two more frequently oaks (*Quercus alba* and *Quercus stellate*) in North America. Our study has demonstrated that the resilience and resistance to drought varied significantly among silver fir provenances.

Silver fir is a species that highlights low genetic variability among populations, but high genetic diversity within populations, even in marginal populations [33,73], which could be a benefit for adapting to climate warming. Heer et al. [74] analyzed dendroecological and genetic data of surviving silver fir trees to the drought episodes of the 1970s and 1980s that caused forest dieback in Central Europe and found fifteen genes associated with the dendrophenotypes, including genes linked to photosynthesis and drought stress. Therefore, besides the so-called "avoidance strategy" of silver fir through bud cessation at the end of July and deep root system [75], there is a genetic basis of adaptation to drought.

The correlations between drought parameters and wood characters of silver fir provenances are positive. The most significant correlations have been obtained between radial growth and resilience. Our results are in accordance with findings from Eilmann et al. [56], while other studies have shown that drought-tolerant provenances were less productive [76]. Correlations between drought parameters and wood density were non-significant, indicating that wood density cannot be used as an indicator of drought sensitivity. Results can be explained by lower genetic variation of WD compared to RW among silver fir provenances at this age. Similar results for silver fir have been obtained by George at al. [57], while for other species like *Picea abies* and *Pseudotsuga menziesii*, correlations between wood density and trees sensitivity to drought have been found to be moderate to strong negative [55,77].

The wood characteristics varied, especially along the longitude, which represents an important gradient of increasing aridity eastward within Romania. In the Bucova trial, located in Banat Mountains with a warmer climate and an increasing deficit in rainfall, the best-performing provenances come from Eastern Carpathians.

The growth response functions revealed that the climatic variables of the trial sites were the significant drivers of the growth performance of the silver fir provenances. The main climatic variables explaining the radial growth of silver fir were MT_{VEG}, $MP_{OCT\text{-}MAR}$, and $MP_{JAN\text{-}MAR}$, while for latewood percentage were MAT, MT_{VEG}, MAP, and MT_{VEG}. The negative correlations between RW and temperature during the growing season and positive correlations with precipitation suggest that warming and water deficit could have a negative impact on silver fir growth in climatic marginal sites, the more so because precipitation patterns are projected to change more than temperature in near future.

5. Conclusions

Even though silver fir experienced the most stressful droughts over the last two decades, it has revealed a plastic response to drought. Results revealed significant genetic variation among silver fir provenances by resistance, recovery, and resilience to drought. The provenance-specific response depended on the climatic conditions of the planting site and drought year. However, there are some local and foreign provenances that combine high radial growths and high drought tolerance.

Silvicultural practices and forest adaptive management should increase and maintain a high genetic diversity and resilience within forest stands. One of the adaptive measures could be selection, transfer, and planting of high-productive and drought resilient forest reproductive material in reforestation programs (assisted migration). Assisted migration may support adaptation process and help to conserve and increase genetic diversity, especially at the species distribution edges.

Finally, we argue that silver fir holds a great potential to thrive under warmer and drier conditions at the eastern limit of its distribution, in the southeastern Carpathians.

Author Contributions: Conceptualization: G.M.; methodology: G.M., A.M.A. and M.V.B.; software: G.M., A.M.A., E.S. and M.V.B.; validation: G.M., A.M.A. and M.V.B.; statistical analysis: G.M. and A.M.A.; resources: G.M. and M.V.B.; writing-original draft preparation: G.M.; writing-review and editing: G.M., A.M.A., E.S. and M.V.B.; project administration: G.M. and A.M.A.; funding acquisition: G.M. All authors have read and agreed to the published version of the manuscript.

Funding: This research was funded by Ministry of Research, Innovation and Digitization, grant number PN 19070303, Nucleu Program.

Institutional Review Board Statement: Not applicable.

Informed Consent Statement: Not applicable.

Data Availability Statement: Datasets generated and/or analyzed during the current study are available from the corresponding author on request.

Acknowledgments: We would like to thank the editor and anonymous reviewers for their useful advice that helped to improve the manuscript.

Conflicts of Interest: The authors declare no conflict of interest.

Appendix A

Table A1. List of silver fir provenances tested in comparative trials.

No. Prov.	Provenance	Country	Location within Distribution Range	Longitude E	Latitude N	Altitude (m)
1	Cheia	Romania	Southeastern edge	25°55′	45°25′	950
2	Azuga I	Romania	Southeastern edge	25°35′	45°25′	1100
3	Ghelinta	Romania	Eastern edge	26°20′	45°55′	880
4	Avrig	Romania	Southeastern edge	24°30′	45°40′	660
5	Valea Motilor	Romania	Core	22°45′	46°30′	750
6	Bucium	Romania	Core	23°10′	46°15′	910
7	Vadul Dobri	Romania	Southeastern edge	22°35′	45°40′	1150
8	Tismana	Romania	Southeastern edge	23°00′	45°05′	950
9	Polovragi	Romania	Southeastern edge	23°48′	45°15′	1100
10	Cozia	Romania	Southeastern edge	24°20′	45°20′	1300
11	Gura Teghii	Romania	Eastern edge	26°20′	45°35′	1100
12	Naruja I	Romania	Eastern edge	26°40′	45°40′	800
13	Soveja	Romania	Eastern edge	26°40′	46°00′	750
14	Asau	Romania	Eastern edge	26°25′	46°25′	1050
15	Tusnad	Romania	Eastern edge	25°50′	46°10′	650
16	Toplita	Romania	Eastern edge	25°25′	46°55′	930
17	Garcin	Romania	Southeastern edge	25°45′	45°35′	1000
18	Rasnov	Romania	Southeastern edge	25°32′	45°35′	700
19	Valiug	Romania	Southeastern edge	22°10′	45°12′	600
20	Rusca Montana	Romania	Southeastern edge	22°28′	45°35′	880

Table A1. *Cont.*

No. Prov.	Provenance	Country	Location within Distribution Range	Longitude E	Latitude N	Altitude (m)
21	Azuga II	Romania	Southeastern edge	25°35′	45°25′	1125
22	Toplita II	Romania	Eastern edge	23°25′	46°55′	900
23	Rakitovo	Bulgaria	Southeastern edge	24°05′	41°59′	1550
24	Devin	Bulgaria	Southeastern edge	24°24′	41°42′	1500
25	Kitilovo	Bulgaria	Southeastern edge	26°13′	42°54′	500
26	St. Dimitrov	Bulgaria	Southeastern edge	23°09′	42°15′	1450
27	Raslog	Bulgaria	Southeastern edge	23°40′	42°01′	1600
29	Vallombrosa	Italy	Southern edge	11°33′	43°45′	960
30	Paularo	Italy	Core	13°30′	46°31′	950
32	San Bruno	Italy	Southern edge	16°20′	38°33′	1250
33	Abeti Soprani	Italy	Southern edge	14°20′	41°52′	800
34	Trieben	Austria	Core	14°30′	47°28′	1125
36	Passail	Austria	Core	15°32′	47°13′	800
37	Liezen	Austria	Core	14°15′	47°31′	800
38	Hohe Wand	Austria	Core	16°04′	47°49′	750
40	Todtmoos	Germany	Western edge	8°05′	47°47′	320
41	Enzklosterle	Germany	Western edge	8°30′	48°16′	280
42	Sulzburg	Germany	Western edge	7°43′	47°51′	560
43	Greseuss	France	Western edge	6°09′	48°28′	400
44	Lepilat	France	Western edge	4°00′	44°40′	340
45	Le Joux	France	Western edge	6°15′	46°40′	260
46	Naruja II	Romania	Eastern edge	26°40′	45°40′	750
47	Moinesti	Romania	Eastern edge	26°25′	46°25′	940
48	Pangarati	Romania	Eastern edge	26°10′	46°53′	860
49	Rasca	Romania	Eastern edge	25°14′	47°20′	560
50	Malini	Romania	Eastern edge	25°56′	47°24′	820
51	Gura Putnei	Romania	Eastern edge	25°33′	47°47′	620
52	Solca	Romania	Eastern edge	24°52′	47°40′	480
53	Botiza	Romania	Core	23°05′	47°40′	970
54	Strambu Baiut	Romania	Core	22°55′	47°35′	760
55	Valea Iadului	Romania	Core	22°40′	46°50′	800
56	Ilisoara Mures	Romania	Eastern edge	25°08′	46°55′	1050
58	Brezno Michalova	Slovakia	Core	20°20′	48°40′	700
59	Banska Bystrica	Slovakia	Core	19°15′	48°40′	850
60	Banska Bystrica	Slovakia	Core	19°15′	48°40′	800
61	Lidečko	Czech Republic	Core	18°02′	49°05′	740
62	Vizovice	Czech Republic	Core	17°52′	49°12′	650
63	Zarovice	Czech Republic	Core	17°01′	49°30′	860
64	Deblin	Czech Republic	Core	16°32′	49°18′	740
65	Skarzysko	Poland	Northern edge	20°50′	51°07′	130

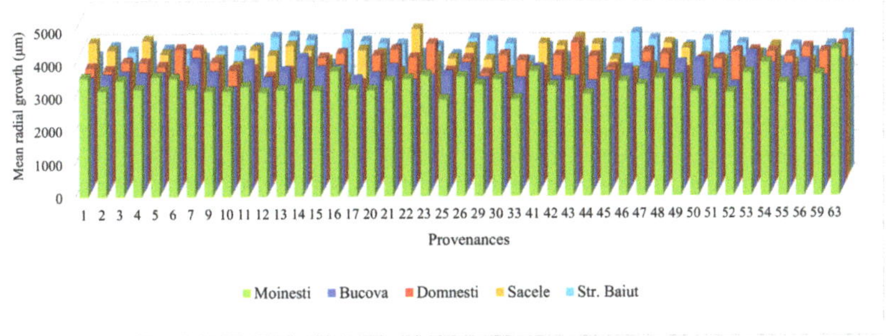

Figure A1. Variation of mean radial growth for period 1997–2018 of silver fir provenances in trial sites.

Table A2. Descriptive statistics of silver fir dendrochronology.

	GLK	RBAR	EPS	SNR	Provenance rbar min	Provenance rbar max
Bucova	0.596	0.659	0.999	989.01	0.702—Prov. 30	0.892—Prov. 17
Domnesti	0.682	0.773	0.999	1700.15	0.752—Prov. 55	0.854—Prov. 16
Moinesti	0.635	0.691	0.999	1096.01	0.607—Prov. 33	0.905—Prov. 55
Sacele	0.612	0.706	0.999	1605.15	0.633—Prov. 22	0.920—Prov. 26
Str. Baiut	0.627	0.689	0.999	1001.36	0.728—Prov. 3	0.918—Prov. 50

Table A3. Phenotypic correlations between the wood characters and geographic coordinates of the origin place of silver fir provenances.

Trial	Trait	LW	EW	LWP	WD	LAT	LONG	ALT
SACELE	RW	0.570 ***	0.870 ***	−0.258 *	0.082	−0.083	−0.172	−0.028
	LW	-	0.098	0.599 ***	0.007	−0.092	0.303 *	−0.123
	EW		-	−0.663 ***	0.090	−0.043	−0.389 **	0.034
	LWP			-	0.059	−0.126	0.522 ***	−0.115
	WD				-	−0.103	0.050	−0.065
DOMNESTI	RW	0.885 ***	0.934 ***	−0.098	−0.189	0.230	−0.128	0.133
	LW	-	0.667 ***	0.310 *	0.011	0.264	−0.189	0.003
	EW		-	−0.424 **	−0.292 *	0.171	−0.096	0.207
	LWP			-	0.179	0.205	−0.029	−0.217
	WD				-	−0.299 *	−0.173	−0.170
BUCOVA	RW	0.495 **	0.801 ***	−0.242	−0.077	0.240	0.292 *	0.166
	LW	-	−0.123	0.709 ***	−0.201	0.103	−0.425 **	−0.176
	EW		-	−0.764 ***	0.051	0.201	0.627 ***	0.300 *
	LWP			-	−0.172	−0.059	−0.723 ***	−0.326 *
	WD				-	−0.240	0.050	−0.031
STRAMBU BAIUT	RW	0.695 ***	0.827 ***	−0.071	−0.304 *	0.081	0.127	0.094
	LW	-	0.172	0.632 ***	−0.137	−0.077	−0.036	−0.115
	EW		-	−0.594 ***	−0.306 *	0.166	0.198	0.230
	LWP			-	0.039	−0.151	−0.224	−0.258
	WD				-	−0.279 *	0.165	0.096
MOINESTI	RW	0.809 ***	0.918 ***	−0.192	−0.362 *	0.486 **	−0.100	0.072
	LW	-	0.508 ***	0.397 **	−0.245	0.266	−0.344 *	−0.005
	EW		-	−0.549 ***	−0.365 *	0.532 ***	0.086	0.113
	LWP			-	0.148	−0.341 *	−0.382 *	−0.036
	WD				-	−0.268	−0.044	−0.099

The level of significance is represented as follows: * $p < 0.05$; ** $p < 0.01$; *** $p < 0.001$.

References

1. Field, C.B.; Barros, T.F.V.; Stocker, Q.D.; Dokken, D.J.; Ebi, K.L.; Mastrandrea, M.D.; Mach, K.J.; Plattner, G.K.; Allen, S.K.; Tignor, M. *IPCC Report on Managing the Risks of Extreme Events and Disasters to Advance Climate Change Adaptation*; Cambridge University Press: Cambridge, UK, 2012; p. 582.
2. Masson-Delmotte, V.; Zhai, P.; Pörtner, H.O.; Roberts, D.; Skea, J.; Shukla, P.R.; Pirani, A.; Moufouma-Okia, W.; Péan, C.; Pidcock, R. *IPCC Special Report on the Impacts of Global Warming of 1.5 °C Above Pre-Industrial Levels and Related Global Greenhouse Gas Emission Pathways, in the Context of Strengthening the Global Response to the Threat of Climate Change, Sustainable Development, and Efforts to Eradicate Poverty*; Intergovernmental Panel on Climate Change, IPCC: Geneva, Switzerland, 2019; p. 26.
3. Mishra, A.K.; Singh, V.P. A review of drought concepts. *J. Hydrol.* **2010**, *391*, 202–216. [CrossRef]
4. Naumann, G.; Spinoni, J.; Vogt, J.V.; Barbosa, P. Assessment of drought damages and their uncertainties in Europe. *Environ. Res. Lett.* **2015**, *10*, 124013. [CrossRef]
5. Allen, C.D.; Macalady, A.K.; Chenchouni, H.; Bachelet, D.; McDowell, N.; Vennetier, M.; Kitzberger, T.; Rigling, A.; Breshears, D.D.; Hogg, E.H.; et al. A global overview of drought and heat-induced tree mortality reveals emerging climate change risks for forests. *For. Ecol. Manag.* **2010**, *259*, 660–684. [CrossRef]

6. Ciais, P.; Reichstein, M.; Viovy, N.; Granier, A.; Ogée, J.; Allard, V.; Aubinet, M.; Buchmann, N.; Bernhofer, C.; Carrara, A.; et al. Europe-wide reduction in primary productivity caused by the heat and drought in 2003. *Nature* **2005**, *437*, 529–533. [CrossRef] [PubMed]
7. Lloret, F.; Escudero, A.; Iriondo, J.M.; Martinez-Vilalta, J.; Valladares, F. Extreme climatic events and vegetation: The role of stabilizing processes. *Glob. Chang. Biol.* **2012**, *18*, 797–805. [CrossRef]
8. Blauhut, V.; Stahl, K.; Stagge, J.H.; Tallaksen, L.M.; de Stefano, L.; Vogt, J. Estimating drought risk across Europe from reported drought impacts, drought indices, and vulnerability factors. *Hydrol. Earth Syst. Sci.* **2016**, *20*, 2779–2800. [CrossRef]
9. Briffa, K.R.; van der Schrier, G.; Jones, P.D. Wet and dry summers in Europe since 1750: Evidence of increasing drought. *Int. J. Clim.* **2009**, *29*, 1894–1905. [CrossRef]
10. Gudmundsson, L.; Seneviratne, S.I. European drought trends. *Proc. Int. Assoc. Hydrol. Sci.* **2015**, *369*, 75–79. [CrossRef]
11. Dumitrescu, A.; Bojariu, R.; Birsan, M.-V.; Marin, L.; Manea, A. Recent climatic changes in Romania from observational data (1961–2013). *Theor. Appl. Clim.* **2014**, *122*, 111–119. [CrossRef]
12. Spinoni, J.; Szalai, S.; Szentimrey, T.; Lakatos, M.; Bihari, Z.; Nagy, A.; Németh, Á.; Kovács, T.; Mihic, D.; Dacic, M.; et al. Climate of the Carpathian Region in the period 1961–2010: Climatologies and trends of 10 variables. *Int. J. Clim.* **2014**, *35*, 1322–1341. [CrossRef]
13. Ionita, M.; Scholz, P.; Chelcea, S. Assessment of droughts in Romania using the Standardized Precipitation Index. *Nat. Hazards* **2016**, *81*, 1483–1498. [CrossRef]
14. Meinshausen, M.; Smit, S.; Calvin, J.H.; Daniel, K.; Kainuma, J.S.; Lamarque, M.L.T.; Matsumoto, J.F.; Montzka, K.; Raper, S.; Riahi, S. The RCP greenhouse gas concentrations and their extensions from 1765 to 2300. *Clim. Chang.* **2011**, *109*, 213–241. [CrossRef]
15. Spinoni, J.; Vogt, J.V.; Naumann, G.; Barbosa, P.; Dosio, A. Will drought events become more frequent and severe in Europe? *Int. J. Clim.* **2018**, *38*, 1718–1736. [CrossRef]
16. Hamrick, J. Response of forest trees to global environmental changes. *For. Ecol. Manag.* **2004**, *197*, 323–335. [CrossRef]
17. Lindner, M. Developing adaptive forest management strategies to cope with climate change. *Tree Physiol.* **2000**, *20*, 299–307. [CrossRef] [PubMed]
18. Bréda, N.; Huc, R.; Granier, A.; Dreyer, E. Temperate forest trees and stands under severe drought: A review of ecophysiological responses, adaptation processes and long-term consequences. *Ann. For. Sci.* **2006**, *63*, 625–644. [CrossRef]
19. Mauri, A.; de Rigo, D.; Caudullo, G. *Abies alba* in Europe: Distribution, habitat, usage and threats. In *European Atlas of Forest Tree Species*; San-Miguel-Ayanz, J., de Rigo, D., Caudullo, G., Houston, D.T., Mauri, A., Eds.; Publication Office of the European Union: Brusels, Belgium, 2016; pp. 48–49.
20. Maiorano, L.; Cheddadi, R.; Zimmermann, N.E.; Pellissier, L.; Petitpierre, B.; Pottier, J.; Laborde, H.; Hurdu, B.I.; Pearman, P.B.; Psomas, A.; et al. Building the niche through time: Using 13,000 years of data to predict the effects of climate change on three tree species in Europe. *Glob. Ecol. Biogeogr.* **2012**, *22*, 302–317. [CrossRef]
21. Dyderski, M.K.; Paz, S.; Frelich, L.E.; Jagodzinski, A.M. How much does climate change threaten European forest tree species distributions? *Glog. Chang. Biol.* **2018**, *24*, 1150–1163. [CrossRef]
22. Vitasse, Y.; Bottero, A.; Rebetez, M.; Conedera, M.; Augustin, S.; Brang, P.; Tinner, W. What is the potential of silver fir to thrive under warmer and drier climate? *Eur. J. For. Res.* **2019**, *138*, 547–560. [CrossRef]
23. Tinner, W.; Colombaroli, D.; Heiri, O.; Henne, P.D.; Steinacher, M.; Untenecker, J.; Vescovi, E.; Allen, J.R.; Carraro, G.; Conedera, M. The past ecology of Abies alba provides new perspectives on future responses of silver fir forests to global warming. *Ecol. Monogr.* **2013**, *83*, 419–439. [CrossRef]
24. Ruosch, M.; Spahni, R.; Joos, F.; Henne, P.D.; Knaap, W.O.; Tinner, W. Past and future evolution of *Abies alba* forests in Eu-rope—Comparison of a dynamic vegetation model with palaeodata and observations. *Glob. Chang. Biol.* **2016**, *22*, 727–740. [CrossRef]
25. Latreille, A.C.; Pichot, C. Local-scale diversity and adaptation along elevational gradients assessed by reciprocal transplant experiments: Lack of local adaptation in silver fir populations. *Ann. For. Sci.* **2017**, *74*, 77. [CrossRef]
26. Becker, M. The role of climate on present and past vitality of silver fir forests in the Vosges mountains of northeastern France. *Can. J. For. Res.* **1989**, *19*, 1110–1117. [CrossRef]
27. Maaten-Theunissen, M.; Kahle, H.P.; Maaten, E. Drought sensitivity of Norway spruce is higher than that of silver fir along an altitudinal gradient in southwestern Germany. *Ann. For. Sci.* **2013**, *70*, 185–193. [CrossRef]
28. Bouriaud, O.; Popa, I. Comparative dendroclimatic study of Scots pine, Norway spruce, and silver fir in the Vrancea Range, Eastern Carpathian Mountains. *Trees* **2008**, *23*, 95–106. [CrossRef]
29. Lévesque, M.; Saurer, M.; Siegwolf, R.T.W.; Eilmann, B.; Brang, P.; Bugmann, H.; Rigling, A. Drought response of five conifer species under contrasting water availability suggests high vulnerability of Norway spruce and European larch. *Glob. Chang. Biol.* **2013**, *19*, 3184–3199. [CrossRef] [PubMed]
30. Zang, C.; Hartl-Meier, C.; Dittmar, C.; Rothe, A.; Menzel, A. Patterns of drought tolerance in major European temperate forest trees: Climatic drivers and levels of variability. *Glob. Chang. Biol.* **2014**, *20*, 3767–3779. [CrossRef]
31. Gazol, A.; Camarero, J.J. Functional diversity enhances silver fir growth resilience to an extreme drought. *J. Ecol.* **2016**, *104*, 1063–1075. [CrossRef]
32. Bosela, M.; Lukac, M.; Castagneri, D.; Sedmák, R.; Biber, P.; Carrer, M.; Konôpka, B.; Nola, P.; Nagel, T.A.; Popa, I.; et al. Contrasting effects of environmental change on the radial growth of co-occurring beech and fir trees across Europe. *Sci. Total Environ.* **2018**, *615*, 1460–1469. [CrossRef] [PubMed]

33. Klisz, M.; Ukalski, K.; Ukalska, J.; Jastrzębowski, S.; Puchałka, R.; Przybylski, P.; Mionskowski, M.; Matras, J. What Can We Learn from an Early Test on the Adaptation of Silver Fir Populations to Marginal Environments? *Forests* **2018**, *9*, 441. [CrossRef]
34. Mihai, G.; Bîrsan, M.-V.; Dumitrescu, A.; Alexandru, A.; Mirancea, I.; Ivanov, P.; Stuparu, E.; Teodosiu, M.; Daia, M. Adaptive genetic potential of European silver fir in Romania in the context of climate change. *Ann. For. Res.* **2018**, *61*, 95–108. [CrossRef]
35. Eriksson, G.; Namkoong, G.; Roberds, J.H. Dynamic gene conservation for uncertain futures. *For. Ecol. Manag.* **1993**, *62*, 15–37. [CrossRef]
36. Kremer, A. How well can existing forests withstand climate change. In *Climate Change and Forest Genetic Diversity: Implications for Sustainable Forest Management in Europe*; Koskela, J., Buck, A., Teissier du Cros, E., Eds.; Bioversity International: Rome, Italy, 2007; pp. 3–17.
37. Lindner, M.; Maroschek, M.; Netherer, S.; Kremer, A.; Barbati, A.; Garcia–Gonzalo, J.; Seidl, R.; Delzon, S.; Corona, P.; Kol-strom, M. Climate change impacts, adaptive capacity and vulnerability of European forest ecosystems. *For. Ecol. Manag.* **2010**, *259*, 698–709. [CrossRef]
38. Mihai, G.; Teodosiu, M.; Birsan, M.-V.; Alexandru, A.-M.; Mirancea, I.; Apostol, E.-N.; Garbacea, P.; Ionita, L. Impact of Climate Change and Adaptive Genetic Potential of Norway Spruce at the South-Eastern Range of Species Distribution. *Agric. For. Meteorol.* **2020**, *291*, 108040. [CrossRef]
39. Rehfeldt, G.E.; Wykoff, W.R.; Ying, C.C. Physiologic Plasticity, Evolution, and Impacts of a Changing Climate on Pinus Contorta. *Clim. Chang.* **2001**, *50*, 355–376. [CrossRef]
40. Wang, T.; O'Neill, G.A.; Aitken, S.N. Integrating environmental and genetic effects to predict responses of tree populations to climate. *Ecol. Appl.* **2010**, *20*, 153–163. [CrossRef]
41. Rehfeldt, G.E.; Ying, C.C.; Spittlehouse, D.L.; Hamilton, D.A. Genetic responses to climate in Pinus contorda: Niche breadth, climate change, and reforestation. *Ecol. Monogr.* **1999**, *69*, 375–407. [CrossRef]
42. Sáenz-Romero, C.; Kremer, A.; Nagy, L.; Újvári-Jármay, É.; Ducousso, A.; Kóczán-Horváth, A.; Hansen, J.K.; Mátyás, C. Common garden comparisons confirm inherited differences in sensitivity to climate change between forest tree species. *PeerJ* **2019**, *7*, e6213. [CrossRef] [PubMed]
43. Buras, A.; Sass-Klaassen, U.; Verbeek, I.; Copini, P. Provenance selection and site conditions determine growth performance of pedunculate oak. *Dendrochronologia* **2020**, *61*, 125705. [CrossRef]
44. Andalo, C.; Beaulieu, J.; Bousquet, J. The impact of climate change on growth of local white spruce populations in Québec, Canada. *For. Ecol. Manag.* **2005**, *205*, 169–182. [CrossRef]
45. Rweyongeza, D.M.; Yang, R.-C.; Dhir, N.K.; Barnhardt, L.K.; Hansen, C. Genetic Variation and Climatic Impacts on Survival and Growth of White Spruce in Alberta, Canada. *Silvae Genet.* **2007**, *56*, 117–127. [CrossRef]
46. St Clair, J.B.; Mandel, N.L.; Vance-Borland, K.W. Genecology of Douglas fir in western Oregon and Washinton. *Ann. Bot.* **2005**, *96*, 1199–1214. [CrossRef]
47. Chakraborty, D.; Wang, T.; Andre, K.; Konnert, M.; Lexer, M.J.; Matulla, C.; Schueler, S. Selecting Populations for Non-Analogous Climate Conditions Using Universal Response Functions: The Case of Douglas-Fir in Central Europe. *PLoS ONE* **2015**, *10*, e0136357. [CrossRef] [PubMed]
48. Schmidtling, R.C. Use of provenance tests to predict response to climate change: Loblolly pine and Norway spruce. *Tree Physiol.* **1994**, *14*, 805–817. [CrossRef] [PubMed]
49. Gömöry, D.; Longauer, R.; Hlásny, T.; Pacalaj, M.; Strmeň, S.; Krajmerová, D. Adaptation to common optimum in different populations of Norway spruce (*Picea abies* Karst.). *Eur. J. For. Res.* **2011**, *131*, 401–411. [CrossRef]
50. Kapeller, S.; Lexer, M.J.; Geburek, T.; Hiebl, J.; Schueler, S. Intraspecific variation in climate response of Norway spruce in the eastern Alpine range: Selecting appropriate provenances for future climate. *For. Ecol. Manag.* **2012**, *271*, 46–57. [CrossRef]
51. Klisz, M.; Ukalska, J.; Koprowski, M.; Tereba, A.; Puchałka, R.; Przybylski, P.; Jastrzębowski, S.; Nabais, C. Effect of provenance and climate on intra-annual density fluctuations of Norway spruce *Picea abies* (L.) Karst. in Poland. *Agric. For. Meteorol.* **2019**, *269*, 145–156. [CrossRef]
52. Vitasse, Y.; Delzon, S.; Bresson, C.C.; Michalet, R.; Kremer, A. Altitudinal differentiation in growth and phenology among populations of temperate-zone tree species growing in a common garden. *Can. J. For. Res.* **2009**, *39*, 1259–1269. [CrossRef]
53. Burczyk, J.; Giertych, M. Response of Norway Spruce (*Picea abies* [L] Karst) annual increments to drought for various prov-enances and locations. *Silvae Genet.* **1991**, *40*, 146–152.
54. Arend, M.; Kuster, T.; Günthardt-Goerg, M.S.; Dobbertin, M. Provenance-specific growth responses to drought and air warming in three European oak species (*Quercus robur*, *Q. petraea* and *Q. pubescens*). *Tree Physiol.* **2011**, *31*, 287–297. [CrossRef] [PubMed]
55. Dalla-Salda, G.; Martinez-Meier, A.; Cochard, H.; Rozenberg, P. Genetic variation of xylem hydraulic properties shows that wood density is involved inadaptation to drought in Douglas-fir (*Pseudotsuga menziesii* (Mirb.)). *Ann. For. Sci.* **2011**, *68*, 747–757. [CrossRef]
56. Eilmann, B.; de Vries, S.M.; Ouden, J.D.; Mohren, G.M.; Sauren, P.; Sass-Klaassen, U. Origin matters! Difference in drought tolerance and productivity of coastal Douglas-fir (*Pseudotsuga menziesii* (Mirb.)) provenances. *For. Ecol. Manag.* **2013**, *302*, 133–143. [CrossRef]
57. George, J.-P.; Schueler, S.; Karanitsch-Ackerl, S.; Mayer, K.; Klumpp, R.T.; Grabner, M. Inter- and intra-specific variation in drought sensitivity in Abies spec. and its relation to wood density and growth traits. *Agric. For. Meteorol.* **2015**, *214*, 430–443. [CrossRef]

58. George, J.-P.; Grabner, M.; Campelo, F.; Karanitsch-Ackerl, S.; Mayer, K.; Klumpp, R.T.; Schüler, S. Intra-specific variation in growth and wood density traits under water-limited conditions: Long-term-, short-term-, and sudden responses of four conifer tree species. *Sci. Total. Environ.* **2019**, *660*, 631–643. [CrossRef]
59. Stokes, M.A.; Smiley, T.L. *An Introduction to Tree-Ring Dating*; University of Chicago Press: Chicago, IL, USA, 1996; p. 73.
60. Grissino-Mayer, H.D. Evaluating crossdating accuracy: A manual and tutorial for the computer program COFECHA. *Tree Ring Res.* **2001**, *57*, 205–221.
61. Fritts, H.C. *Tree Rings and Climate*; Academic Press: San Francisco, CA, USA, 1976; p. 582.
62. Briffa, K.R.; Jones, P.D. Basic chronology statistics and assessment. In *Methods of Dendrochronology*; Cook, E.R., Kairiukstis, L., Eds.; Kluwer Academic: Dordrecht, The Netherlands, 1990; pp. 137–153.
63. Speer, J.H. *Fundamentals of Tree-Ring Research*; Indiana State University: Terre Haute, IN, USA, 2009; p. 508.
64. Bunn, A.G. A dendrochronology program library in R (dplR). *Dendrochronologia* **2008**, *26*, 115–124. [CrossRef]
65. Dumitriu-Tataranu, I.; Ghelmeziu, N.; Florescu, I.; Milea, I.; Mos, V.; Tocan, M. *Estimarea Calitatii Lemnului Prin Metoda Carotelor de Sondaj*; Editura Tehnica: Bucuresti, Romania, 1983; p. 347.
66. Dumitrescu, A.; Birsan, M.-V. ROCADA: A gridded daily climatic dataset over Romania (1961–2013) for nine meteorological variables. *Nat. Hazards* **2015**, *78*, 1045–1063. [CrossRef]
67. McKee, T.B.N.; Doesken, J.; Kleist, J. The relationship of drought frequency and duration to time scales. In Proceedings of the 9th Conference on Applied Climatology, Dallas, TX, USA, 15–20 January 1995; pp. 179–184.
68. Lloret, F.; Keeling, E.G.; Sala, A. Components of tree resilience: Effects of successive low-growth episodes in old ponderosa pine forests. *Oikos* **2011**, *120*, 1909–1920. [CrossRef]
69. Sagnard, F.; Barberot, C.; Fady, B. Structure of Genetic diversity in *Abies alba* Mill. from southwestern Alps: Multivariate analysis of adaptive and non-adaptive traits for conservation in France. *For. Ecol. Manag.* **2002**, *157*, 175–189. [CrossRef]
70. Sindelar, J.; Beran, F. Comparison of some exotic species of *Abies genus* with chosen silver fir provenances on the plots of town Pisek. *Commun. Inst. For. Bohem.* **2008**, *24*, 99–114.
71. Forner, A.; Aranda, I.; Granier, A.; Valladares, F. Differential impact of the most extreme drought event over the last half century on growth and sap flow in two coexisting Mediterranean trees. *Plant Ecol.* **2014**, *215*, 703–719. [CrossRef]
72. Gazol, A.; Camarero, J.J.; Anderegg, W.R.L.; Vicente-Serrano, S.M. Impacts of droughts on the growth resilience of Northern Hemisphere forests. *Glob. Ecol. Biogeogr.* **2017**, *26*, 166–176. [CrossRef]
73. Brousseau, L.; Postolache, D.; Lascoux, M.; Drouzas, A.D.; Källman, T.; Leonarduzzi, C.; Liepelt, S.; Piotti, A.; Popescu, F.; Roschanski, A.M.; et al. Local Adaptation in European Firs Assessed through Extensive Sampling across Altitudinal Gradients in Southern Europe. *PLoS ONE* **2016**, *11*, e0158216. [CrossRef]
74. Heer, K.; Behringer, D.; Piermattei, A.; Bassler, C.; Brandl, R.; Fady, B.; Jehl, H.; Liepelt, S.; Lorch, S.; Piotti, A. Linking dendroecology and association genetics in natural populations: Stress responses archived in tree rings associate with SNP genotypes in silver fir (*Abies alba* Mill.). *Mol. Ecol.* **2018**, *27*, 1428–1438. [CrossRef] [PubMed]
75. Jones, H.G.; Sutherland, R.A. Stomatal control of xylem embolism. *Plant Cell Environ.* **1991**, *14*, 607–612. [CrossRef]
76. St Clair, J.B.; Howe, G.T. Genetic maladaptation of coastal Douglas-fir seedlings to future climates. *Glob. Chang. Biol.* **2007**, *13*, 1441–1454. [CrossRef]
77. Rosner, S.; Světlík, J.; Andreassen, K.; Børja, I.; Dalsgaard, L.; Evans, R.; Karlsson, B.; Tollefsrud, M.; Solberg, S. Wood density as a screening trait for drought sensitivity in Norway spruce. *Can. J. For. Res.* **2014**, *44*, 154–161. [CrossRef]

MDPI

Review

Valuing Forest Ecosystem Services. Why Is an Integrative Approach Needed?

Gabriela Elena Baciu [1,2], Carmen Elena Dobrotă [3,4,*] and Ecaterina Nicoleta Apostol [2,*]

1 Forest Management Planning and Terrestrial Measurements, Department of Forest Engineering, Faculty of Silviculture and Forest Engineering, "Transilvania" University, 1 Ludwig van Beethoven Str., 500123 Brașov, Romania; gabriela.baciu@icas.ro
2 "Marin Drăcea" Romanian National Institute for Research and Development in Forestry, 128 Eroilor Blvd., 077190 Voluntari, Romania
3 Faculty of Business and Administration, University of Bucharest, 4-12 B-dul Regina Elisabeta, 030018 Bucharest, Romania
4 Institute of National Economy, Romanian Academy, 13 Calea 13 Septembrie, 050711 Bucharest, Romania
* Correspondence: dobrotacarmen@yahoo.com (C.E.D.); Ecaterina.apostol@icas.ro (E.N.A.); Tel.: +40-7-2633-2258 (C.E.D.); +40-7-2181-8827 (E.N.A.)

Abstract: Among the many types of terrestrial ecosystems, forests have some of the highest levels of biodiversity; they also have many interdependent economic, ecological and social functions and provide ecosystem services. They supply a range of tangible, marketable goods, as well as a variety of nonmarketable and intangible services derived from various forest functions. These translate into social, cultural, health and scientific benefits for people's quality of life. However, because they cannot be traded on a market, nonmarketable and intangible services are often perceived as free, inexhaustible and, as a result, underestimated. The human–nature interaction has affected both nature (via resource consumption) and society (via development of human welfare and well-being). Decision-makers, both public and private, often manage natural capital for multiple aims. In recent years it has been found that the single, individual approach estimating the value for these goods and services is not able to provide information that generates and supports decisions and policies in complex areas of current relevance such as the constant loss of biodiversity, climate change and global warming in close connection with the need for social development and ensuring an acceptable level of well-being for the greatest part of humanity. An integrated assessment with advanced techniques and methods using a pluralist framework of a heterogeneous set of values is considered a better approach to the valuation of such complex nature of the ecosystem goods and services. This assessment should take into account both costs and benefits trade-off issues among the multiple uses of ecosystem goods and/or services, especially the relationships between them and how they influence or determine the economic, social and cultural development of society. It should also consider the estimation of the complex inverse effect, from society to nature, whose goods and services can be diminished to exhaustion by the extensive and intensive anthropization of natural ecosystems with major impact on the number and quality of goods and services provided by ecosystems. Research has shown that applying an integrative assessment approach that utilizes tools developed by sustainability sciences could be an important component of future environmental policy making.

Keywords: biodiversity; forests; valuing ecosystem services; climate change; policy making

Citation: Baciu, G.E.; Dobrotă, C.E.; Apostol, E.N. Valuing Forest Ecosystem Services. Why Is an Integrative Approach Needed? *Forests* **2021**, *12*, 677. https://doi.org/10.3390/f12060677

Academic Editors: Alessandra De Marco, Pierre Sicard and Mihai A. Tanase

Received: 6 April 2021
Accepted: 19 May 2021
Published: 25 May 2021

1. Introduction

The Earth's population relies on the benefits provided by ecosystems, including ecosystem provisioning, regulation and cultural and support services [1]. Over time, humans have transformed ecosystems to meet their needs and desires. Nowadays, climate change and biodiversity loss are major challenges for both developed and developing countries. According to the Intergovernmental Panel on Climate Change (IPCC, 2018), if global warming exceeds 1.5 °C, climate change will severely affect humanity and ecosystems. An analysis of how people's use and management of natural resources affects ecosystem resilience is necessary because people's daily choices will result in continued biodiversity loss and new social costs [2]. Forests, which cover one-third of Earth's land surface, are an immense and renewable source of ecosystem services (ESs) [3,4]. They represent an extraordinary opportunity to mitigate climate change through carbon sequestration [5,6], soil stabilization and natural disaster mitigation [5]; forest conservation efforts (e.g., establishing protected areas) do not contradict territorial and regional development objectives [3] since changes in land cover and land use are among the major drivers of forest area reduction, biodiversity loss and land and ecosystem degradation at the global, regional and local levels [7–12]. In this respect, all these aspects should be kept together, to establish correlation among services and their impact on communities' development. In addition, the emergence of states of necessity (e.g., economic crises and social, political and military conflicts) could potentially intensify the use of resources and ESs offered by forests [13]. There are a number of less visible services provided by forests that support local development through cultural services [14] or sustainable tourism services [3]. Depending on the goals of the valuation of the ESs, some services should be seen and evaluated in a strict correlation and an integrative manner. Many studies have addressed how cultural services can be integrated into spatial planning methods; they showed that using spatial mapping and integrating information on habitat types, landscape features and land-use methods with information on existing infrastructure, number of visitors to the area and proximity to local communities during stakeholder consultations often led to increased stakeholder involvement in the planning process [15].

Recent research has analyzed ESs in relation to bioeconomic strategy objectives. This trend reflects how ESs and bioeconomy strategy, two key concepts in sustainability science, must be addressed together, especially given the effects of bioeconomy strategies on ESs [16,17]. Recent sustainable development initiatives have embraced the concept of a circular economy; this paradigm challenges the current linear behavioral model of take–do–consume–throw, which produces excessive waste and inefficiently uses natural resources [18]. The new EU Forest Strategy (2021–2027) emphasizes the need to ensure that the multifunctional potential of EU forests and their vital ESs are managed sustainably.

However, when discussing natural capital (NC), ecosystems and ESs, it is important to integrate concepts and methods that give a perceptible expression of their value. Depending on the final purpose of the analysis and evaluation, at least one of the following types of value can be assigned to NC and then calculated or estimated: philosophical value, economic value, social value, aesthetic value, inheritance value (for future generations), altruistic value [19,20], egoistic value [19], biospheric value [19,21] or intangible and cultural value [22]. Previous research has shown that everything is valuable but in different ways. Art objects often have sentimental value, historical value or financial value [23]. Landscapes, mountains and forests can have economic value and recreational value. In addition, great works of art, as well as natural landscapes, possess a distinct noninstrumental and nonutilitarian value, which is a central concern when works of art or landscapes are evaluated. Though some may think the value of art and landscapes comes from their beauty, others may not consider them beautiful. As such, beauty is a particular case of aesthetic value [23,24]. Aesthetic value is defined as the value possessed by an object, event or state of affairs by virtue of its ability to cause pleasure (positive value) or dissatisfaction (negative value) [23]. It is often seen as more subjective than other types of value and is usually of low priority in policy debates [24]. An example of this is the complex relationship between human aesthetic experience and the development of ethical attitudes towards the environment [25,26]. For ESs, their value often reflects contributions to human welfare and well-being, and a distinction can be made between use value derived from direct or indirect use of ESs and nonuse value derived from the intrinsic value of ecosystems and their biodiversity [27]. Currently, macro-indicators such as GDP report the values of goods and services exchanged in the market, but they do not reflect the values of nonmarket ESs, the deterioration of ESs or the loss of biodiversity. The inclusion of ES indicators in national accounts would allow for not only an economic assessment, but also an environmental and social assessment of a country's development [27,28]. Additionally, mapping ESs and establishing assessment indicators [29] is an important and current issue, with the EU Biodiversity Strategy explicitly calling for this action under Action 5 [26,30–32].

A holistic approach, using sustainability science methods and techniques developed for ES valuation, seems to be now the challenge for value pluralism of forest ecosystems, including well-known services and other indirect benefits such as health, education, equality and governance [17,32–34]. There are many complexities that have to be taken into account in order to value ESs. Their resources provide multifaceted benefits, and for some of them, it is difficult to quantify their value. Cost–benefit analysis allows the aggregation of the values of ESs on a single monetary scale of measurement [35]. However, public sector entities are deeply involved in such efforts. A plethora of multinational organizations are involved, including TEEB, WAVES (Wealth Accounting for the Value of Ecosystem Services, a World Bank program) and IGPBES. National governments are more involved in assessing ESs. The United Kingdom conducts an evaluation of national ecosystems that includes the assessment of several ESs. In the United States, all departments and agencies in the executive branch are now directed to "develop and institutionalize policies to promote the consideration of ecosystem services... and, where appropriate, monetary or non-monetary values for those services" [36] (p. 8/32). In Romania, the ES valuation process is at the beginning; up to now, several exploratory studies have been conducted related to the value of ESs in natural protected areas, and a case study on Piatra Craiului National Park has been conducted [37,38]. The studies revealed that even though the Piatra Craiului protected area generates significant ESs, very low economic values are mirrored in the earnings of the park administration. Thus, in-depth studies combining biodiversity aspects with economic evaluations of ESs will be a strong base for decision-makers for promoting sustainable development public policies in this area.

This paper aims to explore why an integrative approach for valuing and assessing forest ESs is needed, taking into account the many interdependent factors involving ESs and their associated values, as well as current challenges people face.

2. Evaluation of Ecosystem Services—Why Is It Necessary?

Natural resources associated with production (such as wood, food and energy resources), as well as services associated with protection (such as air quality), are assets that help increase the efficiency of services provided to people by NC [39]. The exploitation of NC produces social costs and benefits, referred to as externalities [39]. From an economic viewpoint, externalities occur when a variable (not the price) generated by an economic unit influences the production processes of other economic units or of the population. For example, the construction of a slaughterhouse could produce water, land or air pollution, all of which are negative externalities that affect other economic units and the local population. Due to the difficulty in measuring total benefits or already proven multiple benefits, decision-makers are often required to depend on cost-effectiveness analyses of different management options. More importantly, trade-offs of benefits and burden distribution happen between space, time and social groups, and in general, the perceived value of ecosystems has not been accounted for all of the services the ecosystems provide. One study assessed the monetary and nonmonetary values of forest ecosystems in eight Mediterranean countries and found that wood and wood fuel represent less than one-third of the total economic value (TEV) of forests in the countries under study. The other, nontimber services offered by the assessed ecosystems—recreational activities, fishing, protection provided by the river network and carbon sequestration—made up between 25 and 96% of the ecosystems' TEV.

Scientists have long reported the implications of biodiversity loss. In 1872, Yellowstone National Park became the first geographic area defined as a protected area due to the initiative of several scientists [40]. The economic view that people's survival depends on natural resources, which are limited, has been held since the 18th century (Malthus 1888); the concept of ESs, or services offered by nature to people, was developed in the 1960s and 1970s [39–42]. Many natural processes improve human well-being [43] and welfare, but human activity negatively affects ecosystems through ecosystem conversion, habitat fragmentation, landscape alteration and the anthropization of the natural environment over time [26] and biodiversity loss, which ultimately harms human well-being [31,32,43].

Globally, the importance of protecting and sustainably managing forest ESs has been recognized through a series of UN-adopted documents. These include the 'Rio Forest Principles' from the 1992 United Nations Conference on Environment and Development [44]; the United Nations Framework Convention on Climate Change (UNFCCC) [45], which emphasizes the importance of forests in terms of the global greenhouse gas (GHG) balance; the Convention on Biological Diversity [46,47], which addresses forest biodiversity; the United Nations Forum on Forests (UNFF); the UN Convention to Combat Desertification (UNCCD) [36]; and the Paris Agreement [48], which calls for major reforms in order to fight global warming.

However, in recent years, ESs emerged as an important issue on the public agenda through discussions on topics such as biodiversity loss [2,40,41], land-use and spatial planning [7,9,10], climate change [42], circular economies [49,50] and bioeconomies [16,51] and public policies [36,52–54] and strategies [16]. To address all of these challenges requires sound decision-making [55]; the development of a tool for measuring TEV is necessary to support the political decision-making process and to inform both citizens and businesses about the benefits and costs inherent in projects, programs and policies [56]. There is a growing consensus that in spatial planning, land management and other decision-making contexts, the economic valuation of ESs is essential for the development of efficient public policies and strategies [57]. The value of ESs and biodiversity is assigned based on what societies are ready to offer in exchange for nature conservation [25,58] because the valuation of ESs can vary with time and spaces [59], ranging from simply raising awareness to analyzing various policy choices and scenarios in detail [60]. The estimated loss of ESs from 1997 to 2011 due to land-use change is $4.3–20.2 trillion per year [19].

3. Ecosystem Services and Natural, Socioeconomic and Public Policy Challenges

In recent decades, the concept of nature and ESs as capital has gained visibility [61], as society can receive important goods and services, such as clean air and water, flood control and crop pollination, by conserving and restoring natural habitats [56]. These goods and services, if properly considered, may be valuable enough to justify the protection of forest ESs [62]. Public debates on ESs have hit a sensitive chord. For some, the concept of ESs presents an opportunity to include all of the environmental benefits that the market failed to account for in public and private decision-making. For others, the possibility of structuring payments for ESs that assign and respect property rights and bring the power of the market to a bearable level may seem just as attractive [36].

Addressing climate change requires mitigation and action to adapt to new conditions. Forests and the forestry sector play a significant role in mitigating climate change by capturing CO_2 and producing timber products, as well as by substituting materials whose processing requires high energy consumption [63–65]. They also provide services that can help people adapt to both current and future climate risks [42]. While ESs are part of the solution to climate change, they are also affected by climate change. Climate change will impact forests and may impair their ability to provide essential ecosystem services in the decades to come. Addressing this challenge requires adjustments to forest management strategies as of now, but it is still unclear to what extent this is already in progress [66]. An EFI study found that forests and the role of the forestry sector could be significantly enhanced through Climate-Smart Forestry [63]. This approach aims to increase the climate benefits of forests and the forestry sector in a way that creates synergies with other forest-related needs. It is based on three pillars: (1) reducing or eliminating GHG emissions to mitigate climate change, (2) adapting forest management to build resilient forests and

(3) actively managing forests in order to sustainably increase productivity and provide all of the benefits that forests can offer [62–64,67]. The European Environmental Bureau, an international nonprofit association that has assembled over 160 civil society organizations from more than 35 European countries, stated in 2021 that "the global material footprint is already beyond ecological limits, being over 100 billion tonnes per year and, if we continue 'business as usual', is expected to double in the next 40 years. The impact of excessive consumption is significant. In the European Green Deal, the European Commission states that 'resource extraction and processing account for more than 90% of the global impact on biodiversity loss and water quality and about half of global climate change emissions'" [50].

In this context, sustainable development has become a global concept that transcends different sciences with environmental, social, cultural and economic dimensions. A bioeconomy is currently being promoted both for policymakers and businesses as a sustainable action plan for reconciling environmental, social and economic goals [16,68,69]. Human activity has led to the degradation of the natural environment, which has had a far-reaching impact on society and the economy and has created new conceptual frameworks for how people interact with and depend on the environment. A bioeconomy generally involves replacing fossil fuels with bio-based ones, so three main goals—involving resources, biotechnology and agroecology—are becoming more prevalent in the scientific literature [16]. In 2020, a review of 45 documents and articles showed that, although the publications were diverse and the approaches used were still quite new, eight topics were predominant: (a) the technical and economic feasibility of biomass extraction and use; (b) the potential and challenges of a bioeconomy; (c) frames and tools; (d) the sustainability of biology-based processes, products and services; (e) the ecological sustainability of a bioeconomy; (f) the governance of a bioeconomy; (g) biosecurity; and (h) bioremediation [16]. Though both the bioeconomy and NC combine economics and natural sciences and propose new interdisciplinary frameworks for environmental sustainability, the two concepts are rarely applied together [51]. A circular economy would positively impact ecological systems by not exhausting or overburdening them with technological and productive tasks. This is reflected in the environmental benefits of the circular economy. For example, a circular economy would emit less GHGs; the soil, air and water would remain unchanged; and natural reservations would be preserved [18,53]. Forest ecosystems provide services and products such as wood, pollination and clean drinking water. In a linear economy, these services will eventually be depleted by the constant extraction of products from ecosystems or will be affected by the release of toxins from technological processes [53,69]. If the products extracted from an ecosystem are used in a rational and intelligent technological and economic cycle, and the technological processes do not discharge toxic substances into the environment, then the soil, air and water will remain resistant and productive [52,69,70]. Understanding ESs and their economic applications offers a number of environmental and economic advantages because assessing NC and ES flows provides a powerful economic engine for nature conservation and nature-based solutions to current economic challenges, processes and industrial systems [49].

4. Ecosystem Valuation: Utilitarian vs. Nonutilitarian Approaches

The importance of ESs for human society has multiple dimensions: ecological, sociocultural and economic [71]. Over time, concerns related to ES valuation have led to the development of various methods for conducting these assessments, from mapping and modeling supply and demand for ESs to determine their market value (utilitarian approach) to social and environmental assessment techniques to assess their nonmarket value (nonutilitarian approach).

Utilitarian Approach

The utilitarian approach is intrinsically linked with cost–benefit analysis and welfare economics since they approach human well-being in terms of individual satisfaction based on the individual utility of goods and services. At the same time, environmental psychology research confirms that the relevance of ESs for human well-being is more than the satisfaction of individual needs and consists of physical and psychological health, social integration and cultural identity (ACB). While market valuation is relatively simple to perform, challenges arise when estimating the nonmarket value of an ecosystem. From the seminal classification of Krutilla (1967), the utilitarian approach divides the TEV of ESs into two types of value: the use value, which relates to ESs associated with production and protection functions for which market prices usually exist, and the nonuse value, which reflects the satisfaction of knowing that biodiversity and ESs are preserved and that future generations will also benefit from them [58]. Both of these categories have subsequently been disaggregated into multiple components. Use value was broken up into direct use, indirect use, optional, quasi-optional and bequest values; nonuse value was split into existence or intrinsic, aesthetic, altruist, bequest, moral and religious values [21,40,58,59,72,73]. Direct use value is associated with the benefits of using ESs, such as raw materials. Indirect use value is associated with regulating services like water quality regulation. The optional and quasi-optional values are the values of ESs based on the option to use the services at a certain time in the future. Of the nonuse values, existence or intrinsic value is usually presented as the value attributed by an individual to the continued existence of a service or good, regardless of its current or possible uses [58]. Both use and nonuse values are associated with the utilitarian approach, which primarily aims to express the associated values of ESs in monetary terms and takes into account the utility of NC for humans and for the socioeconomic system [71]. This includes ecosystem resources that can be used or are used by the population and by economic units in their daily activities.

In a neoclassical economy, on which environmental economics and assessment methods are based, the nonuse values are defined and measured in monetary units based on a willingness to pay (WTP) or a willingness to accept (WTA) [19,39,58]. Nonuse values such as WTP are estimated by methods of preference declared in questionnaires or interviews, including both the contingent assessment method (CVM) and direct choice experiments (DCEs) [39]. Two assessment approaches are commonly used to estimate nonuse values. The first approach asks how many respondents would be willing to pay for ESs (or their attributes in the case of DCE) if they were absolutely certain they would never use them. In this case, the interviews would be based on nonusers. The second approach asks respondents, including users, to divide the total WTP for ESs into different categories, such as inheritance, existence and own use. Such statement decomposition approaches have been applied in many CVM-related ES applications and have been useful in understanding the relative quotas of value categories in WTP estimates [39,74] or in identifying warm glow effect in willingness to pay (WTP) responses [75]. In most cases, the proportions of nonuse values in WTP are considered to be quite substantial, representing between 40 and 90% of the total WTP [39,74]. Despite its popularity, the approach to decomposition stated in interviews has substantial shortcomings and is highly controversial, mainly due to the cognitive difficulty of addressing the components of an unfamiliar and inseparable value. An individual's total WTP for an ES is usually a consequence of different overlapping and correlated motivations that may be inseparable and, as such, inaccessible to the researcher [76]. In most cases, the ES assessment is completed when a choice must be made among different services.

Over time, the desire to conduct a comprehensive economic assessment of ESs has led to the identification and refinement of various measurement methods. The first significant economic assessment of ESs, including from a nonmonetary perspective, was made by Costanza in 1997 based on the fact that ecosystems provide benefits to populations through ecosystem functions and components (i.e., services). Ecosystems are unique and irreplaceable, which makes them invaluable. Based on this, the author grouped ESs into categories and calculated their unit values, using assessment techniques based mainly on people's WTP. The resulting values were then multiplied by the area occupied by all US ecosystems and totalled $33 trillion per year, more than double the annual GDP, which was estimated at $16 trillion [20]. Fourteen years later, the value of ESs globally was estimated at $18 billion per year, of which 19% came from ES climate regulation and 4% came from raw materials related to productive functions. ES contributions to recreation, protection against extreme phenomena, the water supply, erosion control, nutrient cycling, habitat, genetic resources and nonwood products represent the rest of the value [20].

Costanza's work can be considered pioneering. From other perspectives, however, the proposed methodology was both technically and ethically challenged because ecosystems, as a support for life, are constantly evolving and cannot be measured monetarily. There is skepticism about the association of ecology with the economy; many specialists consider a strong involvement in the economic sector for the conservation of ecosystems dangerous, which could lead to an increase in nature depreciation. For example, developing countries could request and receive financial compensation in accordance with the estimated value of the ESs they provide, as long as they preserve them. Costanza's approach produced much debate and criticism, but it is better to have debate and criticism among scientists, policy-makers and stakeholders than to have nothing. However, despite the interest in making monetary assessments of ESs, these are not the only possible value assessments. In 2010, TEEB, published by The Ecological and Economic Foundations, developed the concept of TEV and presented a classification of TEV components and assessment tools that can be used to assess various components of ESs. The authors hypothesized that the value of ESs and biodiversity is determined by what a society is willing to offer in exchange for nature conservation. Society and policy-makers need to understand that ecosystems are unique and limited resources and that depreciation or degradation involves costs to society. From an economic point of view, when a resource is limited, an opportunity cost exists, representing the value of the best of the sacrificed chances (i.e., the one that is given up when a choice is made). However, the difficulty of conducting a monetary assessment of ESs is due to the fact that the changes to ecosystems are irreversible or are reversible for a prohibitive cost. The estimated economic value is a cumulation of choices of the buyer, which includes a multitude of preferences for ecology, society, health, technology and expectations regarding the future [58]. The modification of any of the factors listed influences the estimated economic value [58,77] and could lead to different scenarios being planned [77].

The evaluation methods identified in the VET methodology fall into three categories: (a) direct market valuation approaches, such as the price-based method, cost-based method and production function-based method; (b) revealed preference approaches, including the travel cost method and hedonic pricing method; and (c) simulated valuation, such as the contingent valuation method, choice modeling and group valuation.

Price-based methods are most often used to calculate the value of provided goods and services. Because they are traded on the market, their value is relatively easy to calculate. Examples include the value of wood, honey or tourist services [58]. Cost-based methods [39] are based on several identified techniques, such as the avoided costs method, which assesses the costs that would have occurred in the absence of the ES. The replacement cost method estimates the costs of replacing ESs with artificial technologies, the restoration cost method assesses the costs of counteracting the effects of ecosystem loss or restoration and the production function-based method estimates how much of the nonmarket ESs contribute to other services or goods traded on the market, noting how much the services contribute to increasing the productivity or price of those goods or services.

The travel cost method is relevant mainly for determining the value of recreational services associated with biodiversity and ESs. The method is based on the principle that recreational experiences can be associated with a cost that consists of direct costs and opportunity costs. In the case of tourism, changing ecosystem biodiversity can influence the demand to visit that location. The hedonic pricing method is based on the added value that a landscape, or location near an ecosystem, can bring to a market, such as the real estate market. Changing the biodiversity of an ecosystem can change the market value of a property. The revealed preference approaches require a large amount of complex data and statistics and so are expensive and time-consuming. In addition, since these methods are based on direct observation of clients, they can provide an image at a certain moment in time [58].

The contingent valuation method uses questionnaires through which respondents provide information on how they would be willing to pay to protect ESs and how much they would be willing to pay to accept ecosystem loss or degradation. The choice modeling method focuses on modeling human behavior in particular contexts; this method starts with the supposition that people must choose from two or more alternatives when making a decision, one of which is the price in money. The group valuation method combines the use of questionnaires with elements of the deliberative process from political science and is becoming a widespread method for collecting values such as the uniqueness of ecosystems and social justice, as well as altruism towards other people and towards future generations compared to the species that live in the ecosystem. These methods should be applied carefully, and their limitations should be considered, especially when evaluating the nonuse value of a service that does not have a corresponding price on the market [36,54].

Extensive research conducted in Europe through the study Operationalisation of Natural Capital and Ecosystem Services Integrated (OpenNESS) [78] classified the methods used for evaluating ESs into the following categories: (i) biophysical methods, which are used for mapping ESs and include matrix approaches, ecosystem modelling with InVEST (Integrated Valuation of Ecosystem Services and Tradeoffs [79], E-Tree [80] or ESTIMAP [81,82]; (ii) integrated mapping-modelling approaches; (iii) land-use scoring [83]; (iv) participatory mapping; (v) sociocultural methods for understanding social preferences or values for ESs, such as deliberative assessment methods, preference prioritization methods, multicriteria analysis methods and photo-elicitation surveys; (vi) monetary methods for estimating the economic value of services, such as preference methods, revealed preference methods and travel cost methods [58] or hedonic pricing methods [58,84]; and (vii) integrative approaches [85]. The selection of a particular method for a specific case can depend on many factors, including the decision-making context; the strengths and limitations of each method; and pragmatic reasons such as available data, resources and expertise. Each method has specific features that inform its relevance or appropriateness to certain decisions or problems in the context of the study. The ability of a method to address a specific purpose may be the primary factor influencing method selection. Most methods are able to characterize the current state of ecosystem service demand or supply, but only a few are able to explore potential future service provision through modeling approaches and participatory scenario development (which was specifically designed to address this purpose). Some methods focus on specific ESs, such as biophysical models of soil erosion, or specific groups of services, such as photo-series analyses of cultural ESs. Other methods attempt to provide a more holistic or strategic overview of multiple ESs, which may be used to assess trade-offs [86] between the supply of different services (e.g., matrix-based approaches) or the demand for services by different stakeholders (e.g., PGIS, preference assessment methods, photo-elicitation or MCDA). The integration of ES assessment with life cycle assessment (LCA) is important for developing decision support tools for environmental sustainability. LCA methods have traditionally been employed as environmental management tools to assess the environmental impacts of production processes from 'cradle to grave' [87]. The method was developed in the 1960s in reaction to the 'Limits to Growth' discourse, which raised concerns about natural resource finiteness. The assessments were initially limited to energy efficiency and emissions and were information for internal use by companies.

After the 1980s, academia and governments began using LCA as well; methodological development progressed and was supported by formal attempts at international standardization [88]. LCA has since become a reference tool for the assessment of sustainability issues in the context of production–consumption systems, obviously bearing both strengths and weaknesses [89,90]. Despite emerging interest in the topic, additional work is needed for tackling the integration of ES issues in LCA approaches [91].

The nonutilitarian approach identifies four types of value: ecological value, sociocultural value, value with direct economic significance and intrinsic value [40,92].

The ecological value is determined by the integrity of the regulation and habitat functions of the ecosystem and by various ecosystem parameters such as complexity, diversity and scarcity (de Groot). The most appropriate methods to evaluate the ecological value are the biophysical methods mentioned above as well as integrated mapping–modeling approaches and land-use scoring [92]. Sociocultural value is mainly related to aspects such as physical and mental health, education, cultural diversity and identity (heritage value), freedom and spiritual values. The most used methods to evaluate it are participatory mapping and sociocultural methods described above [92].

As regards the economic value, the monetary methods, such as direct methods of valuation based on market prices or indirect valuation methods (e.g WTP, WTA, Replacement cost, travel cost, Hedonic pricing), are the most commonly identified. [92].

For determining intrinsic value, the most adequate methods could be preference prioritization methods, multicriteria analysis methods and photo-elicitation surveys, combined with biophysical methods such as ecological models.

In conclusion, the utilitarian approach is in line with the philosophy of environmental economists who are in favor of extension of monetary valuation methods to nonmarket ESs, while the nonutilitarian approach is aligned with the concepts of ecological economists who consider the substitutability and valuation of NC controversial. Boundaries between utilitarian and nonutilitarian approaches (Figure 1) are blurred, and they benefit from an abundant and expanding body of literature [2].

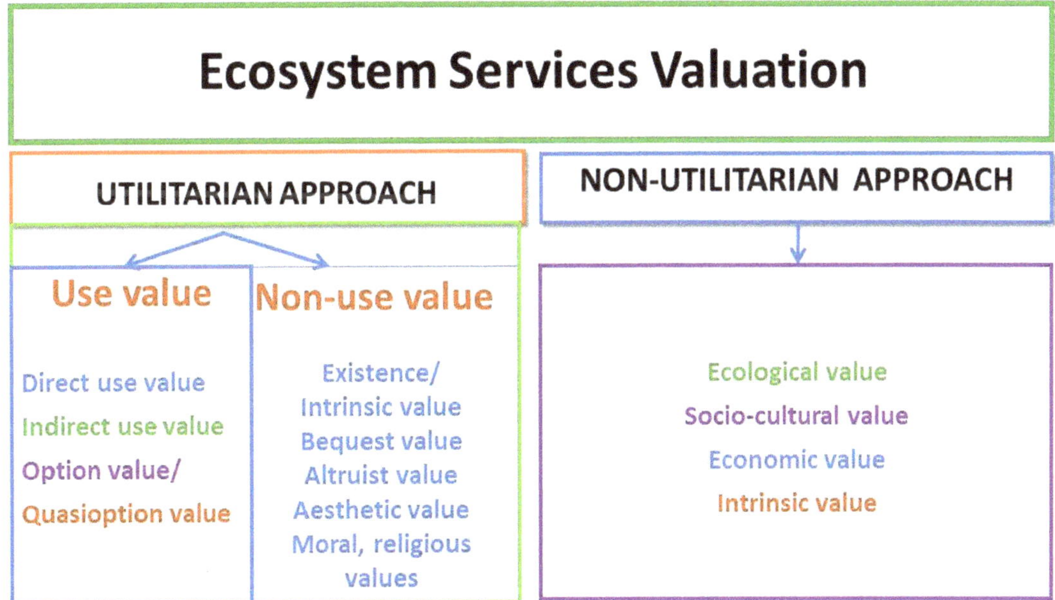

Figure 1. Utilitarian and nonutilitarian frameworks for valuing ESs. (adapted after TEEB).

The nonutilitarian approach is recognized as an important component of the ES valuation and an important motivation for increasing conservation efforts, but using monetary units to raise awareness of policymakers about their importance is a powerful tool [71].

5. Cost–Benefit Analysis of ES

Data on each ecosystem and each service highlighted the need to preserve ecosystems to ensure sustainable development. Even if ecosystems are subject to intensive and extensive exploitation, people must take care of them to ensure continuity. Therefore, in response to the exploitation of resources, plans must be made to conserve ecosystems. An environmental cost–benefit analysis (CBA) is best suited for this purpose [62]. First, because a CBA presents the territorial distribution of benefits and costs and compares this distribution with the distribution of biodiversity, it allows for the identification of important areas for both people and biodiversity (win–win areas), as well as the identification of areas of potential conflict and areas in need of compromises (negotiations). In these areas, the net economic benefits of ecosystem conservation are low, but biodiversity values are high, or vice versa. Second, a CBA highlights which areas have the highest unit cost benefits, thus indicating the most effective places for conservation efforts. Third, maps with ESs could help identify providers and consumers of ESs, enabling the identification of efficient and equitable payment mechanisms for financing conservation projects [62,93]. The core activity in an environmental CBA is estimating monetary values of the environment, especially the economic value of nonmarketable goods and services; the objective of the analysis is to estimate the TEV that arises from a policy proposal [94]. In 1970, CBAs were introduced for use on publicly financed projects with an environmental impact in the US. Since then, CBAs have been continuously adapted and applied to different methods and techniques, such as stated preference methods (which include the contingent valuation method, WTP, WTA, choice experiments, deliberative group valuation and health risk valuation) and revealed preference methods (which include the travel cost and hedonic price methods) [35]. At the same time, an important aspect that has to be taken into consideration when performing CBA is spatiotemporal frames, meaning that ESs are generated at different scales from short-term site level to long-term global level, and any slight change in the spatial or temporal frame approached in CBA can generate different consequences and stakeholders included in the CBA.

6. Ecosystem Service Valuation—What Is Next?

ES approaches and assessment efforts have changed the discourse on issues such as nature conservation, natural resource management and other areas of public policy. It is now accepted that in order to create a win–win situation rather than a compromise between environment and development, strategies for natural resource management and conservation through investment in the conservation, restoration [68] and sustainable use of ecosystems should be based on a combination of all values that occur when estimating the TEV [3,56,95,96] (Figure 2). Nonmarket assessments and methods used for cultural and environmental services have been criticized for their inability to provide values that represent or substantiate the total value of an ecosystem, but economists' efforts to involve interdisciplinary teams and incorporate a variety of methods and information into their research have demonstrated their flexibility, which reinforces the idea that they are effective in the process of diluting public policy decisions [13,57,76,97]. At the same time, actions have to be based on evidence, data and analysis to assess how public policies are beneficial for both people and nature [98], and the valuation methods have to be adapted to the local conditions and stakeholders involved [99]. The moving from conceptual frameworks and theory to practical integration of ESs into credible, replicable, scalable and sustainable public policies will require radical transformations [100] towards systematical integration of the ESs in decision-making at the individual, corporate or governmental level [101].

The ways in which ESs can be included in national accounts have generated a great deal of debate because it is, to some extent, a matter of choice [102,103]. In 2002, the UN's System of Environmental–Economic Accounting—Experimental Ecosystem Accounting (SEEA EEA) showed that the concept of valuation has made a significant difference in attempts to incorporate the generated ES values into national accounts [104,105]. Accounting ESs supposedly quantifies the amount of ESs provided by an ecosystem to socioeconomic systems [25,106]. This can highlight ES contributions to the economy, social well-being, jobs and livelihoods.

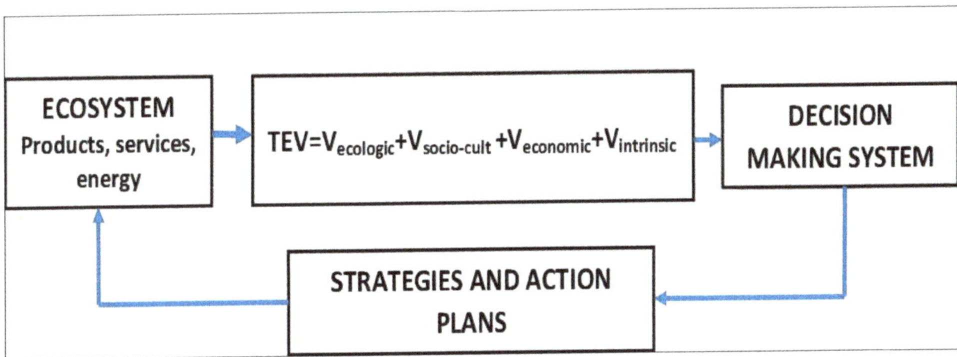

Figure 2. Integrative approach of valuing ESs.

SEEA methodology gave rise to the concept of the information pyramid (Figure 3), which combines basic economic, ecological and sociodemographic data. These data can be collected, centralized, processed and used for the development of analyses and studies that provide evidence for public policies and lead to the development of aggregate key indicators at the macro level.

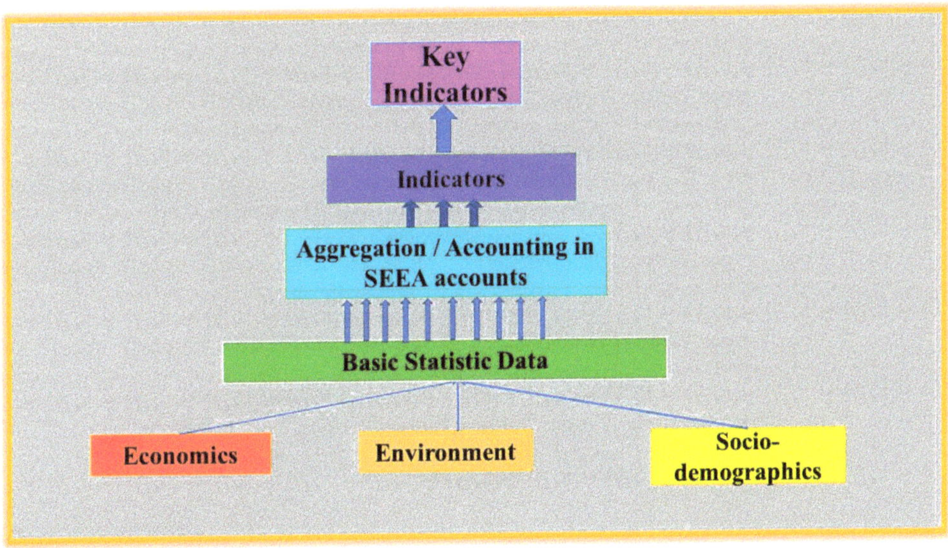

Figure 3. Information pyramid for ES key indicators (from SEEA).

However, creating such key indicators is a challenge. Using exchange value methods based on market techniques to quantify ESs [107] is easier because these are already compatible with the Systems on National Accounts; well-being value-based methods are difficult to translate into exchange value terms [106,108]. This shows that more effort should be put into the development of a pluralistic value-based approach able to capture both monetary and nonmonetary values [105].

In addition, the development of experimental ES accounts revealed the need to develop different indicators for separate ESs since each service has different characteristics. For forest ecosystems, the main indicators are related to timber production, biomass harvesting for energy, wild food provision, climate regulation, fire management, air quality regulation, noise reduction, water purification and recreational and aesthetic values. The accounts developed at the EU level [109] face many challenges, such as a lack of data and a lack of availability at the required spatial resolution [106], because natural, historical and cultural resources do not have an explicit monetary value. A different conclusion is reached if the cost of living with regard to the maintenance of nature in acceptable conditions is compared to conditions in which nature is allowed to degrade [20], showing that the single-value approaches are not an option anymore [110].

7. Conclusions

NC produces multiple ecosystem services with differences in values in human life and measurement requirements. The values vary between time and space. Valuation of an individual service or by a single method may result in the overestimation of values of some of them. At the same time, the exploitation of NC generates costs that translate into negative externalities or trade-offs for the environment and for society.

In real life, people do compromise between them. Policy and management decision-making requires information of different dimensions. Information from integrated valuation methods would provide information from different aspects and help policymakers to make informed and pragmatic decisions.

ES valuation does not aim to establish prices in order to capitalize on ESs through the market. Instead, it highlights how ESs contribute to human well-being and welfare and how they are an essential tool for developing efficient public policies and strategies based on scientific evidence. Utilitarian and nonutilitarian approaches to NC have developed multiple methods and techniques for assessing different types of value for ecosystems. However, there is still a significant lack of reliable evidence on nonuse values of ESs. Many approaches to ES assessment remain controversial because they raise concerns related to the availability and accuracy of data. Establishing accurate methods for calculating VET of ESs, as well as indicators and methods for their modeling and calculation, is a topic that requires further research. Using a pluralist framework composed of a set of decision-making instruments adapted to specific spatial and temporal scales involved, in which CBA is an important component, will allow identifying win–win areas and areas of potential conflicts, both for people and for the environment. Such techniques may be the best solution for supporting the public policy measures needed to mitigate current challenges. In recent years, there has been an increased focus on how climate change affects ecosystems, as well as on how ESs connect to sustainability science topics like environmental economies, bioeconomies and circular economies. Further research that utilizes an integrative approach to connect ES valuation to sustainability science is needed in order to support the decision-making process and public policies.

Author Contributions: Conceptualization, G.E.B., C.E.D. and E.N.A.; methodology, G.E.B., software, G.E.B.; validation, G.E.B.; formal analysis, G.E.B.; investigation, G.E.B.; resources, G.E.B.; data curation, G.E.B. and E.N.A.; writing—original draft preparation, G.E.B.; writing—review and editing, G.E.B., C.E.D. and E.N.A.; visualization, C.E.D.; supervision, E.N.A.; project administration, G.E.B.; funding acquisition, G.E.B. and E.N.A. All authors have read and agreed to the published version of the manuscript.

Funding: This study was funded by the Romanian Ministry of Research and Innovation, within the Nucleu National Programme, the Project PN-19070109, Contract No. 12N/2019.

Acknowledgments: We would like to thank to Bogdan Apostol, the project coordinator, for his support in preparing this paper.

Conflicts of Interest: The authors declare no conflict of interest.

References

1. Millennium Ecosystem Assessment. *Ecosystems and Human Well-Being: Synthesis*; Island Press: Washington, DC, USA, 2005.
2. Salles, J.M. Valuing biodiversity and ecosystem services: Why put economic valuation Nature? *Comptes Rendus Biol.* **2011**, *334*, 469–482. [CrossRef]
3. Loomis, J.J.; Knaus, M.; Dziedzic, M. Integrated quantification of forest total economic value. *Land Use Policy* **2019**, *84*, 335–346. [CrossRef]
4. Azul, A.M.; Brandli, L.; Salvia, A.L.; Wall, T. *Life on Land: Encyclopedia of the UN Sustainable Development Goals*; Springer: Cham, Switzerland, 2020; pp. 1–11. [CrossRef]
5. Pearson, T.R.H. *Measurement Guidelines for Forest Carbon Sequestration*; US Department of Agriculture, Forest Service, Northern Research Station: Newton Square, PA, USA, 2007.
6. Kornatowska, B.; Sienkiewicz, J. Forest ecosystem services—Assessment methods. *For. Ecosyst. Serv. Assess. Methods* **2018**, *60*, 248–260. [CrossRef]
7. Caetano, M. LULC Applications, Advanced Training Course on Land Remote Sensing, D3L1. 5 September 2007. Available online: https://earth.esa.int/landtraining07/D3L1-Caetano.pdf (accessed on 24 February 2021).
8. Nkonya, E.M.; Phillip, D.; Mogues, T.; Pender, J.; Kato, E. Impacts of Community-driven Development Programs on Income and Asset Acquisition in Africa: The Case of Nigeria. *World Dev.* **2012**, *40*, 1824–1838. [CrossRef]
9. Kindu, M.; Schneider, T.; Teketay, D.; Knoke, T. Changes of ecosystem service values in response to land use/land cover dynamics in Munessa–Shashemene landscape of the Ethiopian highlands. *Sci. Total Environ.* **2016**, *547*, 137–147. [CrossRef]
10. Quintas-Soriano, C.; Castro, A.J.; Castro, H.; García-Llorente, M. Impacts of land use change on ecosystem services and implications for human well-being in Spanish drylands. *Land Use Policy* **2016**, *54*, 534–548. [CrossRef]
11. Xu, L.; Saatchi, S.S.; Shapiro, A.; Meyer, V.; Ferraz, A.; Yang, Y.; Bastin, J.-F.; Banks, N.; Boeckx, P.; Verbeeck, H. Spatial Distribution of Carbon Stored in Forests of the Democratic Republic of Congo. *Sci. Rep.* **2017**, *7*, 1–12. [CrossRef]
12. Ligate, E.J.; Chen, C.; Wu, C. Evaluation of tropical coastal land cover and land use changes and their impacts on ecosystem service values. *Ecosyst. Health Sustain.* **2018**, *4*, 188–204. [CrossRef]
13. Giurgiu, V.; Badea, O. *Pădurile și Schimbările de Mediu în Romania. Schimbări Climatice Globale—Grija Pentru Resurse Naturale*; Editura Academiei Române: București, Romania, 2015; pp. 69–92.
14. Chan, K.M.A.; Guerry, A.D.; Balvanera, P.; Klain, S.; Satterfield, T.; Basurto, X.; Bostrom, A.; Chuenpagdee, R.; Gould, R.K.; Halpern, B.S.; et al. Where are Cultural and Social in Ecosystem Services? A Framework for Constructive Engagement. *BioScience* **2012**, *62*, 744–756. [CrossRef]
15. Vasiljevic, N.; Gavrilovic, S. Cultural Ecosystem Services. In *Life on Land*; Leal Filho, W., Azul, A.M., Brandli, L., Lange Salvia, A., Wall, T., Eds.; Springer International Publishing: Cham, Switzerland, 2021; pp. 209–218. ISBN 978-3-319-95981-8.
16. D'Amato, D.; Bartkowski, B.; Droste, N. Reviewing the interface of bioeconomy and ecosystem service research. *Ambio* **2020**, *49*, 1878–1896. [CrossRef]
17. Fang, X.; Zhou, B.; Tu, X.; Ma, Q.; Wu, J. "What Kind of a Science is Sustainability Science?" An Evidence-Based Reexamination. *Sustainability* **2018**, *10*, 1478. [CrossRef]
18. Aarikka-Stenroos, L.; Ritala, P.; Thomas, L.D.W. Circular economy ecosystems: A typology, definitions, and implications. In *Handbook of Sustainability Agency*; Teerikangas, S., Onkila, T., Koistinen, K., Mäkelä, M., Eds.; Edgar: Cheltenham, UK, 2021.
19. Kim, M.-S.; Stepchenkova, S. Altruistic values and environmental knowledge as triggers of pro-environmental behavior among tourists. *Curr. Issues Tour.* **2020**, *23*, 1575–1580. [CrossRef]
20. Costanza, R.; de Groot, R.; Sutton, P.; van der Ploeg, S.; Anderson, S.J.; Kubiszewski, I.; Farber, S.; Turner, R.K. Changes in the global value of ecosystem services. *Glob. Environ. Chang.* **2014**, *26*, 152–158. [CrossRef]
21. Martin, C.; Czellar, S. Where do biospheric values come from? A connectedness to nature perspective. *J. Environ. Psychol.* **2017**, *52*, 56–68. [CrossRef]
22. Daniel, T.C.; Muhar, A.; Arnberger, A.; Aznar, O.; Boyd, J.W.; Chan, K.M.A.; Costanza, R.; Elmqvist, T.; Flint, C.G.; Gobster, P.H.; et al. Contributions of cultural services to the ecosystem services agenda. *Proc. Natl. Acad. Sci. USA* **2012**, *109*, 8812–8819. [CrossRef]
23. Plato, L.; Meskin, A. Aesthetic Value. In *Encyclopedia of Quality of Life and Well-Being Research*; Michalos, A.C., Ed.; Springer International Publishing: Dordrecht, The Netherlands, 2014; pp. 76–78.
24. Brady, E. Aesthetics in Practice: Valuing the Natural World. *Environ. Values* **2006**, *15*, 277–291. [CrossRef]
25. Belmontes, J.A.; Pintor, A.L.; Rodríguez, M.A.; Gómez-Sal, A. On the usefulness of ecosystem services evaluations. *Pirineos* **1997**, *149–150*, 145–152. [CrossRef]
26. Pascual, U.; Muradian, R.; Brander, L.; Gómez-Maes, J.E.; Paracchini, M.P.; Zulian, G.; Alkemade, R. Synergies and trade-offs between ecosystem service supply, biodiversity and habitat conservation status in Europe. *Biol. Conserv.* **2012**, *1*, 155.
27. Coscieme, L.; Stout, J.C. Ecosystem Services Evaluation. In *Encyclopedia of Ecology*, 2nd ed.; Brian, F., Ed.; Elsevier: Oxford, UK, 2019; pp. 288–293.
28. Morelli, F.; Jiguet, F.; Sabatier, R.; Dross, C.; Princé, K.; Tryjanowski, P.; Tichita, M. Spatial covariance between ecosystem services and biodiversity pattern at a national scale (France). *Ecol. Indic.* **2017**, *82*, 574–586. [CrossRef]

29. Czúcz, B.; Haines-Young, R.; Kiss, M.; Bereczki, K.; Kertész, M.; Vária, A.; Potschin-Youngb, M.; Arany, I. Ecosystem service indicators along the cascade: How do assessment and mapping studies position their indicators? *Ecol. Indic.* **2020**, *118*, 106729. [CrossRef]
30. Dick, J.; Al-Assaf, A.; Andrews, C.; Díaz-Delgado, R.; Groner, E.; Halada, L'.; Izakovičová, Z.; Kertész, M.; Khoury, F.; Krašić, D.; et al. Ecosystem services: A rapid assessment method tested at 35 sites of the LTER-Europe network. *Ekológia* **2014**, *33*, 217–231.
31. Maes, J.; Teller, A.; Erhard, M.; Liquete, C.; Braat, L. *Mapping and Assessment of Ecosystems and their Services: An Analytical Framework for Ecosystem Assessments under Action 5 of the EU Biodiversity Strategy to 2020*; Publication Office of the European Union: Luxembourg, 2013; Available online: http://ec.europa.eu/environment/nature/knowledge/ecosystem_assessment/pdf/MAESWorkingPaper2013.pdf (accessed on 24 February 2021).
32. Reyers, B.; O'Farrell, P.J.; Cowling, R.M.; Egoh, B.N.; Le Maitre, D.C.; Vlok, J.H.J. Ecosystem Services, Land-Cover Change, and Stakeholders: Finding a Sustainable Foothold for a Semiarid Biodiversity Hotspot. *Ecol. Soc.* **2009**, *14*. Available online: http://www.ecologyandsociety.org/vol14/iss1/art38/ (accessed on 24 February 2021). [CrossRef]
33. Kates, R.W. What kind of a science is sustainability science? *Proc. Natl. Acad. Sci. USA* **2011**, *108*, 19449–19450. [CrossRef] [PubMed]
34. Turner, B.L.; Kasperson, R.E.; Matson, P.A.; McCarthy, J.J.; Corell, R.W.; Christensen, L.; Eckley, N.; Kasperson, J.X.; Luers, A.; Martello, M.L.; et al. A framework for vulnerability analysis in sustainability science. *Proc. Natl. Acad. Sci. USA* **2003**, *100*, 8074–8079. [CrossRef]
35. Wegner, G.; Pascual, U. Cost-benefit analysis in the context of ecosystem services for human well-being: A multidisciplinary critique. *Glob. Environ. Chang.* **2011**, *21*, 492–504. [CrossRef]
36. Simpson, R.D. Ecosystem Services: What are the Public Policy Implications. PERC POLICY SERIES, No. 55. 2016. Available online: https://www.perc.org/wp-content/uploads/old/pdfs/PS-55-EcosystemServices_Simpson_PERC.pdf (accessed on 25 February 2021).
37. Popa, B.; Coman, C.; Borz, S.A.; Nita, D.M.; Codreanu, C.; Ignea, G.; Marinescu, V.; Ioraş, F.; Ionescu, O. Total Economic Value of Natural Capital—A Case Study of Piatra Craiului National Park. *Not. Bot. Horti Agrobot. Cluj-Napoca* **2013**, *41*, 608–612. [CrossRef]
38. Popa, B.; Pascu, M.; Nita, D.M.; Codreanu, C. *The Value of Forest Ecosystem Services in Romanian Protected Areas—A Comparative Analysis of Management Scenarios*; Bulletin of the Transilvania University of Braşov, Series II: Forestry, Wood Industry, Agricultural Food Engineering, Vol. 6 (55) No.2—2013; Transilvania University of Braşov: Braşov, Romania, 2013.
39. Freeman, A.M., III; Joseph, A.; Catherine, L.K. The Measurement of Environmental and Resource Values: Theory and Methods. In *Resources for the Future*, 3rd ed.; RFF Press: New York, NY, USA, 2002; p. 479.
40. Ghilarov, A.M. Ecosystem functioning and intrinsic value of biodiversity. *Oikos* **2000**, *90*, 408–412. [CrossRef]
41. Balian, E.V.; Drius, L.; Eggermont, H.; Livoreil, B.; Vandewalle, M.; Vandewoestjine, S.; Wittmer, H.; Young, J. Supporting evidence-based policy on biodiversity and ecosystem services: Recommendations for effective policy briefs. *Evid. Policy* **2016**, *12*, 431–451. [CrossRef]
42. Potschin, M.; Haines-Young, R.; Fish, R.R.; Turner, K. Ecosystem Services and Climate Change. In *Routledge Handbook of Ecosystem Services*; Routledge: London, UK, 2018; pp. 481–490. ISBN 978-1-13-802508-0. Available online: https://www.routledge.com/products/9781138025080 (accessed on 26 February 2021).
43. Joachim, H.; Spangenberg, J.S. Value pluralism and economic valuation—defendable if well done. *Ecosyst. Serv.* **2016**, *18*, 100–109.
44. Agenda 21: Programme of action for sustainable development; Rio Declaration on Environment and Development. Statement of Forest Principles. In Proceedings of the Report of the United Nations Conference on Environment and Development, Rio de Janeiro, Brazil, 3–14 June 1992; Volume 3.
45. *United Nations Framework Convention on Climate Change*; UN: New York, NY, USA, 1992.
46. *Convention on Biological Diversity*; UN: New York, NY, USA, 1992. Available online: https://www.cbd.int/convention/text/ (accessed on 2 March 2021).
47. Perrings, C.; Naeem, S.; Ahrestani, F.; Bunker, D.E.; Burkill, P.; Canziani, G.; Elmqvist, T.; Ferrati, R.; Fuhrman, J.; Jaksic, F.; et al. Ecosystem Services for 2020. *Policy Forum Sci.* **2010**, *330*. Available online: https://science.sciencemag.org/content/suppl/2010/10/12/330.6002.323.DC1 (accessed on 3 March 2021). [CrossRef]
48. Paris Agreement. United Nation 2015. Available online: https://unfccc.int/sites/default/files/english_paris_agreement.pdf (accessed on 3 March 2021).
49. Parida, V.; Burström, T.; Visnjic, I.; Wincent, J. Orchestrating industrial ecosystem in circular economy: A two-stage transformation model for large manufacturing companies. *J. Bus. Res.* **2019**, *101*, 715–725. [CrossRef]
50. Bolger, M.; Schweitzer, J.-P.; Arditi, S.; Lutter, S. A circular economy within ecological limits. In *Friends of the Earth Europe*; European Environmental Bureau, Institute for Ecological Economics WU: Wien, Austria, 2019.
51. Neill, A.M.; O'Donoghue, C.; Stout, J.C. A Natural Capital Lens for a Sustainable Bioeconomy: Determining the Unrealised and Unrecognised Services from Nature. *Sustainability* **2020**, *12*, 8033. [CrossRef]
52. Verkerk, P.J.; Costanza, R.; Hetemäki, L.; Kubiszewski, I.; Leskinen, P.; Nabuurs, G.J.; Potočnik, J.; Palahí, M. Climate-Smart Forestry: The missing link. *For. Policy Econ.* **2020**, *115*, 102164. [CrossRef]
53. Furman, E.; Häyhä, T.; Hirvilammi, T. *A Future the Planet can Accommodate*; Finnish Environment Institute: Helsinki, Finland, 2018; pp. 1–4. ISBN 978-952-11-4938-2.

54. Hove, S.V.D. A rationale for science–policy interfaces. *Future* **2007**, *39*, 807–826. [CrossRef]
55. Fisher, B.; Turner, R.K.; Morling, P. Defining and classifying ecosystem services for decision making. *Ecol. Econ.* **2009**, *68*, 643–653. [CrossRef]
56. Commission Staff Working Document. *EU Guidance on Integrating Ecosystems and Their Services into Decision-Making*; SWD (2019) 305 Final PART 1/3; European Commission: Brussels, Belgium, 18 July 2019.
57. Karen, E.; Rebecca, M.A. Moving beyond the exchange value in the non-market valuation of ecosystem services. *Ecosyst. Serv.* **2016**, *18*, 78–86.
58. Baggethun, B.; Martín-López, M.V.; Armsworth, P.; Christie, M.; Cornelissen, H.; Eppink, F.; Farley, J.; Loomis, J.; Pearson, L.; Perrings, C.; et al. Chapter 5: The economics of valuing ecosystem services and biodiversity. In *The Economics of Ecosystems and Biodiversity—The Ecological and Economic Foundations*; Kumar, P., Ed.; Earthscan: London, UK, 2010.
59. David, W.; Pearce, R.; Turner, K. Economics of natural resources and the environment. *Am. J. Agric. Econ.* **1991**. [CrossRef]
60. Jamie, A.; Carr, G.; Petrokofsky, D.; Spracklen, V.; Simon, L.; Lewis, D.; Nicholas, R.; Trull, A.; Vidal, S.; Wicanderi, J.; et al. Anticipated impacts of achieving SDG targets on forests—A review. *For. Policy Econ.* **2021**, *126*, 102423.
61. Schaefer, M.; Goldman, E.; Bartuska, A.M.; Sutton-Grier, A.; Lubchenco, J. Nature as capital: Advancing and incorporating ecosystem services in United States federal policies and programs. *Proc. Natl. Acad. Sci. USA* **2015**, *112*, 7383–7389. [CrossRef]
62. Naidoo, R.; Ricketts, T.H. Mapping the Economic Costs and Benefits of Conservation. *PLoS Biol.* **2006**, *4*, e360. [CrossRef]
63. Nabuurs, G.-J.; Verkerk, P.J.; Schelhaas, M.-J.; Olabarria, J.R.G.; Trasobares, A.; Cienciala, E. Climate-Smart Forestry: Mitigation impacts in three European regions. *Sci. Policy* **2018**. [CrossRef]
64. Fitzgerald, J.; Lindner, M. Adaptive Challenges for European Forests. Adapting forests to climate change Symposium. In Proceedings of the International Conference on Ecological Sciences, Marseille, France, 25 October 2016.
65. Sabbadin, D. *The Elephant is in the Room. Why it Makes Sense Giving Priority to Circular Economy Measures in the Building Industry in the 2020s*; EEB: Brussels, Belgium, 2021.
66. Sousa-Silva, R.; Verbist, B.; Ângela, L.; Valent, P.; Suškevičs, M.; Picard, O.; Hoogstra-Klein, M.A.; Cosofret, V.-C.; Bouriaud, L.; Ponette, Q.; et al. Adapting forest management to climate change in Europe: Linking perceptions to adaptive responses. *For. Policy Econ.* **2018**, *90*, 22–30. [CrossRef]
67. Wijewardana, D. Criteria and indicators for sustainable forest management: The road travelled and the way ahead. *Ecol. Indic.* **2008**, *8*, 115–122. [CrossRef]
68. El-Chichakli, B.; Von Braun, J.; Lang, C.; Barben, D.; Philp, J. Policy: Five cornerstones of a global bioeconomy. *Nat. Cell Biol.* **2016**, *535*, 221–223. [CrossRef]
69. D'Amato, D.; Korhonen, J.; Toppinen, A. Circular, Green, and Bio Economy: How Do Companies in Land-Use Intensive Sectors Align with Sustainability Concepts? *Ecol. Econ.* **2019**, *158*, 116–133. [CrossRef]
70. Dietz, T.; Börner, J.; Förster, J.J.; Von Braun, J. Governance of the Bioeconomy: A Global Comparative Study of National Bioeconomy Strategies. *Sustainability* **2018**, *10*, 3190. [CrossRef]
71. de Groot, R.; Brander, L.; van der Ploeg, S.; Costanza, R.; Bernard, F.; Braat, L.; Christie, M.; Crossman, N.; Ghermandi, A.; Hussain, L.H.S.; et al. Global estimates of the value of ecosystems and their services in monetary units. *Ecosyst. Serv.* **2012**, *1*, 50–61. [CrossRef]
72. Ojea, E.; Loureiro, M.L. Altruistic, egoistic and biospheric values in willingness to pay (WTP) for wildlife. *Ecol. Econ.* **2007**, *63*, 807–814. [CrossRef]
73. Costanza, R.; De Groot, R.; Braat, L.; Kubiszewski, I.; Fioramonti, L.; Sutton, P.; Farber, S.; Grasso, M. Twenty years of ecosystem services: How far have we come and how far do we still need to go? *Ecosyst. Serv.* **2017**, *28*, 1–16. [CrossRef]
74. Kontogianni, A.; Tourkolias, C.; Machleras, A.; Skourtos, M. Service providing units, existence values and the valuation of endangered species: A methodological test. *Ecol. Econ.* **2012**, *79*, 97–104. [CrossRef]
75. Nunes, P.A.; Schokkaert, E. Identifying the warm glow effect in contingent valuation. *J. Environ. Econ. Manag.* **2003**, *45*, 231–245. [CrossRef]
76. Marre, J.-B.; Brander, L.; Thebaud, O.; Boncoeur, J.; Pascoe, S.; Coglan, L.; Pascal, N. Non-market use and non-use values for preserving ecosystem services over time: A choice experiment application to coral reef ecosystems in New Caledonia. *Ocean Coast. Manag.* **2015**, *105*, 1–14. [CrossRef]
77. Hernández-Blanco, M.; Costanza, R.; Anderson, S.; Kubiszewski, I.; Sutton, P. Future scenarios for the value of ecosystem services in Latin America and the Caribbean to 2050. *Curr. Res. Environ. Sustain.* **2020**, *2*, 100008. [CrossRef]
78. Gómez-Baggethun, E.; Martín-López, B.; Barton, D.; Braat, L.; Saarikoski, H.; Kelemen, M.; García-Llorente, E.; van den Bergh, P.J.; Arias, P.; Berry, L.; et al. *State-of-the-Art Report on Integrated Valuation of Ecosystem Services*; Deliverable D.4.1/WP4; European Commision: Brussels, Belgium, July 2014.
79. Sharp, R.; Douglass, J.; Wolny, S.; Arkema, K.; Bernhardt, J.; Bierbower, W.; Chaumont, N.; Denu, D.; Fisher, D.; Glowinski, K.; et al. InVEST 3.9.0.post71+ug.gfb92465 User's Guide. Available online: https://invest-userguide.readthedocs.io/en/latest/ (accessed on 5 March 2021).
80. Zhang, P.; Zhou, C.; Wang, P.; Gao, B.J.; Zhu, X.; Guo, L. E-Tree: An Efficient Indexing Structure for Ensemble Models on Data Streams. *IEEE Trans. Knowl. Data Eng.* **2014**, *27*, 461–474. [CrossRef]
81. Zulian, G.; Polce, C.; Maes, J. ESTIMAP: A GIS-based model to map ecosystem services in the European Union. *Ann. Bot.* **2014**, *4*, 1–7.

82. Zulian, Z.; Paracchini, M.L.; Maes, J.; Liquete, C. *ESTIMAP: Ecosystem Services Mapping at European Scale*; EUR 26474; Publications Office of the European Union: Luxembourg, 2013; ISSN 1831-9424. [CrossRef]
83. Kopperoinen, L.; Itkonen, P.; Niemelä, J. Using expert knowledge in combining green infrastructure and ecosystem services in land use planning: An insight into a new place-based methodology. *Landsc. Ecol.* **2014**, *29*, 1361–1375. [CrossRef]
84. Gibbons, S.; Mourato, S.; Resende, G.M. The Amenity Value of English Nature: A Hedonic Price Approach. *Environ. Resour. Econ.* **2014**, *57*, 175–196. [CrossRef]
85. Mandle, L.; Douglass, J.; Lozano, J.S.; Sharp, R.P.; Vogl, A.L.; Denu, D.; Walschburger, T.; Tallis, H. OPAL: An open-source software tool for integrating biodiversity and ecosystem services into impact assessment and mitigation decisions. *Environ. Model. Softw.* **2016**, *84*, 121–133. [CrossRef]
86. Carpenter, S.R.; Bennett, E.M.; Peterson, G.D. Scenarios for ecosystem services: An overview. Research, part of a Special Feature on Scenarios of global ecosystem services. *Ecol. Soc.* **2006**, *11*, 29. Available online: http://www.ecologyandsociety.org/vol11/iss1/art29/ (accessed on 7 March 2021). [CrossRef]
87. Torabi, F.; Ahmadi, P. Chapter 9—Techno-economic assessment of battery systems. In *Simulation of Battery Systems*; Academic Press: Cambridge, MA, USA, 2020; pp. 311–352. ISBN 978-0-12-816212-5. Available online: https://doi.org/10.1016/B978-0-12-816212-5.00013-1 (accessed on 7 March 2021).
88. Bjørn, A.; Molin, C.; Hauschild, M.Z.; Owsianiak, M. LCA History. In *Life Cycle Assessment: Theory and Practice*; Hauschild, M., Rosenbaum, R.K., Olsen, S., Eds.; Springer: Berlin/Heidelberg, Germany, 2017; pp. 17–30.
89. Curran, M.A. Life Cycle Assessment: A review of the methodology and its application to sustainability. *Curr. Opin. Chem. Eng.* **2013**, *2*, 273–277. [CrossRef]
90. Liu, X.; Bakshi, B.R.; Rugani, B.; de Souza, D.M.; Bare, J.; Johnston, J.M.; Laurent, A.; Verones, F. Quantification and valuation of ecosystem services in life cycle assessment: Application of the cascade framework to rice farming systems. *Sci. Total Environ.* **2020**, *747*, 141278. [CrossRef] [PubMed]
91. Chaplin-Kramer, R.; Sim, S.; Hamel, P.; Bryant, B.; Noe, R.; Mueller, C.; Rigarlsford, G.; Kulak, M.; Kowal, V.; Sharp, R.; et al. Life cycle assessment needs predictive spatial modelling for biodiversity and ecosystem services. *Nat. Commun.* **2017**, *8*, 15065. [CrossRef]
92. De Groot, R.S.; Wilson, M.A.; Boumans, R.M.J. A typology for the classification, description and valuation of ecosystem functions, goods and services. *Ecol. Econ.* **2002**, *41*, 393–408. [CrossRef]
93. Pagiola, S.; Bishop, J.; Landell, N. *Mills-Selling Forest Environmental Services: Market-Based Mechanisms for Conservation and Development*, 1st ed.; Routledge: London, UK, 2002; p. 320.
94. Schmidt, K.; Sachse, R.; Walz, A. Current role of social benefits in ecosystem service assessments. *Landsc. Urban Plan.* **2016**, *149*, 49–64. [CrossRef]
95. Bernetti, I.; Sottini, V.A.; Marinelli, N.; Marone, E. Quantification of the total economic value of forest systems: Spatial analysis application to the region of Tuscany (Italy). *Aestimum* **2013**. [CrossRef]
96. Crook, S.; Levine, A.; Lopez-Carr, D. Perceptions and Application of the Ecosystem Services Approach among Pacific Northwest National Forest Managers. *Sustainability* **2021**, *13*, 1259. [CrossRef]
97. Grainger, D.; Stoeckl, N. The importance of social learning for non-market valuation. *Ecol. Econ.* **2019**, *164*, 106339. [CrossRef]
98. Carpenter, S.R.; Mooney, H.A.; Agard, J.; Capistrano, D.; DeFries, R.S.; Diaz, S.; Dietz, T.; Duraiappah, A.K.; Oteng-Yeboah, A.; Pereira, H.M.; et al. Science for managing ecosystem services: Beyond the Millennium Ecosystem Assessment. *Proc. Natl. Acad. Sci. USA* **2009**, *106*, 1305–1312. [CrossRef] [PubMed]
99. Villegas-Palacio, C.; Berrouet, L.; López, C.; Ruiz, A.; Upegui, A. Lessons from the integrated valuation of ecosystem services in a developing country: Three case studies on ecological, socio-cultural and economic valuation. *Ecosyst. Serv.* **2016**, *22*, 297–308. [CrossRef]
100. Gretchen, C.D.; Pamela, A.M. Ecosystem services: From theory to implementation. *Proc. Natl. Acad. Sci. USA* **2008**, *105*, 9455–9456. Available online: www.pnas.org_cgi_doi_10.1073_pnas.0804960105 (accessed on 12 March 2021).
101. Daily, G.C.; Polasky, S.; Goldstein, J.; Kareiva, P.M.; Mooney, H.A.; Pejchar, L.; Ricketts, T.H.; Salzman, J.; Shallenberger, R. Ecosystem services in decision making: Time to deliver. *Front. Ecol. Environ.* **2009**, *7*, 21–28. [CrossRef]
102. Droste, N.; Bartkowski, B. Ecosystem Service Valuation for National Accounting: A Reply to Obst, Hein and Edens (2016). *Environ. Resour. Econ.* **2018**, *71*, 205–215. [CrossRef]
103. Obst, C.; Hein, L.; Edens, B. National Accounting and the Valuation of Ecosystem Assets and Their Services. *Environ. Resour. Econ.* **2016**, *64*, 1–23. [CrossRef]
104. System of Environmental-Economic Accounting-Ecosystem Accounting, Final Draft. Department of Economic and Social Affairs. Statistics Division United Nations, Version 5 February 2021. Available online: https://seea.un.org/ecosystem-accounting (accessed on 16 March 2021).
105. Turner, K.; Badura, T.; Ferrini, S. Natural capital accounting perspectives: A pragmatic way forward. *Ecosyst. Health. Sustain.* **2019**, *5*, 237–241. [CrossRef]
106. Heckwolf, M.J.; Peterson, A.; Jänes, H.; Horne, P.; Künne, J.; Liversage, K.; Sajeva, M.; Reusch, T.B.; Kotta, J. From ecosystems to socio-economic benefits: A systematic review of coastal ecosystem services in the Baltic Sea. *Sci. Total Environ.* **2021**, *755*, 142565. [CrossRef] [PubMed]

107. Varul, M.Z. Value: Exchange and Use Value. In *Encyclopedia of Consumer Culture*; Southerton, D., Ed.; Project: Capitalist Transcendencies; Sage Publications: Newbury Park, CA, USA, 2011; pp. 1502–1504.
108. Vallecillo, S.; La Notte, A.; Ferrini, S.; Maes, J. How ecosystem services are changing: An accounting application at the EU level. *Ecosyst. Serv.* **2019**, *40*, 101044. [CrossRef]
109. Ecosystem Accounts. Measuring the Contribution of Nature to the Economy and Human Wellbeing. EUROSTAT-Statistics Explained, May 2020; ISSN 2443-8219. Available online: https://ec.europa.eu/eurostat/statistics-explained/index.php/Main_Page (accessed on 18 March 2021).
110. Jacobs, S.; Dendoncker, N.; Martín-López, B.; Barton, D.N.; Gomez-Baggethun, E.; Boeraeve, F.; McGrath, F.L.; Vierikko, K.; Geneletti, D.; Sevecke, K.J.; et al. A new valuation school: Integrating diverse values of nature in resource and land use decisions. *Ecosyst. Serv.* **2016**, *22*, 213–220. [CrossRef]

Article

Applications of TLS and ALS in Evaluating Forest Ecosystem Services: A Southern Carpathians Case Study

Alexandru Claudiu Dobre [1,2], Ionuț-Silviu Pascu [1,2,*], Ștefan Leca [2], Juan Garcia-Duro [2], Carmen-Elena Dobrota [3,4], Gheorghe Marian Tudoran [1] and Ovidiu Badea [1,2]

1 Department of Forest Engineering, Forest Management Planning and Terrestrial Measurements, Faculty of Silviculture and Forest Engineering, "Transilvania" University, 1 Ludwig van Beethoven Str., 500123 Brașov, Romania; dobre.alexandruclaudiu@gmail.com (A.C.D.); tudoran.george@unitbv.ro (G.M.T.); ovidiu.badea63@gmail.com (O.B.)

2 Development in Forestry-Department of Forest Monitoring, "Marin Drăcea" Romanian National Institute for Research, 128 Eroilor Blvd., 077190 Voluntari, Romania; stefan.leca@icas.ro (Ș.L.); juan.garcia.duro@icas.ro (J.G.-D.)

3 Faculty of Business and Administration, University of Bucharest, 4-12 B-dul Regina Elisabeta, County 3, 030018 Bucharest, Romania; dobrotacarmen@yahoo.com

4 Institute of National Economy, Romanian Academy, 13 Calea 13 Septembrie, County 5, 050726 Bucharest, Romania

* Correspondence: ionut.pascu@icas.ro; Tel.: +40-7-2633-2258

Abstract: Forests play an important role in biodiversity conservation, being one of the main providers of ecosystem services, according to the Economics of Ecosystems and Biodiversity. The functions and ecosystem services provided by forests are various concerning the natural capital and the socio-economic systems. Past decades of remote-sensing advances make it possible to address a large set of variables, including both biophysical parameters and ecological indicators, that characterize forest ecosystems and their capacity to supply services. This research aims to identify and implement existing methods that can be used for evaluating ecosystem services by employing airborne and terrestrial stationary laser scanning on plots from the Southern Carpathian mountains. Moreover, this paper discusses the adaptation of field-based approaches for evaluating ecological indicators to automated processing techniques based on airborne and terrestrial stationary laser scanning (ALS and TLS). Forest ecosystem functions, such as provisioning, regulation, and support, and the overall forest condition were assessed through the measurement and analysis of stand-based biomass characteristics (e.g., trees' heights, wood volume), horizontal structure indices (e.g., canopy cover), and recruitment-mortality processes as well as overall health status assessment (e.g., dead trees identification, deadwood volume). The paper, through the implementation of the above-mentioned analyses, facilitates the development of a complex multi-source monitoring approach as a potential solution for assessing ecosystem services provided by the forest, as well as a basis for further monetization approaches.

Keywords: ecosystem services; natural capital; socio-economic system; ecological indicators; terrestrial laser scanning; aerial laser scanning

Citation: Dobre, A.C.; Pascu, I.-S.; Leca, Ș.; Garcia-Duro, J.; Dobrota, C.-E.; Tudoran, G.M.; Badea, O. Applications of TLS and ALS in Evaluating Forest Ecosystem Services: A Southern Carpathians Case Study. *Forests* **2021**, *12*, 1269. https://doi.org/10.3390/f12091269

Academic Editor: Jarosław Socha

Received: 20 August 2021
Accepted: 13 September 2021
Published: 17 September 2021

1. Introduction

Forest is playing a crucial role in biological diversity, local welfare, the balance of carbon emissions, and the global economy [1–3]. In the context of climate change, the understanding of forest ecosystem processes' importance is essential in assuring sustainable management and economic development [4]. Toward this purpose, forest monitoring was established as the main tool for studying the dynamics of forest structure and functioning and its response to anthropogenic influences [3,5]. The necessity of this tool is highlighted by decisional factors' requirements and forest governance [6]. Due to the high complexity of the forest dynamics, a high amount of warranted information is needed in the characterization process.

The primary mechanism of forest monitoring in assuring the data integration is developing forest inventories focused on parameters related to the main dendrometric characteristics of trees (e.g., diameter at breast height (DBH), height-DBH ratio, crown width). Besides these variables, the monitoring also has to take into consideration information regarding the climate (temperature and precipitations) and pollution (atmospheric depositions). However, it is a well-known fact that the traditional forest inventory can be expensive, time-consuming, and requires a large amount of qualified personnel [3]. Moreover, forest inventory is limited to statistically established sample plots, resulting in a weaker representativity at larger scales [7–9].

To overcome the mentioned limitations, alternative solutions and measuring methodologies were sought in the remote-sensing field. In the past decades, remote-sensing systems have evolved, ensuring a large variety of applications [10]. As expected, the remote-sensing portfolio already contains several techniques addressing forest ecology and management [11]. From their beginning, remote-sensing systems were mostly equivalated to satellite imagery. New instruments of interest here, airborne laser scanning, unmanned aerial vehicles, digital photography systems, and terrestrial laser scanners, have more recently captured the researchers' attention, gradually gaining visibility through a large number of scientific studies.

Land cover analysis [12,13], biomass estimation [14–17], hazard identification [18–20], structure assessment [21–26], and ecological indicators are just some of the most frequent applications of remote sensing in forestry. The major advantages of remote sensing are related to its capability of capturing a large amount of data and the possibility of revisiting in relatively short periods, as well as the plurality of the associated analyses [24].

A keen interest in remote sensing was shown toward biophysical parameters, such as DBH, tree height, volume, and implicitly biomass. The majority of these parameters were initially computed employing regression models, with input data derived from crown projections and height measurements from passive sensors [27–29], calibrated with ground samples. New technologies, as is the case of terrestrial laser scanning, propose different approaches for estimating tree characteristics. These provide a more direct method that involves point cloud classification, tree segmentation, and stem reconstruction [30–32]. Besides the biophysical parameters, active remote-sensing systems are used to describe stands' structure through indirect analyses of the number of trees, canopy stratification, and trees distribution. As described in the work of [24,33,34], airborne and terrestrial laser scanning represent optimal solutions in describing forest stands through structural indicators based on point cloud processing.

Regarding this matter, the literature offers a rich variety of active remote-sensing-based forest variables, from foliage indices [24,35,36] (leaf area index—LAI, gap probability—p_{gap}) to trees spatial distribution [37,38] (mostly distance and angles between trees, but also the position itself for marginal trees detection, sampling plot edge effect mitigation, etc.). Satellite imagery also proposes indicators related to the status of forest stand health [39–42], an aspect that will not be detailed here since passive remote sensing does not make the subject of our study.

Disregarding the plethora of variables and its promising evolution, passive remote-sensing technology still demands innovative approaches to address the requirements of ecological relevant indicators [11]. The constant need for ground measurement calibration represents the main disadvantage of most passive remote-sensing systems. Furthermore, the applications based on regression models can lead to important errors due to potentially incorrect assumptions regarding the relationship between forest characteristics [43,44].

In the ecological research field, active remote-sensing data are increasingly being used. Quantifying forest ecosystems information from indices based on active remote-sensing highlights the need for further analysis and adaptation. The processing and uptake of these data are necessary for linking the indicators to the capacity of forest ecosystems to provide benefits. These benefits materialize in what we call ecosystem services and represent the ecosystems' benefits, processes, and assets for providing human well-being [45].

In the field of research, the relationship between ecology and economy has been attributed with great importance, a fact that is corroborated by the very nature of ecosystem services. This has made it possible to develop the concept of natural capital on an environmental basis [46] and led to the idea of value, from a monetary point of view, of the ecosystem services and goods [47]. The need to exploit the benefits of ecosystems derives from their contribution to the human economy [48,49] and their expression in services and commercial goods [50,51].

Nowadays, there is a multitude of methods for evaluating and monetizing services, most of them being subjective. The methods are based on human preferences or physical costs upon which ecosystem services can be integrated [46]. The established methods are based on damaged cost avoided, replacement cost, market price, productivity cost, hedonic pricing, benefits transfer, and contingent evaluation method [51–53].

Despite the difficulties encountered in the process of applying ecosystem evaluation methods, they have an essential role in communicating the value of nature to the decisional factors and policymakers [54]. In this regard, there is an absolute need for objective ecological indicators that can provide information about ecosystem health status and structure.

This paper intends to identify and test several methods and variables applicable to airborne and terrestrial stationary laser scanning to quantify the capacity of the forest ecosystem in providing benefits. The identification of suitable ecosystem services will be performed according to The Economics of Ecosystem and Biodiversity (TEEB) classification [51,55]. Alongside, Millennium Ecosystem Assessment classification (MEA) and Common International Classification of Ecosystem Services (CICES), TEEB represents one of the widely known ecosystem services classification networks. The latter is a global campaign aiming at raising awareness regarding biodiversity's economic benefits and the rising costs of ecosystem degradation. The final purpose of this initiative is to analyze and explain in a mainstream approach the importance of taking action [56]. This classification was adopted because it corresponds faithfully to the functions attributed to the studied stands according to the Romanian forest legislation. The majority of ecosystem functions will be analyzed in relation to the existing indicators, as well as other variables adapted to active remote-sensing sampling. The paper does not intend to calibrate or to validate existing methodologies but to showcase a minimal set of indicators computed through active remote-sensing methods that can offer sufficient information about the ecosystems' capacity to provide services. Furthermore, as mentioned above, the paper aims only at information obtained through the use of ALS and stationary TLS measurements, excluding any other potential data based on satellite imagery or other passive remote-sensing technologies.

2. Materials and Methods

2.1. Study Site

To analyze the identified methods and variables, ten stands were considered in the current study, each of them being designed as a one-hectare rectangular plot with three 15 m-radius circular subplots within them.

The ten one-hectare plots are located in two different areas of the Southern Carpathian mountains, thus covering three of the most representative tree species of Romania. These are sessile oak (*Quercus petraea*) and beech (*Fagus sylvatica*) in the hill region and Norway spruce (*Picea abies*) in the mountainous region (Figure 1).

(a)

(b)

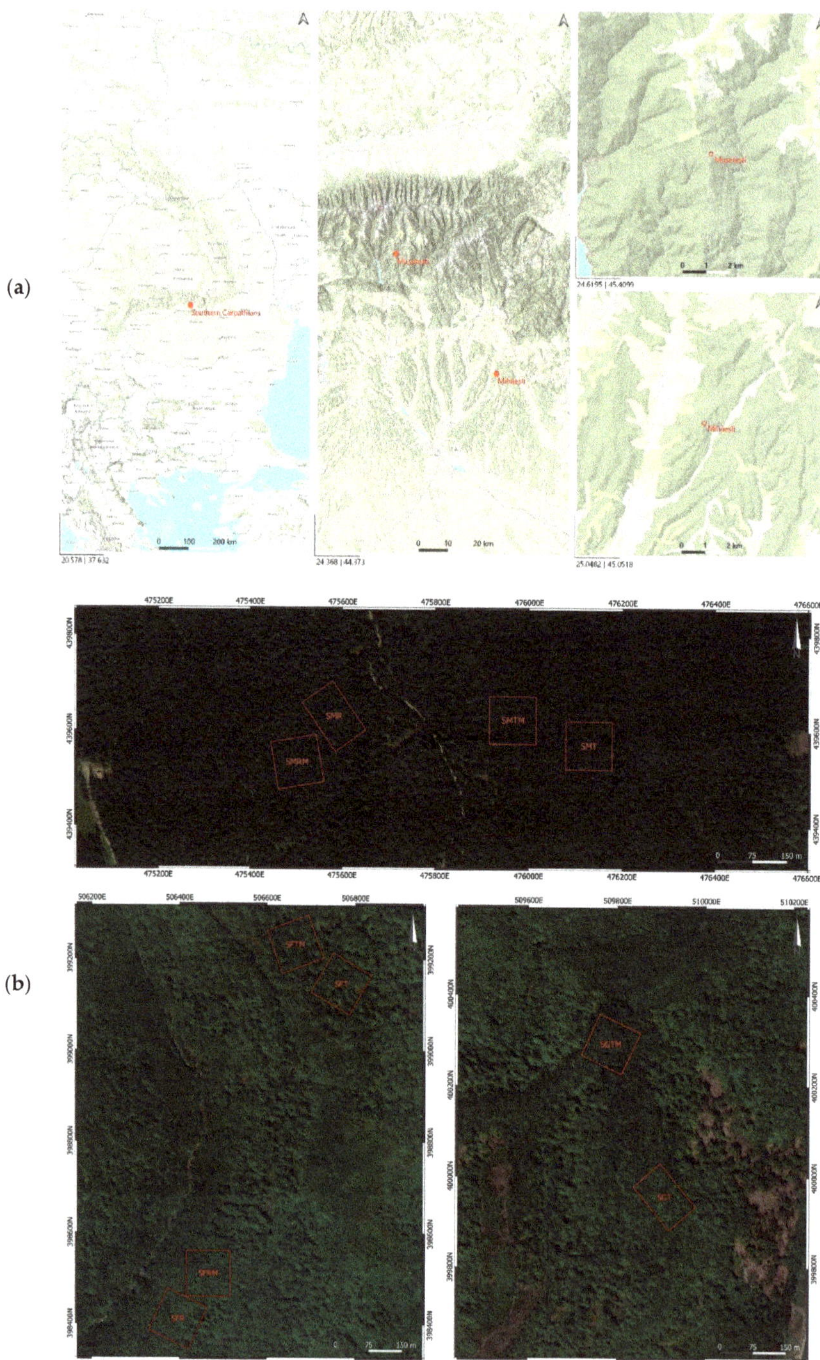

Figure 1. (**a**) Location of the sampled forest stands, (**b**) detailed position (coordinates in WGS 84 projection system).

Both deciduous and coniferous forest plots were considered in the process of assessing the applicability of the studied methods as well as in the evaluation of the different structural characteristics of the plots. Therefore, the plots were chosen in relation to species, age, and applied silvicultural interventions (Table 1).

Table 1. Sample plots characteristics.

Sample Plot	Species	Age [Years]	Silvicultural Interventions	Forest Districts
SGT	Sessile oak	190	Progressive	Mihăești
SGTM	Sessile oak	190	Without interventions	Mihăești
SFR	Beech	40	Thinning	Mihăești
SFRM	Beech	40	Without interventions	Mihăești
SFT	Beech	120	Progressive	Mihăești
SFTM	Beech	120	Without interventions	Mușetești
SMR	Norway spruce	50	Thinning	Mușetești
SMRM	Norway spruce	50	Without interventions	Mușetești
SMT	Norway spruce	150	Progressive	Mușetești
SMTM	Norway spruce	150	Without interventions	Mușetești

2.2. Conventional Field Data Collection

In order to ensure control over the LiDAR data sets, a classical inventory was also carried out in the plots. Field measurements included DBH, tree height, crown height, crown width, and position of each tree (XYZ coordinates) and targeted all the trees with a DBH equal to or greater than 6 cm. To acquire these variables, an integrated GIS field software and electronic mapping and dendrometrics sensors [57] for recording tree positions and canopy characteristics were used.

2.3. Terrestrial Laser Scanner Data

In each 15 m circular subplot, five terrestrial scans were performed accordingly to a cardinal point sampling scheme to compensate for the shadowing (Figure 2a). The scanning process was achieved with a phase shift terrestrial laser scanner [58]. The resulting point clouds were characterized by 8 µs per scan point and over 44 million points per 360° sweep.

(a)

Figure 2. *Cont.*

(**b**)

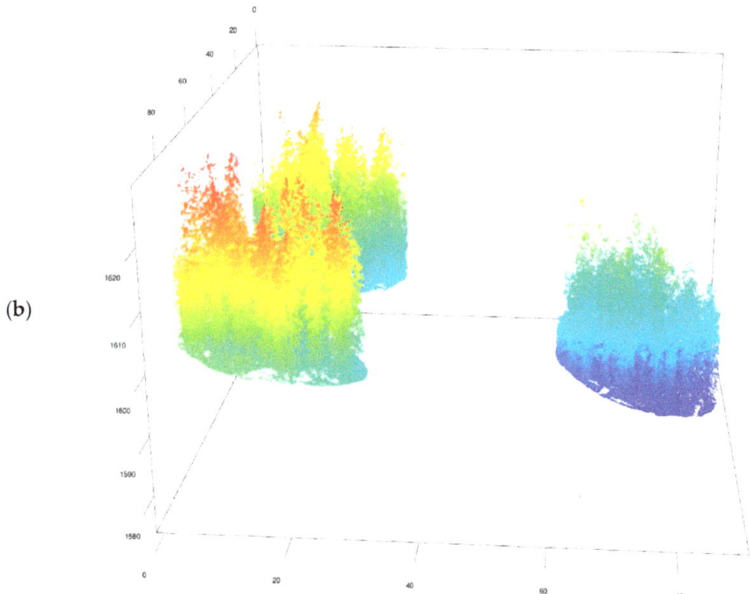

Figure 2. (**a**) TLS ground-based data collection (**b**) terrestrial laser scan of the subplots.

Regarding the TLS pre-processing methods for classification and segmentation of the point cloud prior to obtaining the stems and the foliage, an approach proposed by Pascu et al. was followed [24,30,32,59] (Figure 2b).

2.4. *Airborne Laser Scanner Data*

The airborne LiDAR data for the one-hectare plots were collected through the use of a full-wave airborne laser scanner [60]. The discrete points extraction was conducted by the provider of the data sets, according to the standard processing procedure. Following processing, an average point density of 6 points/m^2 was reached (Figure 3).

Further analyses, such as the ground-non-ground classification, were performed using filtering algorithms by means of dedicated software [61], as shown in the work of [62]. The digital terrain model (DTM) was generated through an inverse distance-weighting interpolation, which ensured a 1×1 m spatial resolution. The DTM was further used as support in the computation of several parameters (e.g., tree height, canopy height).

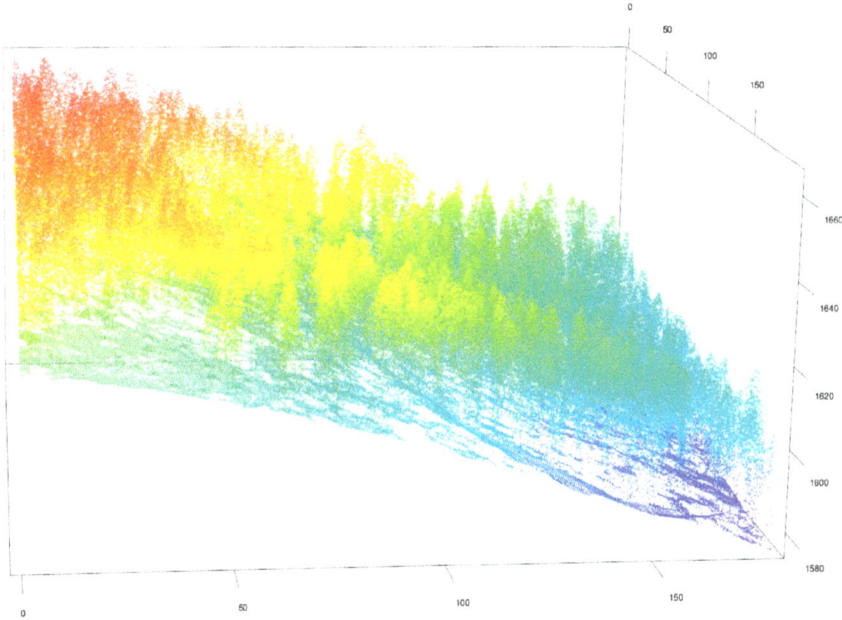

Figure 3. Airborne laser scan of the studied area.

2.5. Ecosystem Services Identification and Evaluation

The literature proposes an entire series of ecosystem functions and services assessment methods (monetary, non-monetary, and integrated methods). In this research, the interest was to gather reliable information needed in applying those evaluation methods. The ecosystem services identification is presented according to the ecosystem functions stated by TEEB, and the paper intends to cover the majority of the functions.

3. Results

3.1. Provisioning Services

Wood products are one of the most prominent resources provided by forest ecosystems [63], being a direct economic benefit that can be easily assessed from a monetary point of view. In literature, wood products, equated to above-ground biomass, are an important variable that can be estimated through remote-sensing techniques. Between the implementation of biophysical parameters relationships [64,65] to allometric models and direct measurements [30,66–68], the above-ground biomass estimation gained impressive interest in research due to the associated accuracy.

In our study, the applied methodology was the one proposed by Pascu et al. in the work of [30]. Therefore, the above-ground estimation implied the use of stand volume derived from number of trees, DBH, and tree height (Figure 4). Even though other studies [69,70] show volume underestimation when based on terrestrial laser scanning data, this was due to low stand heterogeneity. The accuracy presented by Pascu et al. in what concerns the number of trees and DBH is more than satisfactory (errors under 5%). Moreover, the use of terrestrial laser scanning proved to be an adequate approach for the above-ground volume computation [71].

Height values computed through this active remote-sensing technology show biases and errors, also highlighted by several research papers [30,69,70,72,73]. To overcome this limitation, compensations were applied based on airborne laser scanning.

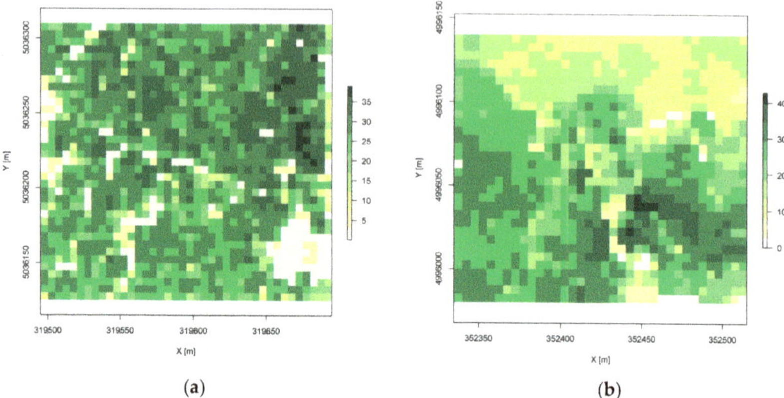

Figure 4. Canopy height model from airborne laser scanning (**a**) SMTM (**b**) SGTM.

For computing above-ground tree volume, the logarithmic regression equation (Equation (1)) described in the work of [74] was used:

$$\log v = a_0 + a_1 \log d + a_2 log^2 d + a_3 \log h + a_4 log^2 h \quad (1)$$

where:

d—tree diameter at breast height;

v—tree volume;

h—tree height;

a_0, a_1, a_2, a_3, a_4—species-specific regression coefficients.

The above-ground volumes for each plot are presented in Table 2. When compared to the field measurements based on the same methodology (Equation (1)), the errors are between 4.6% and 13.3%.

Table 2. Above-ground volume and mean stand characteristics.

Plot	V [m^3 ha^{-1}]	d_m [cm]	h_m [m]	v_m [m^3]
SGT	444.1	21.91	20.7	0.97
SGTM	646.4	24.15	22.36	0.90
SFR	434.6	17.81	21.9	0.37
SFRM	509.8	18.25	26.26	0.48
SFT	457.2	24.83	18.9	1.07
SFTM	622.3	25.45	19.4	1.05
SMR	345.7	17.3	17.8	0.28
SMRM	420.1	17.45	15.8	0.29
SMT	409.5	29.29	21.6	0.90
SMTM	558.3	33.01	26.6	0.93

V—stand volume; d_m—stand mean diameter; h_m—stand mean height; v_m—tree mean volume.

Considering the biophysical parameters, differences in mean stand volume can be observed between the plots where silvicultural interventions were applied and those without interventions. The reduced volume, specific to the young forest stands and to those targeted by interventions, confirms the viability of the methods and results and makes it possible to compare them in terms of wood product provisioning.

3.2. Regulating Services

At the moment, the ecosystem services specific to regulating functions represent a great challenge in the evaluating processes [54]. This function includes services for air quality regulation, moderation of extreme events, erosion prevention, and carbon sequestration [51,55,75]. In the context of evaluating the related services, specific indicators were developed in the field of ecological research.

The assessment methods tend to use indirect measurements and quantify the relationship between different variables. Tree canopy cover, canopy structure indices (e.g., leaf area index), and trees distribution are the most used parameters in the majority of the evaluating approaches [76–78].

3.2.1. Structural Indices

As previously mentioned, forest structure characteristics and biodiversity are the main sources of information for the assessment of ecosystem services. To establish the capacity of ecosystems in supplying regulating services, indices such as Clark-Evans nearest neighbor index (CE), uniform angle index (*UAI*), and relative dominance diameter index were computed at the subplot level.

Clark-Evans nearest neighbor index (CE) describes the horizontal trees distribution by using the mean distance between a reference tree and the nearest neighbors and the mean distance defined by a Poison distribution [79]. CE can range from 0, when the stand is characterized by tree clustering, to 2.1491 [79] in the case of regular distribution.

Uniform angle index (*UAI*) describes the uniform distribution of the nearest neighboring trees in relation to the reference tree [38]. The method is based on the angles between trees, compared to a uniform dispersion angle of 72° (Equation (2)). The interpretation of these values is made according to the confidence interval of 0.475–0.517 [38], describing a random distribution.

$$UAI = \frac{1}{n}\sum_{i=1}^{n} UAI_i = \frac{1}{4n}\sum_{i=1}^{n}\sum_{j=1}^{4} z_{ij} \tag{2}$$

where:

n—number of reference trees

z_{ij}—angle coefficients in relation to the reference (72°), 1 if <72°, 0 if > 72°

UAI_i—uniform angle index

The relative dominance diameter index (IDR) is defined as the ratio between the number of trees with a diameter greater than the reference tree. The value of this indicator reaches values in the range (0–1) and is interpreted in relation to five default thresholds. Thus, in relation to the number of trees with a diameter larger than the reference, the indicator falls into the following categories: shade tolerant, dominated, co-dominant, dominant, predominant. These categories correspond to the Kraft classes, a method used for validating the obtained values. The variable considered in the evaluation of the dominancy indicator may be substituted by other tree characteristics such as height or species.

In the structural indices computation process, the edge effect was removed in order to ensure accurate results. This was performed by selecting only the trees within an inner buffer, defining an area smaller than that of the circular subplots (Figure 5).

The interpretation of these indices made it possible to identify the supplied services and the level to which they could be quantified. CE values greater than 1 suggested that the studied subplots were characterized by a more uniform horizontal structure. An exception was identified in the SFTM-3 subplot, which was characterized by a mean value of 0.4. This could be explained by the smaller number of trees clustered together and by the fact that this circle is crossed by a forest harvesting road. Based on the calculated t-values for the CE, according to the work of [80], the subplots that overpass 1.96 can be described as having a regular distribution (Table 3).

Figure 5. Nearest neighbors identification and reference trees selection; **green**—reference tree; **red**—marginal tree.

Table 3. Horizontal structure indices in the 15 m-radius subplot.

Plots	Subplot	N_{ref}	CE	*t*-Value *	W
SMTM	1	17	1.715	1.19	0.456
	2	11	1.829	2.19	0.432
	3	7	1.689	2.59	0.393
SGTM	1	20	1.558	3.12	0.563
	2	19	1.558	3.14	0.526
	3	25	1.825	2.68	0.510
SFTM	1	31	1.074	0.56	0.547
	2	37	1.272	1.59	0.574
	3	20	0.405	−8.75	0.55
SFRM	1	64	1.423	1.09	0.553
	2	55	1.283	0.91	0.515
	3	37	1.166	0.97	0.5
SMRM	1	92	1.269	0.4	0.532
	2	106	1.171	0.21	0.709
	3	87	1.025	0.04	0.548
SMT	1	46	1.751	3.17	0.531
	2	55	1.604	1.95	0.524
	3	38	1.429	2.41	0.561
SGT	1	39	1.321	2.7	0.545
	2	38	1.556	2.13	0.59
	3	35	1.575	3.65	0.558
SFT	1	26	1.551	4.46	0.635
	2	42	1.781	3.77	0.642
	3	36	1.549	3.34	0.643
SFR	1	71	1.309	0.68	0.715
	2	99	1.866	1.16	0.707
	3	76	1.628	0.95	0.725
SMR	1	102	1.301	0.38	0.719
	2	131	1.244	0.21	0.722
	3	147	1.252	0.37	0.723

N_{ref}—number of reference trees, * *t*—CE value.

When uniform angle index values were analyzed, differences between plots could be observed, thus detecting structural differences between the corresponding stands. The uniform angle index values ranged between 0.393 and 0.725, values covering the entire interpretation interval. Within the old sessile oak stand, without interventions (SGTM), the corresponding subplots reached values equivalent to a rather random distribution. This was the case with the 2.3 (0.510) subplot, reaching values quite different than its counterpart, subplots 2.1 and 2.2, characterized by a clustered structure (Table 2). In the case of the Norway spruce (SMTM), the reached values defined a uniform structure, while the young beech stand with interventions (SFR) was characterized by a clustered structure in all subplots.

Also, from this analysis, the difference between the plots with interventions and those without could be observed. The plots covered with silvicultural treatments tend to describe more clustered structures, an effect caused by the increased distance between trees after harvesting.

Figure 6 facilitates the interpretation of the structure and conditions similarity within a plot. As stated before, discrepancies appeared in SFTM for the CE index and in SMRM for the uniform angle index. The latter was a consequence of a windthrow event that had affected the SMRM-3 subplot.

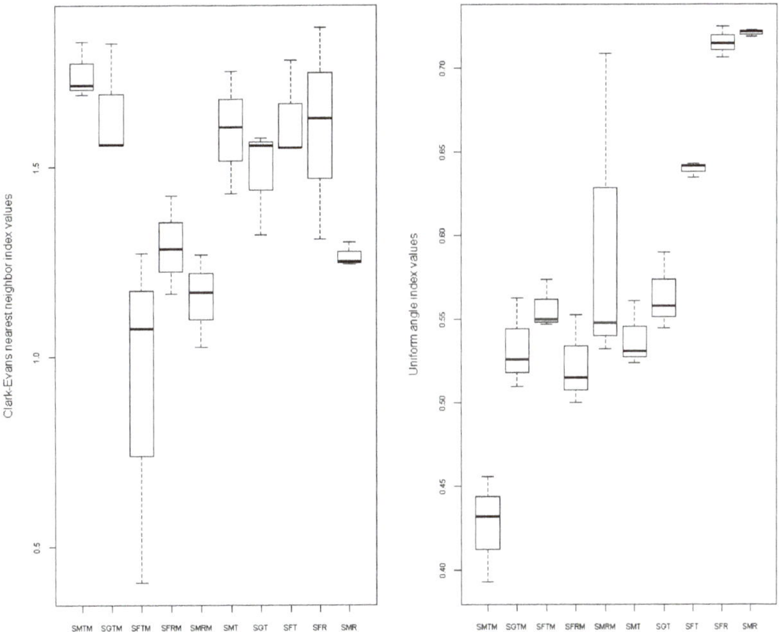

Figure 6. Clark-Evans nearest neighbors index and uniform angle index.

Analyses of relative dominance diameter index described the vertical stand structure at the subplot level. A similarity could be observed between old Norway spruce (SMTM) subplots (Figure 7), indicating a uniform structure within the stand and a uniform tree distribution between classes. The sessile oak stand is characterized by a lower degree of heterogeneity and more unevenness between classes.

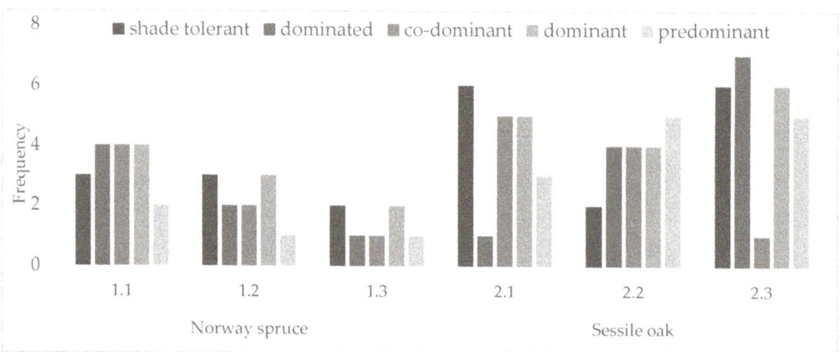

Figure 7. Vertical stand structure (Kraft classes) based on IDR.

3.2.2. Carbon Storage

Carbon is stocked in forest stands in the following five pools: above and below-ground living biomass, soil, litter, and deadwood [81,82]. Apart from the variables used for the above-ground volume, carbon stock evaluation (Table 4) required another set of parameters, namely the theoretical number of trees per hectare, wood density, root-to-shoots ratio, and biomass expansion factor (Equation (3)). These were retrieved from specific yield tables and international guides [83,84].

$$C_{stock} = \sum V * D * (1 + R) * BEF * CF \quad (3)$$

where:

C_{stock}—carbon stock [tC]
V—tree volume [m^3]
D—wood density [t/m^3]
R—root-to-shoot ratio
BEF—biomass expansion factor
CF—carbon fraction

Table 4. Carbon stock required variables.

Plot	V [m^3 ha^{-1}]	D [1] [$kg\ m^{-3}$]	R [1]	BEF [2]	CF [3]	Carbon Stock [$tC{\cdot}ha^{-1}$]
SGT	444.1	584	0.22	1.4	0.48	151.88
SGTM	646.4	584	0.22	1.4	0.48	221.06
SFR	434.6	545	0.19	1.4	0.46	129.66
SFRM	509.8	545	0.19	1.4	0.46	152.09
SFT	457.2	545	0.19	1.4	0.46	136.40
SFTM	622.3	545	0.19	1.4	0.46	185.65
SMR	345.7	353	0.2	1.3	0.51	74.68
SMRM	420.1	353	0.2	1.3	0.51	90.76
SMT	409.5	353	0.2	1.3	0.51	88.47
SMTM	558.3	353	0.2	1.3	0.51	120.61

[1] [83] [2] [84] [3] [85].

Due to the methodology for the above-ground volume, for the sessile oak and beech species, the biomass expansion factor (BEF) was omitted, as the regression equation for volume already took into consideration the branches' volume. Including BEF would have led to biased results.

The obtained carbon stock values ranged between 74.68 $tC{\cdot}ha^{-1}$ (273.82 $tCO_2{\cdot}ha^{-1}$) in the case of the young Norway spruce plot covered with silvicultural intervention and 221.06 $tC{\cdot}ha^{-1}$ (810.55 $tCO_2{\cdot}ha^{-1}$) in the case of the old sessile oak plot. The upper values of the storage capacity interval of the studied plots are in accordance with those stated in the work of [86]. The lower values are a consequence of age and species characteristics (wood density, root-to-shoot ratio, and carbon fraction).

3.2.3. Foliage Indices

Active remote-sensing technology advances allowed for the development of multiple applications addressing the canopy structure, crown dynamics, and phenology [21,24,29,87–91]. These applications based on active remote-sensing data are a powerful tool in the decisional process associated with forestry and ecology sectors. From the variety of indices computed through remote sensing, in the research field, the leaf area index (LAI) is the most commonly used. Furthermore, along with the LAI, an important role in improving the canopy description is held by leaf area density (*LAD*), which offers detailed information regarding the stand vertical structure. Leaf area index estimation as the ratio between leaves (single-faced) area and area of the studied plot, was measured over time through various indirect methods (orbital sensors, hemispherical photography, and light intensity attenuation) [92–94], and still require improvement in what concerns the stability and robustness of their results. Alternatively, airborne laser scanning, despite its limitations related to penetration capability, has promising results in forestry indices and parameters computation, including those above-mentioned [70,95].

In this study, LAI and *LAD* were estimated through the MacArthur and Horn equation [96] developed on the principle of the Beer–Lambert law [97,98] and following methodologies proposed in other related research papers [95,99–102]. Thus, to each voxel from the processed point cloud (voxel—5 × 5 × 1 m), the following proposed equation was applied [102]:

$$LAD_{i-1,i} = \ln\left(\frac{S_e}{S_t}\right)\frac{1}{k\Delta z} \quad (4)$$

where:

S_e—number of pulses entering the voxel;

S_t—number of pulses exiting the voxel;

k—Beer–Lambert law extinction coefficient;

z—voxel height (1 m).

From the variety of estimated indices resulting when applying derivatives of the above-mentioned methodology, of most interest to our study were the total LAI values, the height of the mean *LAD*, and standard deviation corresponding to each voxel (cell of a three-dimensional grid) column taken into consideration.

As shown in the case of the IDR, sessile oak (SGTM) is characterized by an uneven structure, a fact also illustrated in the LAI and *LAD* values. In the northwest part of the plot, the higher density of smaller trees impacted the LAI and height of the mean *LAD*, reaching values in the range 1–3, respectively, 5–10 m.

As expected, the Norway spruce plot is characterized by smaller standard deviation values, suggesting a constant horizontal structure throughout the plot (Figure 8). In the case of the sessile oak plot, the standard deviation trend highlights a generation individualization through higher variation within the upper levels of the canopy.

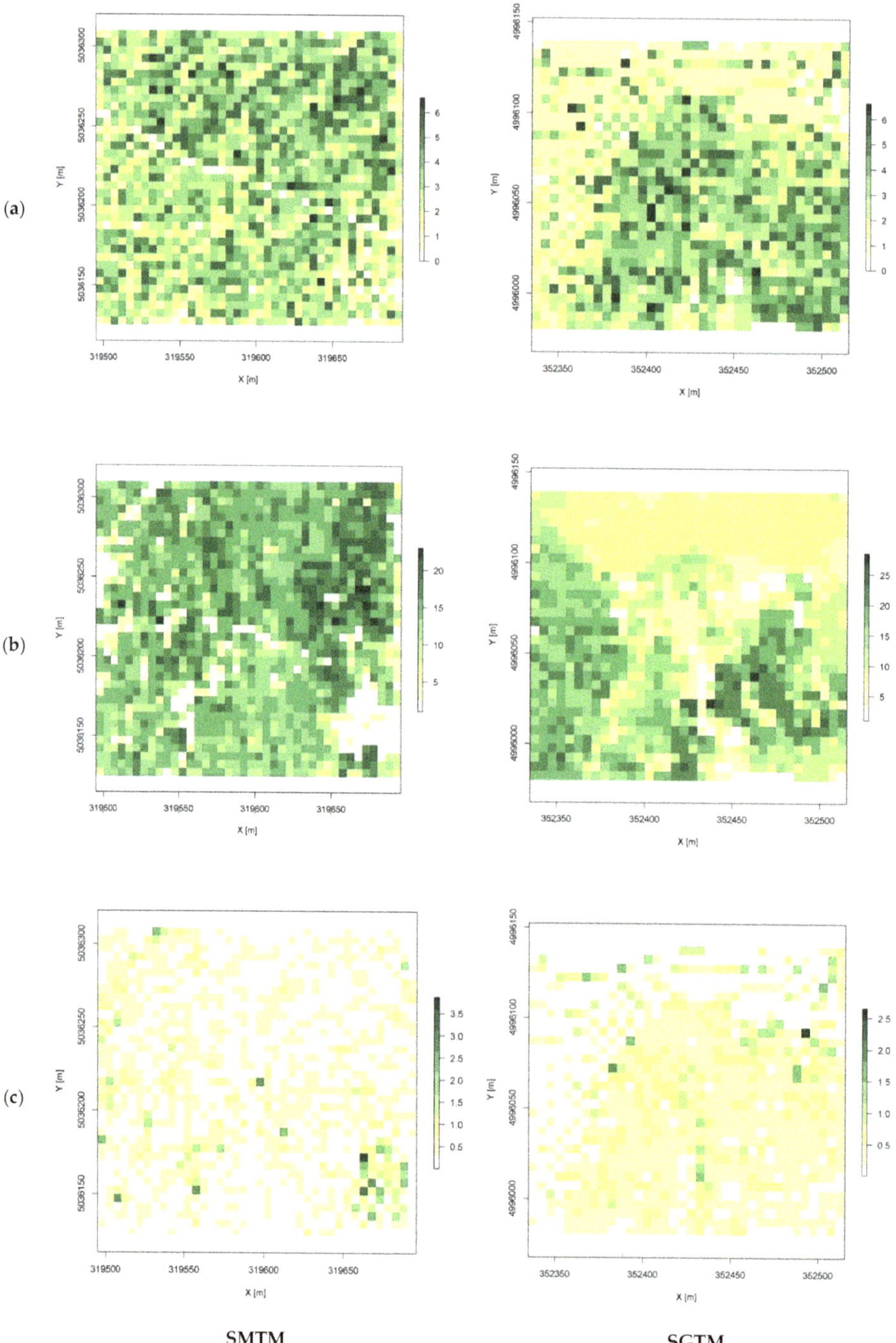

Figure 8. Mapped values of (**a**) LAI; (**b**) height of the mean *LAD*; (**c**) standard deviation corresponding to each voxel column.

3.3. Supporting Services

In the majority of the research papers, this function is not a self-contained one. Millennium Ecosystem Assessment classification (MEA) [103] presents the support function as integration between provisioning, regulating, and cultural functions, quantifying benefits that ensure the rest of the services. In the Common International Classification of Ecosystem Services (CICES), the support function is not promoted as one and is considered an underlying structure that provides indirect outputs [104,105].

Understory biomass has an important ecological significance in forest ecosystem stability and in assessing the relationships between wildlife and their habitat. Despite the low proportion in above-ground volume, the understory biomass represents a tool for the researchers in evaluating the food provisioning and the quality of the environment [106–114].

The understory biomass computation implies a complex and expensive forest inventory due to multiple variables that should be taken into consideration. Active remote-sensing applications that aim at assessing understory biomass were proposed. Terrestrial and airborne laser scanning data were analyzed in order to estimate the understory, following [106,112,115].

This study addressed the methodology proposed by the authors of [116] that aims to predict the presence of shrub layers from aerial-based point clouds. In the mentioned thesis, two indices were computed: (a) undergrowth return fraction and (b) undergrowth cover density. For our case, of most interest was the undergrowth return fraction, expressed as the ratio between the number of points in the 0.5–5 m range and the total number of points (Figure 9).

The old Norway spruce plot, in comparison with the sessile oak, is characterized by a sparse distribution of the shrub layer of lower intensities, with no understory clusters identified. In the case of the sessile oak plot, a central area with a high density of understory vegetation could be observed, mirrored in the northern part, by the lower values of the canopy height model. Overall, the sessile oak plot recorded a value of 0.20, which according to the work of [116], is indicative of a medium-to-high shrub cover intensity (Figure 10).

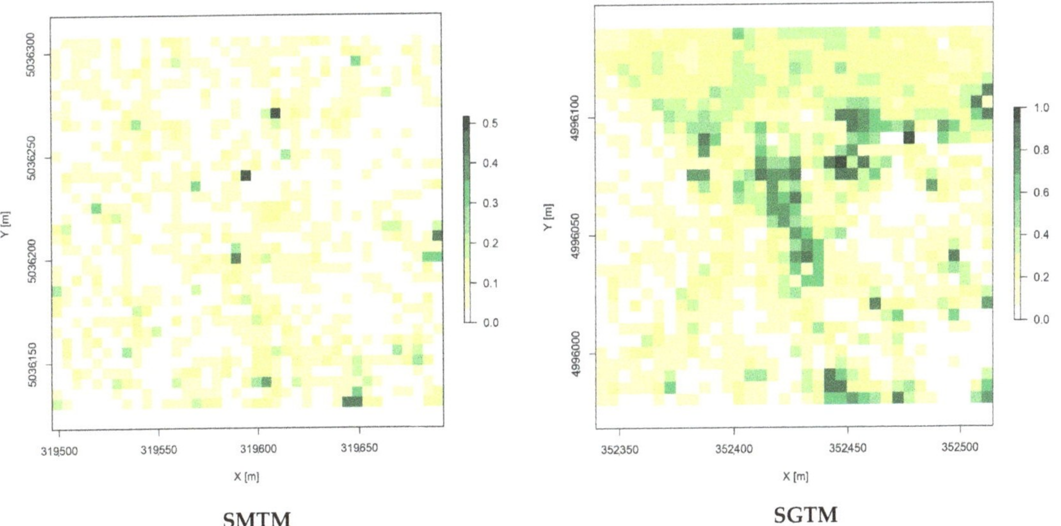

Figure 9. Mapping of the shrub layer (5 × 5 m pixel)—undergrowth return fraction.

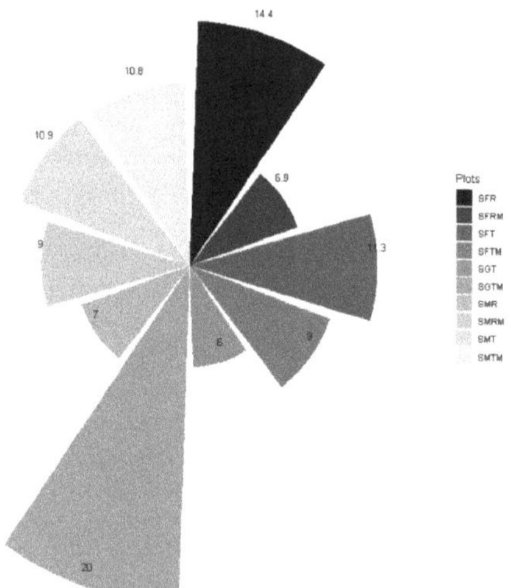

Figure 10. Undergrowth return fraction values.

The majority of the rest of the plots have a low-to-minim shrub cover, covering 6% to 10%. An issue identified through field observations was in the case of SMRM. The plot is characterized by a high shrub coverage, but due to the lower age and high tree density, the laser beams could not penetrate the canopy layer. This resulted in a small number of points near the ground and an underestimation of the shrub layer. This shortcoming can be compensated by using TLS data to complete the ALS points cloud.

3.4. *Structure Analysis for Cultural Services Assessment*

The cultural services are the most problematic in what concerns the evaluation processes. The services provided by the one-hectare plots are not traded on the market, and therefore the methods of valuation applied tend to be more subjective. In addition, the evaluation of the forest ecosystem's capacity in providing these benefits is a challenging one due to public preferences and the number of variables involved.

The forest structure indices computed under the regulating function section and part of the health status information can be used in quantifying the human preferences regarding the ideal distribution and biodiversity. Tree clusters, number of trees, sparse distribution, higher canopy density, light penetration, visibility, understory volume, and snags volume can all be indirectly assessed through tree distribution characteristics and mortality analysis.

The snags identification and mortality characteristics were analyzed based on airborne laser scanning data according to the work of [117] methodology (Figure 11).

After processing, the point clouds were classified into four classes, namely live trees, small snags, live crown edge snags, and higher canopy snags. Due to the small proportion of dead trees in the studied plots, not all classes were well represented. Moreover, following the analysis, none of the snag classes were identified in the Norway spruce plots, apart from sparse, unrepresentative small snags in the understory. By way of comparison, the sessile oak plot presents a higher proportion corresponding to the live crown edge snags class.

A crucial role is attributed to the higher canopy snags class, which makes possible the identification of dead treetops. A higher proportion of snags would have allowed for the evaluation of a ratio between deadwood and the above-ground biomass. This information could have then been used in the carbon sequestration estimation or the mortality rate of the forest stand.

Figure 11. Snags identification and classification (SGTM).

4. Discussion

Forest ecosystems are characterized by various structures and complex processes defined by a plethora of intra- and inter-plot relationships. The assessment of all services provided by these ecosystems' characteristics is still a challenging subject for the research field [118]. Therefore, this study aimed to highlight some of the most important and quantifiable services employing the latest applications of active remote-sensing technology [119].

Taking advantage of the terrestrial stationary laser scanning, the obtained values for above-ground volume, at tree and stand level, was within the characteristic tolerances [30]. Moreover, we compensated the height-specific bias caused by the terrestrial laser scanner's inability to penetrate dense canopies, a well know feature relevant to Romanian forests [24], by deriving a canopy height model from aerial laser scanning.

Compared to the rest of the functions, provisioning could be evaluated most straightforward [76,120]. By only knowing the above-ground volume and market price, this service could be monetized. To better understand this service, a technical approach can further be used by classifying the wood in relation to the type of final product or the quality of the timber, information that can be extracted from management plans.

The assessment of carbon sequestration is a more complex process, and it partially uses wood volume calculations. The results are influenced by multiple biophysical parameters (wood density, the ratio between above and below-ground biomass, or the biomass expansion factor) [84]. All these parameters depend on the species composition, a feature that cannot be presently assessed on a large scale by means of close-range active remote sensing [121,122]. Furthermore, as described in the IPCC Guideline [84,123], forest ecosystems stock a large amount of carbon not only in living biomass, so other pools (soil and dead organic matter) should also be taken into consideration for a real carbon emission/removals analysis. On the other hand, assumptions are made even within the country-level estimation. Pools as soil, deadwood, or litter are considered to be neutral in the carbon emission and removals balance. Therefore, only considering the living biomass can represent a viable solution for carbon sequestration and stock assessment.

The utility of structural indices was highlighted in a long list of studies [24,119,124–126], and the indirect quantification made by means of these indices could be considered a proper method for evaluating forest ecosystem capacity to provide services. Forest structure represents a valuable source of information, and relating structure characteristics to specific services is the approach used in this study.

Soil and water regulation services assured by the forest ecosystem are quantifiable through trees distribution, LAI, *LAD*, and canopy projection [127,128]. The distances, the angles, and the relationship between different individuals define the rate of success of the forest to ensure the regulation function. Indices such as uniform angle index, Clark-Evans nearest neighbor index, and relative dominance diameter index can describe the ideal structure, which prevents gaps or corridors from occurring. The resulting values, corresponding to the studied plots, describe a uniform tree distribution. According to the uniform angle index values, the Norway spruce plot has a better capacity to assure the regulation function if we consider the ideal structure being a random one.

When analyzing the CE, the obtained values tend to reach the upper half of the range, close to a perfectly regular hexagonal distribution (2.1491). There are some dissimilarities between CE and *UAI* regarding, in particular, the sessile oak plot due to the high number of trees in the sampled area. However, of great importance in soil and water regulation services is the fact that none of the plots is characterized by clustered trees that would facilitate soil erosion and low water retention.

Air regulation function was evaluated in this study through the use of LAI and *LAD* [129]. The capacity of the forest ecosystem to provide these services is directly proportional to LAI values. For the studied plots, LAI values are within the 0–6 range, with a considerable proportion in the upper classes throughout the entire old Norway spruce plot. Due to lower values in the canopy height model and implicit smaller crown volumes, the sessile oak recorded lower intensities in the northern part of the plot.

The support function assessment was evaluated based on the shrub cover, indicating the capacity of the studied forests to provide various species' habitat requirements. Important differences were observed between the studied areas. The sessile oak plot (SGTM) recorded a larger area covered by understory vegetation. This analysis can also provide additional information on the forest structure, information that can be used in further biodiversity assessments.

The results regarding mortality have not allowed any further analyses regarding the dead matter stratification. However, it offered enough information to assess the overall health status of the forest ecosystem [117,123,130]. By following the presented structural indices, as well as the ones addressing foliage, while simultaneously considering the human preferences toward the ideal forest structure, a suitable evaluation of the cultural services could be deployed.

Given the results corresponding to our studied plots, the lack of clustered trees, large gaps, and overall canopy structure, an appropriate scale for referencing forest ecosystems' capability to provide cultural benefits could have also been developed.

5. Conclusions

In order to emphasize and maximize the ecological, social, and economic benefits of forests, suitable assessment methods are required. Active remote-sensing technology, with the proven advantages and characteristic limitations, can represent the foundation for multiple approaches aiming to quantify the capacity of the forest ecosystem to provide services. This study highlighted the possibility of using two different active remote-sensing data sets and several techniques to assess the main ecosystem functions according to TEEB classification.

To estimate the key biophysical parameters of a tree, terrestrial laser scanning point clouds proved to be a viable solution. The processing of this data source led to errors associated with DBH of below 1 cm [30] at the subplot level when analyzing the mean tree. Precisions associated with tree coordinates are comparable to those obtained through the electronic field mapping system. Using the TLS-based variables as well as the airborne laser scanning data, the provisioning function, particularly wood products, was evaluated.

This was performed by means of above-ground volume, characterized by errors smaller than 6%.

Combining the terrestrial and aerial laser scans, the evaluation of regulating function was also possible. The indices computed by processing the above-mentioned data sources proved to be a suitable basis for acquiring the forest's horizontal structure and the distribution of trees. CE, *UAI*, and IDR implementations through active remote-sensing approaches can represent the link between ecosystem services and human preferences, but also the qualitative parameters for assessing the degree of ensuring certain services, namely soil stability, air quality, and water regulation.

Challenges still exist in applying active remote-sensing techniques due to the complex ecosystems' intra- and inter-plot relationships. However, the development of tools to address the environmental assessment requirements is encouraged by the stakeholders and decisional factors. Thus, it can be stated that active remote-sensing applications have a significant role in forestry, a role that translates to an overall improvement of human well-being.

Therefore, implementing the described methodologies highlighted the necessity of developing custom reference scales relevant in the assessment processes of the relative capacity of forest ecosystems to supply benefits. To achieve this, the study should be extended to address further stands of different structures, species compositions, and microclimates.

Author Contributions: Conceptualization, A.C.D., I.-S.P., G.M.T. and O.B.; Formal analysis, A.C.D. and I.-S.P.; Investigation, A.C.D., I.-S.P., Ş.L. and J.G.-D.; Methodology, A.C.D. and I.-S.P.; Software, I.-S.P.; Validation, Ş.L., C.-E.D., G.M.T. and O.B.; Visualization, I.-S.P.; Writing—original draft, A.C.D. and I.-S.P.; Writing—review and editing, A.C.D., J.G.-D., G.M.T., I.-S.P. and O.B. All authors have read and agreed to the published version of the manuscript.

Funding: This study was conducted under the project CRESFORLIFE (SMIS 105506), subsidiary contract no. 18/2020, co-financed by the European Regional Development Fund through the 2014–2020 Competitiveness Operational Program.

Conflicts of Interest: The authors declare no conflict of interest.

References

1. Sasaki, N.; Asner, G.; Knorr, W.; Durst, P.; Priyadi, H.; Putz, F. Approaches to classifying and restoring degraded tropical forests for the anticipated REDD+ climate change mitigation mechanism. *iForest Biogeosci. For.* **2011**, *4*, 1–6. [CrossRef]
2. Pan, Y.; Birdsey, R.A.; Fang, J.; Houghton, R.; Kauppi, P.E.; Kurz, W.A.; Phillips, O.L.; Shvidenko, A.; Lewis, S.L.; Canadell, J.G.; et al. A large and persistent carbon sink in the World's forests. *Science* **2011**, *333*, 988–993. [CrossRef]
3. Fu, X.; Zhang, Z.; Cao, L.; Coops, N.C.; Goodbody, T.R.; Liu, H.; Shen, X.; Wu, X. Assessment of approaches for monitoring forest structure dynamics using bi-temporal digital aerial photogrammetry point clouds. *Remote Sens. Environ.* **2021**, *255*, 112300. [CrossRef]
4. Badea, O.; Neagu, S.; Bytnerowicz, A.; Silaghi, D.; Barbu, I.; Iacoban, C.; Popescu, F.; Andrei, M.; Preda, E.; Iacob, C.; et al. Long-term monitoring of air pollution effects on selected forest ecosystems in the Bucegi-Piatra Craiului and Retezat Mountains, southern Carpathians (Romania). *iForest Biogeosci For.* **2011**, *4*, 49–60. [CrossRef]
5. Lausch, A.; Borg, E.; Bumberger, J.; Dietrich, P.; Heurich, M.; Huth, A.; Jung, A.; Klenke, R.; Knapp, S.; Mollenhauer, H.; et al. Understanding forest health with remote sensing, part III: Requirements for a scalable multi-source forest health monitoring network based on data science approaches. *Remote Sens.* **2018**, *10*, 1120. [CrossRef]
6. Rasmussen, L.V.; Jepsen, M.R. Monitoring systems to improve forest conditions. *Curr. Opin. Environ. Sustain.* **2018**, *32*, 29–37. [CrossRef]
7. Grafström, A.; Zhao, X.; Nylander, M.; Petersson, H. A new sampling strategy for forest inventories applied to the temporary clusters of the Swedish national forest inventory. *Can. J. For. Res.* **2017**, *47*, 1161–1167. [CrossRef]
8. Ginzler, C.; Hobi, M.L. Countrywide stereo-image matching for updating digital surface models in the framework of the swiss national forest inventory. *Remote Sens.* **2015**, *7*, 4343–4370. [CrossRef]
9. Masek, J.G.; Hayes, D.; Hughes, M.J.; Healey, S.P.; Turner, D.P. The role of remote sensing in process-scaling studies of managed forest ecosystems. *For. Ecol. Manag.* **2015**, *355*, 109–123. [CrossRef]
10. Lechner, A.M.; Foody, G.M.; Boyd, D.S. Applications in remote sensing to forest ecology and management. *One Earth* **2020**, *2*, 405–412. [CrossRef]

11. Boyd, D.; Foody, G. An overview of recent remote sensing and GIS based research in ecological informatics. *Ecol. Inform.* **2011**, *6*, 25–36. [CrossRef]
12. Antonarakis, A.; Richards, K.; Brasington, J. Object-based land cover classification using airborne LiDAR. *Remote Sens. Environ.* **2008**, *112*, 2988–2998. [CrossRef]
13. Yan, W.Y.; Shaker, A.; El-Ashmawy, N. Urban land cover classification using airborne LiDAR data: A review. *Remote Sens. Environ.* **2015**, *158*, 295–310. [CrossRef]
14. García, M.; Riaño, D.; Chuvieco, E.; Danson, F. Estimating biomass carbon stocks for a Mediterranean forest in central Spain using LiDAR height and intensity data. *Remote Sens. Environ.* **2010**, *114*, 816–830. [CrossRef]
15. Popescu, S.C. Estimating biomass of individual pine trees using airborne lidar. *Biomass Bioenergy* **2007**, *31*, 646–655. [CrossRef]
16. Næsset, E. Estimating tree height and tree crown properties using airborne scanning laser in a boreal nature reserve. *Remote Sens. Environ.* **2002**, *79*, 105–115. [CrossRef]
17. Lim, K.S.; Treitz, P.M. Estimation of above ground forest biomass from airborne discrete return laser scanner data using canopy-based quantile estimators. *Scand. J. For. Res.* **2004**, *19*, 558–570. [CrossRef]
18. Chuvieco, E.; Congalton, R.G. Application of remote sensing and geographic information systems to forest fire hazard mapping. *Remote Sens. Environ.* **1989**, *29*, 147–159. [CrossRef]
19. Chuvieco, E.; Kasischke, E.S. Remote sensing information for fire management and fire effects assessment. *J. Geophys. Res. Space Phys.* **2007**, *112*. [CrossRef]
20. Jaboyedoff, M.; Oppikofer, T.; Abellan, A.; Derron, M.-H.; Loye, A.; Metzger, R.; Pedrazzini, A. Use of LIDAR in landslide investigations: A review. *Nat. Hazards* **2010**, *61*, 5–28. [CrossRef]
21. Lim, K.; Treitz, P.; Wulder, M.; St-Onge, B.; Flood, M. LiDAR remote sensing of forest structure. *Prog. Phys. Geogr. Earth Environ.* **2003**, *27*, 88–106. [CrossRef]
22. Jayathunga, S.; Owari, T.; Tsuyuki, S. Evaluating the performance of photogrammetric products using fixed-wing UAV imagery over a mixed conifer-broadleaf forest: Comparison with airborne laser scanning. *Remote Sens.* **2018**, *10*, 187. [CrossRef]
23. Palace, M.W.; Sullivan, F.B.; Ducey, M.J.; Treuhaft, R.N.; Herrick, C.; Shimbo, J.Z.; Mota-E-Silva, J. Estimating forest structure in a tropical forest using field measurements, a synthetic model and discrete return lidar data. *Remote Sens. Environ.* **2015**, *161*, 1–11. [CrossRef]
24. Pascu, I.-S.; Dobre, A.-C.; Badea, O.; Tănase, M.A. Estimating forest stand structure attributes from terrestrial laser scans. *Sci. Total. Environ.* **2019**, *691*, 205–215. [CrossRef] [PubMed]
25. Wulder, M.; Hall, R.J.; Coops, N.; Franklin, S. High spatial resolution remotely sensed data for ecosystem characterization. *BioScience* **2004**, *54*, 511–521. [CrossRef]
26. Lefsky, M.A.; Cohen, W.B.; Parker, G.G.; Harding, D.J. Lidar remote sensing for ecosystem studies. *Bioscience* **2002**, *52*, 19–30. [CrossRef]
27. Holmgren, J.; Nilsson, M.; Olsson, H. Estimation of tree height and stem volume on plots using airborne laser scanning. *For. Sci.* **2003**, *49*, 419–428.
28. McRoberts, R.E.; Tomppo, E.O. Remote sensing support for national forest inventories. *Remote Sens. Environ.* **2007**, *110*, 412–419. [CrossRef]
29. Lefsky, M.; Cohen, W.; Acker, S.; Parker, G.; Spies, T.; Harding, D. Lidar remote sensing of the canopy structure and biophysical properties of Douglas-Fir Western Hemlock Forests. *Remote Sens. Environ.* **1999**, *70*, 339–361. [CrossRef]
30. Pascu, I.-S.; Dobre, A.-C.; Badea, O.; Tanase, M.A. Retrieval of forest structural parameters from terrestrial laser scanning: A Romanian case study. *Forests* **2020**, *11*, 392. [CrossRef]
31. Cabo, C.; Ordóñez, C.; López-Sánchez, C.A.; Armesto, J. Automatic dendrometry: Tree detection, tree height and diameter estimation using terrestrial laser scanning. *Int. J. Appl. Earth Obs. Geoinf.* **2018**, *69*, 164–174. [CrossRef]
32. Othmani, A.; Voon, L.F.L.Y.; Stolz, C.; Piboule, A. Single tree species classification from Terrestrial Laser Scanning data for forest inventory. *Pattern Recognit. Lett.* **2013**, *34*, 2144–2150. [CrossRef]
33. Wallace, L.; Lucieer, A.; Malenovský, Z.; Turner, D.; Vopěnka, P. Assessment of forest structure using two UAV techniques: A comparison of airborne laser scanning and structure from motion (SfM) point clouds. *Forests* **2016**, *7*, 62. [CrossRef]
34. Jarron, L.R.; Coops, N.C.; MacKenzie, W.H.; Tompalski, P.; Dykstra, P. Detection of sub-canopy forest structure using airborne LiDAR. *Remote Sens. Environ.* **2020**, *244*, 111770. [CrossRef]
35. Zheng, G.; Moskal, L.M. Retrieving Leaf Area Index (LAI) using remote sensing: Theories, methods and sensors. *Sensors* **2009**, *9*, 2719–2745. [CrossRef] [PubMed]
36. Hosoi, F.; Omasa, K. Voxel-based 3-D modeling of individual trees for estimating leaf area density using high-resolution portable scanning lidar. *IEEE Trans. Geosci. Remote Sens.* **2006**, *44*, 3610–3618. [CrossRef]
37. Zhang, G.; Hui, G.; Zhao, Z.; Hu, Y.; Wang, H.; Liu, W.; Zang, R. Composition of basal area in natural forests based on the uniform angle index. *Ecol. Inform.* **2018**, *45*, 1–8. [CrossRef]
38. Zhao, Z.; Hui, G.; Hu, Y.; Wang, H.; Zhang, G.; Von Gadow, K. Testing the significance of different tree spatial distribution patterns based on the uniform angle index. *Can. J. For. Res.* **2014**, *44*, 1419–1425. [CrossRef]
39. Lausch, A.; Erasmi, S.; King, D.J.; Magdon, P.; Heurich, M. Understanding forest health with remote sensing, Part I: A review of spectral traits, processes and remote-sensing characteristics. *Remote Sens.* **2016**, *8*, 1029. [CrossRef]

40. Listopad, C.M.; Masters, R.E.; Drake, J.; Weishampel, J.; Branquinho, C. Structural diversity indices based on airborne LiDAR as ecological indicators for managing highly dynamic landscapes. *Ecol. Indic.* **2015**, *57*, 268–279. [CrossRef]
41. Wang, K.; Franklin, S.E.; Guo, X.; Cattet, M. Remote sensing of ecology, biodiversity and conservation: A review from the perspective of remote sensing specialists. *Sensors* **2010**, *10*, 9647–9667. [CrossRef] [PubMed]
42. Kerr, J.T.; Ostrovsky, M. From space to species: Ecological applications for remote sensing. *Trends Ecol. Evol.* **2003**, *18*, 299–305. [CrossRef]
43. Duncanson, L.; Niemann, K.; Wulder, M. Estimating forest canopy height and terrain relief from GLAS waveform metrics. *Remote Sens. Environ.* **2010**, *114*, 138–154. [CrossRef]
44. Lefsky, M.A.; Cohen, W.B.; Acker, S.A.; Spies, T.A.; Parker, G.G.; Harding, D. Lidar remote sensing of forest canopy structure and related biophysical parameters at H.J. Andrews experimental forest, Oregon, USA. In Proceedings of the International Geoscience and Remote Sensing Symposium (IGARSS), Seattle, WA, USA, 6–10 July 1998.
45. Daily, G.C.; Matson, P.A. Ecosystem services: From theory to implementation. *Proc. Natl. Acad. Sci. USA* **2008**, *105*, 9455–9456. [CrossRef] [PubMed]
46. Gómez-Baggethun, E.; de Groot, R. Natural capital and ecosystem services: The ecological foundation of human society. *Ecosyst. Serv.* **2010**, *30*, 105–121. [CrossRef]
47. Wyatt, T.D.; de Groot, R.S. Valuing nature. *Glob. Ecol. Biogeogr. Lett.* **1993**, *3*, 90. [CrossRef]
48. Costanza, R.; D'Arge, R.; De Groot, R.; Farber, S.; Grasso, M.; Hannon, B.; Limburg, K.; Naeem, S.; O'Neill, R.V.; Paruelo, J.; et al. The value of the world's ecosystem services and natural capital. *Nature* **1997**, *387*, 253–260. [CrossRef]
49. Balmford, A.; Bruner, A.; Cooper, P.; Costanza, R.; Farber, S.; Green, R.E.; Jenkins, M.; Jefferiss, P.; Jessamy, V.; Madden, J.; et al. Ecology: Economic reasons for conserving wild nature. *Science* **2002**, *297*, 950–953. [CrossRef]
50. De Groot, R.; Brander, L.; van der Ploeg, S.; Costanza, R.; Bernard, F.; Braat, L.; Christie, M.; Crossman, N.; Ghermandi, A.; Hein, L.; et al. Global estimates of the value of ecosystems and their services in monetary units. *Ecosyst. Serv.* **2012**, *1*, 50–61. [CrossRef]
51. Kumar, P. *The Economics of Ecosystems and Biodiversity: Ecological and Economic Foundations*; Earthscan Publications Ltd.: London, UK, 2013.
52. Häyhä, T.; Franzese, P.P. Ecosystem services assessment: A review under an ecological-economic and systems perspective. *Ecol. Model.* **2014**, *289*, 124–132. [CrossRef]
53. Bradbeer, J.; Pearce, D. Economic values and the natural world. *Geogr. J.* **1995**, *161*, 335. [CrossRef]
54. Kornatowska, B.; Sienkiewicz, J. Forest ecosystem services-assessment methods. *Folia For. Pol. A For.* **2018**, *60*, 248–260. [CrossRef]
55. Wittmer, H.; Gundimeda, H. *The Economics of Ecosystems and Biodiversity for Local and Regional Policy Makers*; Routledge: Abingdon-on-Thames, UK, 2011.
56. Sukhdev, P. *The Economics of Ecosystem and Biodiversity*; Yale School of Forestry and Environmental Studies: New Heaven, CT, USA, 2011.
57. IFER. Monitoring and Mapping Solutions. Ltd. FieldMap. 2016. Available online: https://www.youtube.com/watch?v=edBBWh0JyIU&ab_channel=YaleCampus (accessed on 1 March 2021).
58. FARO Technologies Inc. *Faro Scene*; FARO: Lake Mary, FL, USA, 2018.
59. Hackenberg, J.; Spiecker, H.; Calders, K.; Disney, M.; Raumonen, P. SimpleTree—An efficient open source tool to build tree models from TLS clouds. *Forests* **2015**, *6*, 4245–4294. [CrossRef]
60. RIEGL. *LMS-Q680i*; RIEGL Laser Measurement Systems GmbH: Horn, Austria, 2012.
61. Terrasolid Ltd. *TerraScan*; Terrasolid v021 Ltd.: Helsinki, Finland, 2021.
62. Axelsson, P. DEM generation from laser scanner data using adaptive TIN models. *Int. Arch. Photogramm. Remote Sens.* **2000**, *33*, 110–117.
63. Brandt, P.; Abson, D.J.; DellaSala, D.A.; Feller, R.; von Wehrden, H. Multifunctionality and biodiversity: Ecosystem services in temperate rainforests of the Pacific Northwest, USA. *Biol. Conserv.* **2014**, *169*, 362–371. [CrossRef]
64. Basuki, T.; van Laake, P.; Skidmore, A.; Hussin, Y. Allometric equations for estimating the above-ground biomass in tropical lowland Dipterocarp forests. *For. Ecol. Manag.* **2009**, *257*, 1684–1694. [CrossRef]
65. Patenaude, G.; Hill, R.; Milne, R.; Gaveau, D.; Briggs, B.; Dawson, T. Quantifying forest above ground carbon content using LiDAR remote sensing. *Remote Sens. Environ.* **2004**, *93*, 368–380. [CrossRef]
66. De Tanago, J.G.; Lau, A.; Bartholomeus, H.; Herold, M.; Avitabile, V.; Raumonen, P.; Martius, C.; Goodman, R.C.; Disney, M.; Manuri, S.; et al. Estimation of above-ground biomass of large tropical trees with terrestrial LiDAR. *Methods Ecol. Evol.* **2017**, *9*, 223–234. [CrossRef]
67. Calders, K.; Newnham, G.; Burt, A.; Murphy, S.; Raumonen, P.; Herold, M.; Culvenor, D.S.; Avitabile, V.; Disney, M.; Armston, J.D.; et al. Nondestructive estimates of above-ground biomass using terrestrial laser scanning. *Methods Ecol. Evol.* **2014**, *6*, 198–208. [CrossRef]
68. He, Q.; Chen, E.; An, R.; Li, Y. Above-ground biomass and biomass components estimation using LiDAR data in a coniferous forest. *Forests* **2013**, *4*, 984–1002. [CrossRef]
69. Brack, C.; Schaefer, M.; Jovanovic, T.; Crawford, D. Comparing terrestrial laser scanners' ability to measure tree height and diameter in a managed forest environment. *Aust. For.* **2020**, *83*, 161–171. [CrossRef]
70. Roussel, J.R.; Caspersen, J.; Béland, M.; Thomas, S.; Achim, A. Removing bias from LiDAR-based estimates of canopy height: Accounting for the effects of pulse density and footprint size. *Remote Sens. Environ.* **2017**, *198*, 1–16. [CrossRef]

71. Dassot, M.; Colin, A.; Santenoise, P.; Fournier, M.; Constant, T. Terrestrial laser scanning for measuring the solid wood volume, including branches, of adult standing trees in the forest environment. *Comput. Electron. Agric.* **2012**, *89*, 86–93. [CrossRef]
72. Astrup, R.; Ducey, M.J.; Granhus, A.; Ritter, T.; von Lüpke, N. Approaches for estimating stand-level volume using terrestrial laser scanning in a single-scan mode. *Can. J. For. Res.* **2014**, *44*, 666–676. [CrossRef]
73. Ducey, M.J.; Astrup, R.; Seifert, S.; Pretzsch, H.; Larson, B.C.; Coates, K.D. Comparison of forest attributes derived from two terrestrial lidar systems. *Photogramm. Eng. Remote Sens.* **2013**, *79*, 245–257. [CrossRef]
74. Giurgiu, V. *Dendrometrie și Auxologie Forestieră*; Ceres: Bucharest, Romania, 1979.
75. McVittie, A.; Hussain, S. The Economics of Ecosystems and Biodiversity—Valuation Database Manual. Available online: http://doc.teebweb.org/wp-content/uploads/2014/03/TEEB-Database-and-Valuation-Manual_2013.pdf (accessed on 1 March 2021).
76. Alamgir, M.; Turton, S.M.; Macgregor, C.J.; Pert, P.L. Assessing regulating and provisioning ecosystem services in a contrasting tropical forest landscape. *Ecol. Indic.* **2016**, *64*, 319–334. [CrossRef]
77. Burkhard, B.; Kroll, F.; Nedkov, S.; Müller, F. Mapping ecosystem service supply, demand and budgets. *Ecol. Indic.* **2012**, *21*, 17–29. [CrossRef]
78. De Groot, R.; Alkemade, R.; Braat, L.; Hein, L.; Willemen, L. Challenges in integrating the concept of ecosystem services and values in landscape planning, management and decision making. *Ecol. Complex.* **2010**, *7*, 260–272. [CrossRef]
79. Clark, P.J.; Evans, F.C. Distance to nearest neighbor as a measure of spatial relationships in populations. *Ecology* **1954**, *35*, 445–453. [CrossRef]
80. Vorčák, J.; Merganič, J.; Saniga, M. Structural diversity change and regeneration processes of the Norway spruce natural forest in Babia hora NNR in relation to altitude. *J. For. Sci.* **2012**, *52*, 399–409. [CrossRef]
81. Ajrhough, S.; Maanan, M.; Mharzi Alaoui, H.; Rhinane, H.; El Arabi, E.H. *Mapping Forest Ecosystem Services: A Review*; International Archives of the Photogrammetry, Remote Sensing and Spatial Information Science—ISPRS Archives; ISPRS: Hannover, Germany, 2019.
82. Houghton, R.A.; House, J.I.; Pongratz, J.; Van Der Werf, G.R.; Defries, R.S.; Hansen, M.C.; Le Quéré, C.; Ramankutty, N. Carbon emissions from land use and land-cover change. *Biogeosciences* **2012**, *9*, 5125–5142. [CrossRef]
83. Giurgiu, V.; Decei, I.; Draghiciu, D. *Metode si Tabele Dendrometrice*; Ceres: Bucharest, Romania, 2004.
84. Eggleston, S.; Buendia, L.; Miwa, K.; Ngara, T.; Tanabe, K. *Guidelines for National Greenhouse Gas Inventories: Agriculture, Forestry and Other Land Use*; IPCC: Hayama, Japan, 2006.
85. Lamlom, S.; Savidge, R. A reassessment of carbon content in wood: Variation within and between 41 North American species. *Biomass Bioenergy* **2003**, *25*, 381–388. [CrossRef]
86. Justine, M.F.; Yang, W.; Wu, F.; Tan, B.; Khan, M.N.; Zhao, Y. Biomass stock and carbon sequestration in a chronosequence of *pinus massoniana* plantations in the upper reaches of the Yangtze River. *Forests* **2015**, *6*, 3665–3682. [CrossRef]
87. Calders, K.; Schenkels, T.; Bartholomeus, H.; Armston, J.D.; Verbesselt, J.; Herold, M. Monitoring spring phenology with high temporal resolution terrestrial LiDAR measurements. *Agric. For. Meteorol.* **2015**, *203*, 158–168. [CrossRef]
88. Danson, F.; Hetherington, D.; Morsdorf, F.; Koetz, B.; Allgower, B. Forest canopy gap fraction from terrestrial laser scanning. *IEEE Geosci. Remote Sens. Lett.* **2007**, *4*, 157–160. [CrossRef]
89. Danson, F.M.; Morsdorf, F.; Koetz, B. Airborne and terrestrial laser scanning for measuring vegetation canopy structure. *Laser Scanning Environ. Sci.* **2009**, 201–219. [CrossRef]
90. Pascu, I.S.; Dobre, A.-C.; Zamfira, V.; Apostol, E.; Leca, Ș.; Pitar, D.; Apostol, B.; Chivulescu, Ș.; Ciceu, A.; Duro, J.G.; et al. Phenological analysis through the use of multitemporal TLS observations. *Rev. Silvic. Cineg.* **2020**, *XXV*, 38–45.
91. Jenkins, R.B. Airborne laser scanning for vegetation structure quantification in a south east Australian scrubby forest-woodland. *Austral. Ecol.* **2011**, *37*, 44–55. [CrossRef]
92. Savastru, D.M.; Zoran, M.A.; Savastru, R.S. Geospatial information for assessment of climate change impact on forest phenology. In *Seventh International Conference on Remote Sensing and Geoinformation of the Environment (RSCy2019)*; International Society for Optics and Photonics: Bellingham, WA, USA, 2019; Volume 11174, p. 1117402. [CrossRef]
93. Nezval, O.; Krejza, J.; Světlík, J.; Šigut, L.; Horáček, P. Comparison of traditional ground-based observations and digital remote sensing of phenological transitions in a floodplain forest. *Agric. For. Meteorol.* **2020**, *291*, 108079. [CrossRef]
94. Alivernini, A.; Fares, S.; Ferrara, C.; Chianucci, F. An objective image analysis method for estimation of canopy attributes from digital cover photography. *Trees* **2018**, *32*, 713–723. [CrossRef]
95. Stark, S.C.; Leitold, V.; Wu, J.; Hunter, M.; De Castilho, C.V.; Costa, F.R.C.; McMahon, S.M.; Parker, G.; Shimabukuro, M.T.; Lefsky, M.A.; et al. Amazon forest carbon dynamics predicted by profiles of canopy leaf area and light environment. *Ecol. Lett.* **2012**, *15*, 1406–1414. [CrossRef]
96. MacArthur, R.H.; Horn, H.S. Foliage profile by vertical measurements. *Ecology* **1969**, *50*, 802–804. [CrossRef]
97. Swinehart, D.F. The beer-lambert law. *J. Chem. Educ.* **1962**, *39*, 333. [CrossRef]
98. Calloway, D. Beer-lambert law. *J. Chem. Educ.* **1997**, *74*, 744. [CrossRef]
99. Sumida, A.; Nakai, T.; Yamada, M.; Ono, K.; Uemura, S.; Hara, T. Ground-based estimation of leaf area index and vertical distribution of leaf area density in a Betula ermanii forest. *Silva Fenn.* **2009**, *43*. [CrossRef]
100. Stark, S.C.; Enquist, B.; Saleska, S.R.; Leitold, V.; Schietti, J.; Longo, M.; Alves, L.; de Camargo, P.B.; Oliveira, R.C. Linking canopy leaf area and light environments with tree size distributions to explain Amazon forest demography. *Ecol. Lett.* **2015**, *18*, 636–645. [CrossRef]

101. Parker, G.G.; Harding, D.J.; Berger, M.L. A portable LIDAR system for rapid determination of forest canopy structure. *J. Appl. Ecol.* **2004**, *41*, 755–767. [CrossRef]
102. Kamoske, A.G.; Dahlin, K.M.; Stark, S.C.; Serbin, S.P. Leaf area density from airborne LiDAR: Comparing sensors and resolutions in a temperate broadleaf forest ecosystem. *For. Ecol. Manag.* **2018**, *433*, 364–375. [CrossRef]
103. Reid, W.V.; Mooney, H.A.; Cropper, A.; Capistrano, D.; Carpenter, S.R.; Chopra, K.; Zurek, M.B. *Ecosystems and Human Well-being-Synthesis: A Report of the Millennium Ecosystem Assessment*; Island Press: Washington, DC, USA, 2005.
104. Haines-Young, R.; Potschin, M. *CICES V5. 1. Guidance on the Application of the Revised Structure*; Fabis Consulting: Nottingham, UK, 2018.
105. Haines-Young, R.; Potschin, M. Common international classification of ecosystem services (CICES, Version 4.1). *Eur. Environ. Agency* **2012**, *33*, 107.
106. Li, S.; Wang, T.; Hou, Z.; Gong, Y.; Feng, L.; Ge, J. Harnessing terrestrial laser scanning to predict understory biomass in temperate mixed forests. *Ecol. Indic.* **2020**, *121*, 107011. [CrossRef]
107. Martire, S.; Castellani, V.; Sala, S. Carrying capacity assessment of forest resources: Enhancing environmental sustainability in energy production at local scale. *Resour. Conserv. Recycl.* **2015**, *94*, 11–20. [CrossRef]
108. Street, G.M.; Rodgers, A.R.; Avgar, T.; Fryxell, J.M. Characterizing demographic parameters across environmental gradients: A case study with Ontario moose (*Alces alces*). *Ecosphere* **2015**, *6*, art138. [CrossRef]
109. Pringle, R.M.; Fox-Dobbs, K. Coupling of canopy and understory food webs by ground-dwelling predators. *Ecol. Lett.* **2008**, *11*, 1328–1337. [CrossRef]
110. Arnold, J.M.; Gerhardt, P.; Steyaert, S.M.; Hochbichler, E.; Hackländer, K. Diversionary feeding can reduce red deer habitat selection pressure on vulnerable forest stands, but is not a panacea for red deer damage. *For. Ecol. Manag.* **2018**, *407*, 166–173. [CrossRef]
111. Ewald, M.; Dupke, C.; Heurich, M.; Muller, J.P.; Reineking, B. LiDAR remote sensing of forest structure and GPS telemetry data provide insights on winter habitat selection of european roe deer. *Forests* **2014**, *5*, 1374–1390. [CrossRef]
112. Martinuzzi, S.; Vierling, L.A.; Gould, W.A.; Falkowski, M.J.; Evans, J.S.; Hudak, A.T.; Vierling, K.T. Mapping snags and understory shrubs for a LiDAR-based assessment of wildlife habitat suitability. *Remote Sens. Environ.* **2009**, *113*, 2533–2546. [CrossRef]
113. Nilsson, M.-C.; Wardle, D.A. Understory vegetation as a forest ecosystem driver: Evidence from the northern Swedish boreal forest. *Front. Ecol. Environ.* **2005**, *3*, 421–428. [CrossRef]
114. Gilliam, F. The ecological significance of the herbaceous layer in temperate forest ecosystems. *BioScience* **2007**, *57*, 845–858. [CrossRef]
115. Hill, R.; Broughton, R. Mapping the understorey of deciduous woodland from leaf-on and leaf-off airborne LiDAR data: A case study in lowland Britain. *ISPRS J. Photogramm. Remote Sens.* **2009**, *64*, 223–233. [CrossRef]
116. Gersom, Z. *Mapping the Shrub Layer in a Forest Using LiDAR*; Wageningen University and Research Centre: Wageningen, The Netherlands, 2018.
117. Wing, B.M.; Ritchie, M.W.; Boston, K.; Cohen, W.B.; Olsen, M.J. Individual snag detection using neighborhood attribute filtered airborne lidar data. *Remote Sens. Environ.* **2015**, *163*, 165–179. [CrossRef]
118. Tallis, H.; Ricketts, T.H.; Daily, G.C.; Polasky, S. Natural Capital: Theory and Practice of Mapping Ecosystem Services—Oxford Scholarship. Available online: https://www.amazon.com/Natural-Capital-Practice-Ecosystem-Services/dp/0199589003 (accessed on 1 March 2021).
119. Maes, J.; Crossman, N.D.; Burkhard, B. Mapping ecosystem services. In *Handbook of Ecosystem Services*; Routledge: London, UK, 2018.
120. Lautenbach, S.; Kugel, C.; Lausch, A.; Seppelt, R. Analysis of historic changes in regional ecosystem service provisioning using land use data. *Ecol. Indic.* **2011**, *11*, 676–687. [CrossRef]
121. Martin, M.; Newman, S.; Aber, J.; Congalton, R. Determining forest species composition using high spectral resolution remote sensing data. *Remote Sens. Environ.* **1998**, *65*, 249–254. [CrossRef]
122. Ørka, H.O.; Dalponte, M.; Gobakken, T.; Næsset, E.; Ene, L.T. Characterizing forest species composition using multiple remote sensing data sources and inventory approaches. *Scand. J. For. Res.* **2013**, *28*, 677–688. [CrossRef]
123. Penman, J.; Gytarsky, M.; Hiraishi, T.; Irving, W.; Krug, T. *Guidelines for National Greenhouse Gas Inventories*; IPCC: Geneva, Switzerland, 2006.
124. Pandey, R. Indices for measuring forest ecosystem goods and services contribution to the rural community: A tool for informed decisions. *J. Environ. Prof. Sri Lanka* **2013**, *1*, 58. [CrossRef]
125. Czúcz, B.; Haines-Young, R.; Kiss, M.; Bereczki, K.; Kertész, M.; Vári, Á.; Potschin-Young, M.; Arany, I. Ecosystem service indicators along the cascade: How do assessment and mapping studies position their indicators? *Ecol. Indic.* **2020**, *118*, 106729. [CrossRef]
126. Burkhard, B.; Santos-Martín, F.; Nedkov, S.; Maes, J. An operational framework for integrated Mapping and Assessment of Ecosystems and their Services (MAES). *One Ecosyst.* **2018**, *3*, e22831. [CrossRef]
127. Song, Z.; Seitz, S.; Li, J.; Goebes, P.; Schmidt, K.; Kühn, P.; Shi, X.; Scholten, T. Tree diversity reduced soil erosion by affecting tree canopy and biological soil crust development in a subtropical forest experiment. *For. Ecol. Manag.* **2019**, *444*, 69–77. [CrossRef]
128. Loustau, D.; Granier, A.; Bréda, N. A generic model of forest canopy conductance dependent on climate, soil water availability and leaf area index. *Ann. For. Sci.* **2000**, *57*, 755–765. [CrossRef]

129. Manes, F.; Marando, F.; Capotorti, G.; Blasi, C.; Salvatori, E.; Fusaro, L.; Ciancarella, L.; Mircea, M.; Marchetti, M.; Chirici, G.; et al. Regulating ecosystem services of forests in ten Italian metropolitan cities: Air quality improvement by PM10 and O3 removal. *Ecol. Indic.* **2016**, *67*, 425–440. [CrossRef]
130. Bobiec, A. Living stands and dead wood in the Białowieża forest: Suggestions for restoration management. *For. Ecol. Manag.* **2002**, *165*, 125–140. [CrossRef]

MDPI
St. Alban-Anlage 66
4052 Basel
Switzerland
Tel. +41 61 683 77 34
Fax +41 61 302 89 18
www.mdpi.com

Forests Editorial Office
E-mail: forests@mdpi.com
www.mdpi.com/journal/forests

www.ingramcontent.com/pod-product-compliance
Lightning Source LLC
LaVergne TN
LVHW071001170726
843515LV00006B/1048